# ITALIAN SCIENTISTS IN THE LOW COUNTRIES IN THE XVIIth AND XVIIIth CENTURIES

**Nieuwe Nederlandse Bijdragen tot de Geschiedenis der Geneeskunde en der Natuurwetenschappen**

**No. 34**

Redactie: M.J. van Lieburg, L.C. Palm, H.A.M. Snelders en R.P.W. Visser

# ITALIAN SCIENTISTS IN THE LOW COUNTRIES IN THE XVIIth AND XVIIIth CENTURIES

Invited Papers from the Congress held in Utrecht on 25-27 May 1988 to commemorate the 350th anniversary of the publication of Galileo Galilei's 'Discorsi e dimostrazioni matematiche intorno a due nuove scienze' (Leyden, 1638)

Edited by

C.S. Maffioli & L.C. Palm

Rodopi

Amsterdam - Atlanta, GA 1989

Published with the financial assistance
of the 'Ministero degli Affari Esteri' of the Republic of Italy
and
of the 'Prins Bernhard Fonds'

CIP-GEGEVENS KONINKLIJKE BIBLIOTHEEK, DEN HAAG

Italian

Italian scientists in the low countries in the XVIIth and XVIIIth centuries / ed. by C.S. Maffioli and L.C. Palm. — Amsterdam : Rodopi. — Ill., fig. — (Nieuwe Nederlandse bijdragen tot de geschiedenis der geneeskunde en der natuurwetenschappen, ISSN 0168—9827 34)
Invited papers from the congress held in Utrecht on 25-27 May 1988 to commemorate the 350th anniversary of the publication of Galilei's 'Discorsi e dimostrazioni matematiche intorno a due nuove scienze' (Leyden, 1638). — Met index.
ISBN 90—5183—121—8
SISO 506 UDC 501(=1.45)(492)"16/17"
Trefw.: natuurwetenschappen ; Nederland ; geschiedenis ; 17e/18e eeuw / natuurwetenschappen ; Italië ; geschiedenis ; 17e/18e eeuw.

Printed in The Netherlands

**Fig. 1 Outline drawing by Galileo, showing Jupiter with the four 'Medicean Stars' revolving around it, in the *Album Amicorum* of Ernst Brinck (Koninklijke Bibliotheek, 's-Gravenhage)**

## CONTENTS

# FOREWORD

When dealing with cultural relations between Italy and the Netherlands what comes first to mind is the relationship which existed in the seventeenth and eighteenth centuries between such artists as painters, engravers, humanists and musicians who often paid visits to each other in their respective countries.

The fact that over the same period there were even more frequent and relevant interchanges between the scientists of our two countries is, on the other hand, far from common knowledge.

The publication at Leyden of Galileo's work *Discorsi e dimostrazioni matematiche intorno a due nuove scienze*, the visit paid by Italian scientists such as Marsili, Barnaba Oriani and Volta, the letters exchanged between university professors and other individual scientists from both countries, all bear witness to the fact that the relationships which were to bring such outstanding results were both deep-rooted and continuous.

The importance of this relationship was clearly stressed by the International Congress *Italian Scientists in the Low Countries in the XVIIth and XVIIIth Centuries* promoted by the University of Utrecht with the participation of many renowned scholars from both countries.

It is our wish to express here our deepest gratitude both to the organizers and to the participants for the dedication shown on this occasion which allowed a large interested audience to understand the results of their research in the field of sciences ranging from mechanics to hydrostatics, biology to mathematics, physics to medicine, as they were viewed in those bygone centuries, in spite of the short time at their disposal.

Thanks to their contributions, even the layman became acquainted with the names of outstanding scholars who, besides having used "doctrinal and technical terms as pertaining to exact sciences", wrote works which had a true philosophical and literary value thus proving, in the words of Croce, that "science accepts association with literature".

Piero Ferraboschi
Ambassador of the Republic
of Italy in the Netherlands

# INTRODUCTION

When Lodewijk Elzevier stepped up towards Galileo's house in Arcetri in May 1636 he was probably aware that the business he was going to discuss with the old mathematician and philosopher of the Grand Duke of Tuscany concerned first of all, according to the author's own words as expressed in a letter to Matthias Bernegger of 15 July 1636, "la ricolta più stimata da me degli studii di tutta la mia vita". That the study of motion was the field par excellence of Galileo was an opinion largely shared, and particularly emphasized by the Servite Paolo Sarpi and Fulgenzio Micanzio, the latter being responsible for the delivery to Elzevier and the mailing to Holland of the manuscript parts of the *Discorsi* that Galileo sent to Venice. With the Strasbourg editions in 1635 and 1636, "Impensis Elzeviriorum", of the Latin translation of the *Dialogo* of 1632 and of the Letter to Christina of Lorraine of 1615, unpublished until then, the Elzeviers had certainly already contributed to the European diffusion of Galileo's thought. However, in the case of the printing of the *Discorsi* the subject-matter was apparently more neutral, and the ideological support had a less obvious Protestant flavour.

Galileo's connections with Flanders and the United Provinces dated back a long while and had been indirectly affected by the changing European equilibrium. Although the universalistic age of Erasmus seemed definitively over, centres like Venice, Florence and Rome remained remarkably attractive to ultramontanes. Sea trade in the Mediterranean favoured intellectual exchanges with the North and even the warfare against Spain of the young Dutch Republic, although partially blocking the 'peregrinatio academica' of Dutch protestants, put a friend of Galileo like Daniello Antonini in contact with the scientific milieu of the Low Countries. The international organization of the Jesuits maintained a certain scientific exchange between Italy and Flanders through its network of colleges, and attracted to the Collegio Romano many distinguished mathematicians from the North. Among them we find e.g. the Brussels Jesuit Odo van Maelcote who in May 1611 read the oration *Nuncius Sidereus Collegii Romani* praising Galileo's astronomical discoveries. However, the most important contact took place during the last years of Galileo's life, almost contemporaneously with the printing of the *Discorsi*. It concerned a possible practical application of Galileo's discovery of the 'Medicean Stars' revolving around Jupiter, namely that of determining longitude at sea and of reforming cartography. After uselessly trying to interest the Spanish crown in his finding, Galileo turned his eyes to a small but powerful Republic, with a feeling of confidence in the republican virtues which we also perceive in his later sorrow for having left Padua and its freedom for a Court in 1610.

The beginning of the Thirty Years War marked the definitive end of the Renaissance, but not of its historical influence. So among the Dutch interlocutors of Elia Diodati, who supported in Paris Galileo's project for measuring the longitude at sea, we find Constantijn Huygens in his capacity as secretary of the Stadholder. But Diodati's interlocutor was also a patron of science, who received Descartes among his guests and favoured the printing of the *Discours de la Méthode*; he was a cultured intellectual, a poet, musician and polyglot, a lover of Italian classicism in architecture who designed his own house at The Hague, inspired by Vitruvius and Palladio. He had studied the Italian language, as did his sons Constantijn and Christiaan. The latter read Galileo's *Discorsi* very early and was to become a major promotor, in his own way, of the Galilean science of motion. In spite of the religious wars the thread was not broken, even though science became more and more the protagonist of the Republic of the literati.

The *Discorsi* are and will remain a central issue and a touchstone of any attempt to evaluate the Scientific Revolution of the XVIIth century, and we should not confine such a unique contribution to the subspeciality of Galileo's studies. At stake there is indeed something more than assessing the degree of experimentalism of Galileo, at least if we think that our main target is to incorporate Galileo's achievement into a synthetic conception — which meets contemporary standards of scope and scholarship — of the emergence on the European scene between 1500 and 1800 of early modern science. A grand challenge indeed for a new generation of historians of science, rendered more arduous by the various directions of research followed in recent years, which have rarely been integrated into a coherent vision. Our historical stage is becoming more and more crowded with forgotten contributions to science and to its institutionalisation, with research approaches and fields rescued from the oblivion put upon them by the XIXth century re-shaping of areas and disciplines, with characters whose dealings with scientific enterprise are not yet well focused, so as to render premature in the present state of affairs any pretension to a new general assessment of the Scientific Revolution. From this point of view a comparative study between different areas, the examination of the paths along which knowledge was transmitted from one country to another, the individualization of possible influences exerted on the most important scientists by traditions far away in space, time or content, the tracing of the web of relationships which substantiated the European circulation of ideas, all these seem necessary preliminary steps to produce a more balanced general picture. The case of Italy and the Low Countries illustrates well our point. Indeed the two areas were certainly among the most advanced in developing the New Science but their polycentric structure and a lower degree of institutionalisation, at least in comparison with centres like Paris and London, have indirectly contributed to focus the attention of historians primarily on the most prominent characters — on giants like Galileo and Huygens first of all — and on the relations of Italian or Dutch scientists with those main centres. This attitude is quite comprehensible and, as e.g. is shown by the contacts of Marcello Malpighi and Antoni van Leeuwenhoek with the Royal Society, it emphasizes a major trend. However, both Malpighi in Bologna and van Leeuwenhoek in Delft were not, though in a quite different manner, isolated figures and they had several other channels of international communication besides sending their works to the Royal Society. The books printed in Holland of Malpighi and many other scientists more or less connected with Bologna's medical school show, for instance, an interest that cannot be reduced to the international bookselling trade of Dutchmen, and reveal the existence of a framework of relationships of which, for the greater part, wait to be dug out and properly understood by historians.

The beginning of scholarly historical work devoted to the analysis of the relations between Italy and the Netherlands in the fields of science and technology goes back at least to the end of the nineteenth century and may be first of all ascribed to the editors of the monumental editions of the works of Galileo and Huygens, one of them also being responsible for a fine edition of the late eighteenth-century correspondence exchanged between Alessandro Volta and Martinus van Marum. These pioneering studies were, however, limited in scope and did not give rise to any consistent tradition of research. Its development was certainly not favoured by the two wars, by the difficulty of cultural exchange between the two countries, by the linguistic barrier and, last but not least, by a more or less implicit historiographical tenet which ended by focusing the attention in both countries on the contacts with England and France (or possibly with Germany, at least in the period from the late eighteenth century onwards). Although we are far from claiming any priority to our theme, which will very likely continue to attract only a limited number of scholars, we think it deserves more attention and consideration than it has got until now. The same papers presented at the congress of Utrecht are far from covering all the main relationships studied in past years by historians (e.g. no words are devoted to the contacts of Christiaan Huygens with the Florentine Accademia del Cimento or to the influence exerted by Giovanni Alfonso Borelli's iatromechanics on Herman Boerhaave) and should be considered as a starting point of new and wider directions of research rather than as more or less definitive issues. The influence exerted by the Netherlanders in Italy should in particular be studied with care. On the other hand, our message has already been positively echoed in *Incontri*, a Dutch journal devoted to the study of cultural relationships between Italy and the Netherlands, which is dedicating a special issue — where the emphasis will be put on the publication and inventory of primary sources — to eighteenth-century scientific contacts.

Many of the special themes dealt with in the congress *Italian Scientists in the Low Countries in the XVIIth and XVIIIth Centuries* held in Utrecht on 25-27 May 1988, were fascinating and we leave the appreciation of their flavour to the reader. However, a few more general themes ask for some introductory remarks. The physical analogy between the area of the main rivers (Rhine, Meuse and their different branches) in Holland and the Po plain, particularly the region between Venice and Bologna, was for instance an important indirect source of intellectual influences and contacts until now largely neglected by historians. The theoretical standpoint of the Italian hydraulicians was quite respected and followed in Holland, while conversely the practical skill of the Dutch and Flemish technicians was highly appreciated in Italy. Only the first of the two parts of this theme is analyzed here, given that the main emphasis of the congress was on the influence exerted by the Italians in the Netherlands. However, the second is at least as promising as the first, particularly for the historian of technology, and even if a figure like Cornelis Meijer has already attracted a certain amount of attention, much more work should be done in this direction.

If we pass from the morphology of the territory to that of the institutions, we perceive a greater affinity between Italy and the Low Countries in the field of universities than in that of scientific societies. The first scientific society in the Republic of the United Provinces, the Hollandsche Maatschappij der Wetenschappen founded at Haarlem in 1752, followed almost a century after the Accademia del Cimento and was largely different in scope and organization from the established great foreign models. It was not the outcome of the will of a prince but of the patricians of a province having, besides ordinary members, a large body of directors belonging to the city regents who sustained its image and finance and who were positively influenced by the so-called physico-theology, a typical feature of the Dutch Enlightenment largely absent in Italy. Furthermore the Hollandsche Maatschappij did not have its own research activity but stimulated useful knowledge by the awarding of prizes and the publication of annual *Verhandelingen*. Even if the work of the Cimento may have indirectly stimulated the constitution in 1675 of a "Theatrum Physicum" at the University of Leyden, and although the Florentine academy was appreciated enough by Petrus van Musschenbroek to publish in 1731 a Latin translation of the *Saggi di naturali esperienze*, its influence was very modest and the same may all the more be said for the other Italian scientific academies. Different, however, is the case of the old Italian universities, particularly Bologna and Padua, that perhaps constituted the main model for the establishment at Leyden in 1575 of the first university in the Northern Netherlands. In fact, although its structure followed the Paris and Louvain pattern, Leyden pursued the intellectual trends in science and, above all, in medicine of the Northern Italian universities. But while the Dutch universities were to be in the late seventeenth century at the forefront in the introduction of experimental physics in the curricula, in a city like Bologna Luigi Ferdinando Marsili (or Marsigli) was obliged to found in 1711-1714 a separate institution, the Istituto e Accademia delle Scienze, because the Bologna Studium was closed to experimental philosophy; Giovanni Poleni, at the University of Padua, even had to wait until 1740 to see the realization of his dream, namely the establishment of a "Teatro di Filosofia Sperimentale" which from the beginning was one of the best in Europe, thanks also to the several apparatus purchased in Leyden. In addition, the structure of the Dutch universities, provided with strong theological faculties, was somewhat different from that of the Northern Italian universities, in which law and medicine largely prevailed over theology. This may be the structural basis for the many philosophical quarrels in the Dutch universities where theologians were often in the front row, a phenomenon scarcely present in Italy where ideological control in the universities was normally exerted by political or religious authorities.

This leads us to another fascinating theme where a comparison between Italy and the Netherlands may be of great interest, namely the question of the influence of religion on science as well as to the different ways in which such influence was exerted in Catholic and Protestant countries. As is well known, it is difficult to find in the history of science a theme on which scholarly opinion is more divided. While the traditional tenet focused on the freedom of thought may still find supporters in the realm of intellectual history, it is conversely possible to maintain from a sociological standpoint that the many religious orders in Catholic Italy gave enough leisure to their members to devote themselves to science while in Protestant Holland trade and not learning absorbed the best energies. Whatever it may be, it is clear that the epoch of dichotomic analyses is definitively over. Nevertheless the issue remains central for an evaluation of the Scientific Revolution of the XVIIth century, and perhaps it will gain still more relevance if the emergence of early modern science is to be regarded from those privileged observatories that were the Italian States and the Low Countries.

If we just glance at the contents of the papers here proposed we see that purely intellectual influences have not been neglected, and that many disciplinary areas have been taken into consideration. Contacts between two scientists which had a primary importance on their research work, like those between Volta and van Marum in the realm of electricity, were, however, an exception. More frequently do we meet a circulation of ideas, with several characters and subject-matters involved. So e.g. Giovanni Battista Benedetti's mechanics of free fall was discussed, through Taisnier's plagiarism, by Simon Stevin and conversely the latter's research in hydrostatics was probably indirectly known by Galileo. It is, on the other hand, certain that Isaac Beeckman read many of Galileo's works, including the *Dialogo* of 1632 which he lent for a couple of days to Descartes in Amsterdam in August 1634. Other examples of such circulation of ideas include the influence of the works of Francesco Redi and Malpighi on van Leeuwenhoek and vice versa of the research on the figures of salts of the Delft microscopist on Domenico Guglielmini, perhaps the best pupil of Malpighi at Bologna. Although Huygens' relationships with Italians are not analyzed, the journey through Italy of a mathematical theme proposed by the great Dutch scientist is discussed through a study of the works of Poleni, Jacopo and Vincenzo Riccati. The already mentioned influence in the field of hydraulics has in its turn been pointed out with reference to the works of Italian and Dutch scientists or engineers like Guglielmini, Willem Jacob 's Gravesande and Johan Lulofs, Poleni and Christiaan Brunings, Paolo Frisi and Johan Frederik Hennert. Finally, in the field of medicine some aspects of the influence of Giambattista Morgagni's pathological anatomy and that of his successor at Padua, Leopoldo Marc'Antonio Caldani, on Eduard Sandifort have been discussed.

The transmission of scientific ideas from one country to another was of course largely determined by the facility of the circulation of books and journals between the two countries. In the case of Italy and Holland, given the difficulties often met with mailing, a decisive factor was the translation or the reprint of books in the other country, at least until papers published or abstracted in different scientific journals became the normal method of communication. The papers presented here which are devoted to the editorial fortune of Bolognese scientists in Holland and to the reprint of the Latin textbooks of 's Gravesande and van Musschenbroek in Italy show how effective this factor was in favouring the mutual diffusion of scientific matters. However, Holland in the XVIIth and XVIIIth centuries was much more than a great printing house, it was a great carrefour of ideas, a meeting point of people coming from the whole of Europe. So in one of the papers here presented we find the young Maltese scholar Giovanni Francesco Buonamico and the Dane Nicolaus Steno (Niels Stensen), in another we see Marsili visiting Boerhaave and 's Gravesande in Leyden, attending the lessons of the first and being invited to the wedding of the sister-in-law of the second. An important task of Marsili's visit to Holland was the acquisition of natural history specimens for the Istituto of Bologna, a business that, together with the purchase of physical instruments, stimulated many relations of Italians with Dutchmen like that between Felice Fontana and Hendrik van Swinden on behalf of the Reale Gabinetto di Fisica e Storia naturale of Florence.

The Dutch milieu was largely a Protestant one, the centre of a widespread network helping religious dissenters coming from Catholic countries. Within this network Switzerland normally represented the first landing-place for Italian refugees like Giovan Francesco Salvemini da Castiglione who was later called to the University of Utrecht where he taught mathematics and astronomy in the 1750's. Switzerland, through the Basel group led by Johann Bernoulli and the scientific milieu of Geneva, had already in the 1720's played an intermediary role between Padua and Leyden when Poleni and 's Gravesande happened to be on the same side in the 'vis viva' controversy, a conflict in which the kind of influence Newton's thought was to exercise in Continental Europe was at stake.

While the eighteenth-century Grand Tour of the Netherlanders in Italy was primarily an educational journey with arts, letters and ancient monuments as the focal cultural points, the several Italian scientists who visited Holland indicate that matters like naturalistic collections, musea, laboratories and workshops of instrument makers were of particular interest for the travellers from the South. Some of them, like the astronomer Barnaba Oriani as well as other Lombard visitors of the second half of the century like Frisi, Volta and Marsilio Landriani, were supported in various ways during their journeys by the government of Austrian Lombardy. Such official support to science was typical of the Italian Enlightenment and bears witness to a new dynamism of the Italian States. However, it was also the symptom of a new age in which the European Republic of Letters was to disappear to make way for new kinds of transmission and organization of science.

*

When at the beginning of 1984 we started with the preparation of our meeting, we proceeded very cautiously by proposing a one-day conference with a limited number of speakers. Thanks to the scientific and organizing support of the Institute for the History of Science of the University of Utrecht, helped by an advising Dutch-Italian Scientific Committee of historians of science as well as by the Dutch Society for the History of Science (GeWiNa), the idea grew into a three-day congress in which, in addition to the sixteen invited papers here published, two research works were also presented by Mrs drs M. Fournier of the Museum Boerhaave of Leyden and by Dr A. Touwaide of the Catholic University of Louvain-la-Neuve which will be published elsewhere The congress was held under the patronage of His Excellency Piero Ferraboschi, Ambassador of the Republic of Italy in the Netherlands, to whom we express our sincere thanks, and was sponsored by the University of Utrecht, by the National Research Council of Italy (C.N.R.), by the San Paolo Bank, by the Eni International Holding b.v. and by the British Council. To all of these bodies, as well as to the Italian Institute of Culture of Amsterdam for its general support, our grateful thanks.

We would also like to express our gratitude to the Utrecht University Museum for the kind hospitality accorded to the official reception of the congress, to Mrs L.M. Boersma-Broekmeyer for her invaluable help in organizing the congress and in many other respects, as well as to Mrs Frances Bulwer for her revision and checking of the English text of some of the papers here proposed. It is finally a pleasure to thank our colleagues in the Scientific Committee of the Congress Prof. V. Cappelletti of the University of Rome, Prof. P. Galluzzi of the University of Siena, Mrs Prof. A.M. Luyendijk-Elshout of the University of Leyden, Prof. C. Maccagni of the University of Genoa, Prof. H.A.M. Snelders of the University of Utrecht, Prof. G. Tabarroni of the University of Bologna, Dr R.P.W. Visser of the University of Utrecht, for their most valuable advice.

C.S. Maffioli — L.C. Palm

# GALILEO'S UPS AND DOWNS IN THE HISTORIOGRAPHY OF THE SCIENTIFIC REVOLUTION

*H. Floris Cohen*

When discussing the historiography of the Scientific Revolution, three notions should be distinguished with some care. I mean the concept of scientific revolution*s*; the *idea* of *The* Scientific Revolution, and the *concept* of The Scientific Revolution, respectively. The first, the concept of scientific revolutions, denotes a general assumption, coined and employed by those who uphold that science progresses over time through revolutionary convulsions rather than in a continuous, gradual fashion. The term has gained wide popularity through Thomas Kuhn's use of it in the particular sense explained in his *The Structure of Scientific Revolutions*, but, as I. Bernard Cohen has shown, both the term and the general idea behind it go back to the late 17th century.[1] As such, however, the concept should be distinguished from The Scientific Revolution, not as a *generic* but rather as a *specific* term employed to conceptualize the unique event of the birth of early modern science in the 17th century. Now, as an overall idea the view that early modern science was born during the period covered roughly by the generations of Copernicus through Newton, is virtually as old as the event itself — its protagonists were well aware that they were bringing about a major transformation in science. The term 'The Scientific Revolution', however, embodying a distinct, analytical tool in the historiography of science, is of far more recent origin. Used in its specific sense, I found a first, rather fleeting occurrence in Martha Ornstein's 1913 discussion of 17th century scientific societies.[2] But the term acquired a far

NB. Wherever these notes appear deficient, their shortcomings will be made good in my forthcoming book (mentioned in note 3), from which substantial chunks of the present paper have been derived.

1. Cohen, I.B. (1976): 'The eighteenth-century origins of the concept of scientific revolution'. In: *Journal of the History of Ideas* 37, 1976, pp. 257-288.

2. Ornstein, M., *The Rôle of Scientific Societies in the Seventeenth Century*. Chicago (UP), 1928, p. 39 (ed. used: facs. reprint, New York (Arno Press), 1975; author's name given as Martha Ornstein [Bronfenbrenner], the latter being the name of her husband, whom she married in 1914, one year after she wrote the dissertation).

more distinct meaning in the works of Alexandre Koyré of the thirties through fifties and subsequently gained wide currency through its use (in a somewhat wider and vaguer sense) by Herbert Butterfield in 1949.[3]

I will not pursue further the intricacies involved in the coinage of this, at present so seemingly self-evident, concept of The Scientific Revolution, nor follow in any detail its vagaries through half a century of subsequent writing by professional historians on the subject of the birth of early modern science. Rather, my job here is to outline briefly what role Galileo was assigned by a variety of historians who, over the past hundred years or so, struggled to give meaning to the idea, and later to the concept, of the Scientific Revolution. I shall use the formation of this concept as a general framework for my account, and divide the process into three stages: its preparation, roughly from Whewell and Comte up to Duhem and Olschki; its actual formation in the works of Koyré and Dijksterhuis, and its further elaboration up to the present time. Thus I am *not* going to present you with an outline of how, ever since Galileo died, historians have been interpreting his contribution to science, however exciting an enterprise this would form in its own right. Rather, my purpose, at once more modest and more topical, is to sketch in the briefest of outlines what role Galileo has been assigned by a number of historians of science engaged in the process of defining what the birth of early modern science essentially came down to. I derive this account from my overall investigations into the historiography of the Scientific Revolution, now well under way and hopefully to appear next year under the title *The Banquet of Truth. An Historiographical Inquiry into Nature and Causes of the 17th Century Scientific Revolution.*

### *Galileo and the Scientific Revolution*

Although the choice of Galileo is not, of course, foreign to the general subject of the present conference, it is clear that no other 16th century or 17th century scientist is so closely tied up with the birth of early modern science as Galileo. It could be argued, as it has been in fact, that such men as Copernicus, or Kepler, or Descartes, played a less than decisive part in the historical process that led to, or constituted, the birth of early modern science. It may, for example, be pointed out that Copernicus was really an unvoluntary revo-

3. An account of how the concept of the Scientific Revolution took shape and has subsequently been employed runs through my forthcoming book *The Banquet of Truth. An Historiographical Inquiry into Nature and Causes of the 17th Century Scientific Revolution.* See also R. Porter, 'The scientific revolution: a spoke in the wheel?' In: R. Porter & M. Teich (eds.) *Revolution in History.* Cambridge (UP), 1986, pp. 290-316.

lutionary, or that Kepler was too much of a mystic to fit in the coming about of early modern science, or that Descartes' system-building was really foreign to the spirit of the new science. In the case of Galileo, however, such a relative disparagement vis-à-vis the process of the Scientific Revolution is virtually inconceivable. For in whatever way one wishes to define the core features of the Scientific Revolution — and there have been scores of ways, as we shall see presently — Galileo holds pride of place in it. Thus there is a quite remarkable entanglement to be detected between the interpretation of Galileo's life's work and historians' views of what the decisively new features of early modern science are to be taken to be. It is the changing interaction pattern between these two closely related yet distinct, because non-identical, lines of historical interpretation that constitutes the real theme of this paper.

I said 'non-identical', because, with one exception only, Galileo has not by historians been taken as the *single-handed* instigator of early modern science — he effected absolutely indispensable transformations in scientific thought, yet it is hard to conceive of a Scientific Revolution which consisted of Galileo only. The process was wider than himself, yet it can scarcely even be thought of but for his unique contribution. This, as noted, cannot be said equally confidently of any other scientist of the 16th/17th centuries, and thus we find here one major source of the peculiar historiographical vagaries over time of the linkage between Galileo and the Scientific Revolution. To another such source — the remarkable malleability of Galileo's work, which appears to allow such a great variety of radically distinct historical interpretations to which it can more or less plausibly be subjected — I shall come back at the end of this paper.

### *Comte and Whewell*

Although in the 18th century there was certainly a distinct awareness that something decisive had happened to science in the preceding century, no attempt was made to pursue the nature of the transformation in any historical depth. At least two indispensable ingredients for such an enterprise can by hindsight be identified as lacking before let us say 1830 — a genuine sense of historical development, and a unitary view of the process of science. The first two studies which might have yielded an overall interpretation of what we are now used to call the Scientific Revolution are Comte's *Cours de Philosophie Positive* and Whewell's twin studies of the *History* and the *Philosophy of the Inductive Sciences*. However, although Comte attempted indeed to hammer out of history a developmental line of science conceived as a whole, his *a priori* division of this line into three inexorably successive stages, coupled with a remarkably unhistorical bent of mind, turned his effort, if viewed from the historiographical point of view, into little more than a Procrustean bed on which the supposedly historical facts were frequently squeezed to death, or at least to scarcely recognizable caricatures. Thus, what remarkably little Comte has to say on Galileo provides quite as gross an anachronistic account as any to be found in the *Cours* in connection with 17th century scientists — nowhere does Comte betray any familiarity with their original works.

Such positivist distortion of the facts of history was exposed at once by the immensely erudite and far less dogmatic philosopher and historian of science, William Whewell, who could not yet foresee how deeply the historiography of science was to be tinged over the next 100 years or so by positivist disfiguring. In his own work, which continues to deserve our admiration for its scope, precision, and fairness, some core features of its overall organization and general approach nevertheless block a treatment of Galileo in his role of prime instigator of the birth of early modern science. Although Whewell's assessment of Galileo's published works is generally fair enough, it is, firstly, split up over his respective contributions to mechanics, to astronomy, and to the other sciences Whewell successively and in mutual separation dealt with. And secondly, Whewell's mode of history writing lacked a feature we are not in fact to encounter at all prior to the twenties of the present century — it is almost never contextual in the sense of inquiring into the thought of a given scientist of the past as one unified whole, to be understood in the framework of the contemporary state of science. Thus, Galileo's contribution to the science of dynamics is discussed by Whewell in the format of a step-by-step retracing of Newton's successive Laws of Motion, so that at the end we are left with a range of separate statements regarding the question of to what extent Galileo, as part of a chain of 16th and 17th century scientists, participated in the discovery of 1) the principle of inertia; 2) the principle that forces accelerate rather than cause motion, and so on. Moreover, in Whewell's treatment the period of the emergence of early modern science ultimately dissolves into a range of essentially separate 'Inductive Epochs', of which Galileo occupied just one. Finally, in Whewell's view the truly revolutionary figure at the birth of early modern science was neither Galileo, nor Kepler or any other scientist of the time, but rather Francis Bacon — it was chiefly through his prophetic awareness of the new era in science and through his formulation of the true way of how to do science that "he [...] converted the Insurrection into a Revolution."[4]

4. Whewell, W., *History of the Inductive Sciences, from the Earliest to the Present Time* (3 vols). London (Parker), 1857 (3rd ed.). The present quotation from vol. II, p. 41.

*Mach's picture - Galileo as an intuitive empiricist*

All this is how, in Comte's and Whewell's comprehensive studies, a treatment of Galileo's work as a major sign-post in the emergence of early modern science continues to elude us. The first book in which such a potentially fruitful combination is in more or less embryonic fashion to be encountered is Ernst Mach's *Science of Mechanics.* Just as was the case with Comte and with Whewell, Mach's study was not primarily concerned with the past for its own sake. Rather, it is particular views of his own on the philosophy of science which led him to its history and colored throughout his interpretation of the past. His basic aim was to eradicate metaphysics from science by showing that the foundations of science ultimately go back to some basic notions of scientists which are characterized by the following two features: 1) these notions represent an 'economy of thought' in that they are the simplest and most self-evident ones conceivable; 2) they derive their simplicity from an unprejudiced, even 'instinctive', look at the world around us — from an intuitive capability of distinguishing the essential *facts* of nature. This means that Mach's conception of science comes down to a fairly extreme form of empiricism. It is this empiricism which he went on to impose on the history of mechanics in his 1883 book *Die Mechanik in ihrer Entwickelung historisch-kritisch dargestellt* (translated in 1893 as *The Science of Mechanics. A Critical and Historical Account of its Development*).

However, it was the first-mentioned idea — Mach's principle of the 'economy of thought' — that made him turn to history in the first place. Mach's basic interest in mechanics was its axiomatics — somewhat crudely formulated, the simpler one made one's axioms the better. Here is where history could come in usefully: it could provide us with alternative axiomatizations.

From his delvings into the early axiomatics of mechanics, Mach came up with Galileo as the central figure. Whereas Whewell had constructed a very careful balance between elements of continuity and of discontinuity in the history of 17th century science, Mach appears as a clear propounder of discontinuity. Dynamics, he claims,

> is entirely a modern science ... Dynamics was founded by Galileo. We shall readily recognise the correctness of this assertion if we but consider a moment a few propositions held by the Aristotelians of Galileo's time.[5]

5. Mach, E., *Die Mechanik in ihrer Entwickelung historisch-kritisch dargestellt.* Leipzig (Brockhaus), 1901 (4th ed.) (translated by Th.J. McCormack as *The Science of Mechanics. A Critical and Historical Account of its Development.* Chicago (Open Court), 1893). The translation was reedited twice so as to bring up to date with current German reeditions. My page

And similarly, on the discovery of the relation of time and space in free fall,

> We must first remark that no part of the knowledge and concepts with which we are now so familiar, existed in Galileo's time, but that Galileo had to develop this knowledge and these concepts for us.[6]

And again:

> An entirely new notion to which Galileo was led is the idea of *acceleration*.[7]

Now what was it that enabled Galileo to achieve all these exploits basically on his own? Evidently, his keen application of the true scientific method. As noted, this was a topic on which Mach entertained views of his own:

> We see thus, ... that Galileo did not supply us with a *theory* of the falling of bodies, but investigated and established, wholly without prejudice, the *actual facts* of falling.[8]

Of Galileo's method it can be said that

> there is no method of procedure more surely calculated to lead to that comprehension of all natural phenomena which is the *simplest* and also attainable with the least expenditure of both feeling and the intellect.[9]

In Mach's view, Galileo's greatness lies above all in his intuitive grasp of the essential features of the motion of bodies, assisted by experiments. Mach was thus well prepared to take Galileo's own lively accounts of how he arrived at his results quite literally. Mathematics, in Mach's view, comes in only through the backdoor, as it were, of scientific discovery: it is just part of the formalization process *after* the discovery has been made.

The overall picture that emerges from Mach's *Science of Mechanics* is one of a complete break, in the fundamental science of dynamics, with the past. This break was single-handedly effected by Galileo, who was enabled to do so by his pioneer application of the true, empiricist method of scientific discovery. This method relies heavily on unprejudiced observation, on discerning the essential elements among the facts nature displays before us, on experimental testing,

---

numbering conforms to the 6th English edition (1960), which comprises all alterations made by Mach up to 1912. The passage quoted in the text is from p. 151. At various places in this and the following notes I have changed somewhat the wording of McCormack's generally excellent, and in fact authorized translation.

Dudley Shapere, in his book *Galileo. A Philosophical Study*. Chicago UP, 1974, has some interesting points on Mach and Galileo (pp. 4-9).

6. Op. cit., in note 5, p. 159.
7. Op. cit., in note 5, p. 174.
8. Op. cit., in note 5, p. 167.
9. Op. cit., in note 5. p. 168.

and also, in a somewhat undefined manner, on the application of mathematics. It was not, however, given to Mach to adhere to the letter to this picture throughout the seven editions his book went through in between the years 1883 and 1912. For in the meantime an increasing number of facts about the history of mechanics came to light that seemed to clash with the simple but highly persuasive picture of discontinuity as sketched by Mach.

Insofar as these facts pertained to Galileo himself, they derived chiefly from the *Edizione Nazionale* then in process of being published, with all its hitherto unknown manuscript materials. More and more was being published as well about other, earlier figures in the history of science, such as Benedetti and other Italian mathematicians of the 16th century. But a radical break with Mach's picture of the origins of early modern science did not occur until an altogether new collection of archival findings, this time emerging from the darkest recesses of the Middle Ages, led Pierre Duhem to give an altogether different account of how early modern science emerged.

### *Duhem's picture - Galileo finishing the Parisian job*

As is well-known, Duhem, in his *Études sur Léonard de Vinci*, defended the view that early modern science was really born in the 14th century school of the Parisian Terminists — Buridan, Oresme, and a host of other scholastics whose manuscript works Duhem single-handedly returned to the world. In his conception, the crucial feat of demolishing Aristotle's natural philosophy had been accomplished by these highly unorthodox thinkers. In anticipating the principle of inertia; in deriving the law of free fall from graphic representations; in favorably discussing the possibility of the daily rotation of the Earth around its own axis, and in underpinning this hypothesis with a plausible dynamics, these men appeared truly to be "Galileo's Parisian forerunners", as Duhem subtitled the third and most revolutionary volume of his Leonardo Studies (1913).

I cannot enter here into a discussion of the strong and weak points of Duhem's thesis. Suffice it to say that it leaves a fairly small role for Galileo. Even though Duhem continued to regard the classic achievements of Galileo, Descartes, Kepler, Newton, and their contemporaries, as the yardstick against which to measure the performance of his Parisian heroes, true novelty was, in his interpretation, almost exclusively attributed to the latter. Time and again we find that, in Duhem's view, the chief contribution Galileo and his contemporaries made to science is that they elaborated, in a more precise fashion, insights that had been arrived at earlier in a purely qualitative vein. Thus Duhem could state that for two and a half centuries nothing essential was added to Buridan's impetus conception, after which period the theory "was divested of its purely qualitative form in order to be clothed now in a more precise, quantitative form."[10] For Duhem mathematics comes in after the fundamental discovery has been made; it helps to make it more precise, to explicate its meaning, to bring out into the open what is potentially there; yet the core of the discovery lies in the qualitative statement itself. It appears from a careful perusal of the 'Leonardo Studies' that for Duhem scientific discovery is made in the format of separate statements containing loose and basically unconnected insights. To combine them, to make them mutually coherent and consistent, to quantify them, to axiomatize them — all such operations, although certainly relevant to science, are at bottom of a secondary nature.

Duhem's thesis, precisely because he blatantly overstated a radical and beautiful case, gave rise to a major debate over the origins of early modern science, in the course of which the concept of the Scientific Revolution emerged as one by-product. It is during this debate, from the twenties through the early sixties, that most of our present-day key notions of the Scientific Revolution have taken shape. Among the historians who took up the challenge involved in the Duhem thesis, I select Anneliese Maier, E.J. Dijksterhuis, and Alexandre Koyré; among those who developed an influential conception but remained outside this budding tradition, I single out Leonardo Olschki. I shall first devote a few words to Olschki.

### *Olschki's picture – Galileo using and transcending the tradition of technological writing in the Italian vernacular*

In between 1919 and 1927, Leonardo Olschki published a 3-volume *Geschichte der neusprachlichen wissenschaftlichen Literatur* ('History of Scientific Literature in the Vernacular'). These highly detailed, dense, and far from closely argued yet captivatingly written volumes aim at providing at once a literary and an historical analysis of two centuries of writing on mixed scientific and technical matters, such as fortification, gunnery, hydraulics, and the like, in the Italian vernacular. Among the authors he discusses are Alberti, Tartaglia, and Bruno, and the entire effort is made to culminate in Galileo. Although these volumes did not aim to provide a history of the Scientific Revolution, together they constitute a particular thesis on the causes of the emergence of early modern science. It is Olschki's contention that it was this very tradition of mixed scientific/technological writing which, by infusing the blood of daily

10. Duhem, P., *Études sur Léonard de Vinci. Ceux qu'il a lus et ceux qui l'ont lu.* 3 Series, the third one subtitled 'Les précurseurs Parisiens de Galilée'. Edition used: Paris (De Nobele), 1955. Originals: Paris (Hermann), 1906, 1909, 1913. The present quotation is from vol. 3, p. 55.

life into the utterly dried-out and other-worldly disquisitions of the scholastics, made early modern science possible. Thus, in the third and final volume, which is almost entirely devoted to Galileo (with whose works and letters Olschki was intimately familiar), Olschki was concerned to show Galileo's background in matters of contemporary technology — the well-known invocation of the Venetian arsenal right at the beginning of the *Discorsi* being just one example. This is *not* to say that Olschki put forward his thesis in the simplified guise given to it by later, Marxist followers — he did not simply hold that early modern science arose exclusively through a concern with practical matters characteristic of early commercial capitalism. Rather, Olschki stated that the tradition of technological writing, although vital for bringing to science the turn towards the empirical it sorely needed, was nevertheless insufficient for inaugurating the new science, which involved lifting those new, empirical notions to the level of theoretical abstraction. Here is how carefully Olschki qualified his much-misunderstood thesis:

> The problems of the economy of power and of how much machines can achieve; of the accuracy of guns; of the resistance of fortifications, are the very same ones that already through two centuries had found treatment in the technical literature. Galileo, however, considered the tradition of the workshops, with which he had become acquainted through his teachers, primarily as the area of experience and observation, as suitable for drawing the preliminary lines of the theoretical foundations of the mechanical arts. This is why the formulation of those questions is nevertheless fundamentally different, and why their solution is fully independent from this direct tradition of the workshops and of the theorists, even though his attention is drawn time and again towards the possibilities of applying in practice the theories he discovered through speculation and experiment.[11]

Insofar as Olschki's conception of the importance of the arts and techniques of the Renaissance for the emergence of early modern science relies on the characterization of the latter as empirical and observational, he quite consciously followed the footsteps of Ernst Mach. Yet, even though more than a modicum of the latter's positivist approach to the history of science can here and there be found in Olschki's work, Olschki's far more refined historical sensibilities allowed him to a considerable extent to transcend the limitations inherent in Machian historiography. It is remarkable that he did so quite

11. Olschki, L., *Geschichte der neusprachlichen wissenschaftlichen Literatur. Erster Band: Die Literatur der Technik und der angewandten Wissenschaften vom Mittelalter bis zur Renaissance.* Heidelberg (Winter), 1919. *Zweiter Band: Bildung und Wissenschaft im Zeitalter der Renaissance in Italien.* Leipzig/Firenze (Olschki), 1922. *Dritter Band: Galilei und seine Zeit.* Halle (Niemeyer), 1927. (All three volumes were reprinted together by Kraus Reprint, Vaduz, 1965). The present quotation is from vol. 3, pp. 156-7.

independently from Duhem's findings, which were to influence so deeply the formation of historians' views on the Scientific Revolution during the very period when Olschki was writing his volumes — apparently in considerable isolation. For, as a critical admirer of Galileo, Olschki was put off by what he regarded as Duhem's disparagement of Galileo's independent merits through the attribution of prime originality to the Paris Terminists. Thus he could not bring himself to attaching any importance to what the Middle Ages had contributed, beyond sheer erudition, to the origins of early modern science. In Olschki's view, Galileo was primarily a mathematical physicist in the Archimedean tradition, yet what enabled Galileo to overstep the limits of classical tradition and scholastic erudition was precisely the creative elaboration of practical matters for which the newly emerged tradition of the scientific and technical literature in the vernacular had done so much to sharpen his eyes.

To the extent that Olschki aimed at making a general point about the emergence of early modern science, it follows as a natural defect of his thesis that he confined it to Italy. Only if he had made good on his announcement, in the Preface to the second volume, to treat French and, possibly, German scientific and technical literature in a similar fashion, something might have emerged that would have come close to making up an argument about the emergence of early modern science in its full setting. This, however, is a shortcoming that Olschki was to share, as we shall see presently, with more than one later historian of the birth of early modern science.

### *Maier's picture – Galileo inadvertently starting the second revolution*

We return now to the Duhem thesis, and to those historians who, rather than rejecting it out of hand, picked it up and used its fruitful elements for framing accounts of the birth of early modern science all their own. Anneliese Maier's chief contribution to the debate over the Duhem thesis was a thorough-going review of the sources. In a far more critical vein than Duhem she established, not more or less superficial lines of resemblance between Parisian and 17th century statements, but truly contextual accounts of the problems addressed by late scholastic thinkers. In so doing, she in effect became the pioneer of what has since become the subspecialty of 'studies in medieval science'. The overall view which emerged from Maier's erudite studies was that of a double revolution. The entire history of the sciences in Christendom, right from its inception in the 13th century up to the 18th century, should be seen as

> ... a history of the gradual overcoming of Aristotelianism. This overcoming did not take place in one single, grand revolution, as people have thought for a long time.

> On the other side, neither was it a stable and continuous emancipation process evenly spread over the centuries. Rather, it displays a development which was executed in two grand stages, of which the first finds its culmination point in the 14th, and the second in the 17th century.[12]

Yet the two stages were not of quite equal importance. Seen as a whole, Maier admitted, it is certainly permitted to regard the new conception of nature embodied in late scholasticism as a precursor to early modern physics, "that is, in the sense of the first stage of a grand developmental process of which the second *and decisive* [my italics. HFC] stage is to be found in the 17th century."[13]

What happened, Maier suggested, is essentially this. Buridan had developed his impetus theory after he had become fully aware of the shortcomings in Aristotle's account of projectile motion. Similarly, Galileo had not begun to develop an alternative approach to kinematics until he had realized a point Buridan might have become aware of but never had, namely, that the continuing motion of a spherical body rolling over a smooth, horizontal surface cannot be accounted for through impetus kinematics. Reflections on this seemingly trivial phenomenon had made Galileo sort of stumble into the principle of inertia which, in its turn, had introduced him into the idealized sphere of a mathematized physics and thus ushered in his drastic revolution.[14]

*Koyré's picture – Galileo's revolutionary vindication of Platonism*

Alexandre Koyré's view of what defined the birth of early modern science and of Galileo's role in it, although utilizing many of the same ingredients, was nevertheless quite incompatible with Maier's. The core of Koyré's ideas on the subject was that far more had been at hand than just a transition from certain scientific tenets to certain others with a better claim to represent reality. In his view, the emergence of early modern science signified no less than a mutation of human thought. And this mutation, he felt sure, had taken place in the early seventeenth century, when it had been effected by Galileo and Descartes. Between them, these two embodied what, in 1939, Koyré was the first to call the 'Scientific Revolution of the 17th century', a concept somewhat later expanded by him so as to encompass the achievement of Copernicus on the one side, and Newton on the other as well.

Now this revolution came down to a decisive change in world-view all by

12. Maier, A., *Die Vorläufer Galileis im 14. Jahrhundert. Studien zur Naturphilosophie der Spätscholastik*. Roma (Edizioni di 'Storia e Letteratura'), 1949, pp. 1-2.
13. Op. cit., in note 12, p. 5.
14. Op. cit., in note 12, pp. 151-154.

itself, "of which modern physical science is at once the expression and the fruit."[15] Earlier concepts and theories lost their meaning because they no longer made sense in the context of the new world-view; the new concepts and theories at once began to look seductively self-evident for the same reason. Thus the birth of early modern science was not just the emergence of a number of new statements about nature; not even such fundamental propositions as the principle of inertia, or Newton's Second Law, marked the transition in their own right. Their very discovery and subsequent adoption could only be accounted for in the framework of this larger transition, which Koyré described as a fundamentally new overall conception of *motion*. And this transition, in its turn, could only take place in the even wider framework of a new conception of the universe at large. Movement ceased to be regarded as a goal-oriented *process* in our heterogeneous, finite Cosmos — largely through the work of Galileo and Descartes, movement henceforth came to be regarded as a value-neutral *state* of bodies on their way through the homogeneous infinity of Euclidean, geometrized space. This sentence sums up the core of Koyré's conception, to which virtually all his later work on the history of 16th/17th century science was to remain subsidiary.

The revolution, Koyré insisted, was no less a revolution for having been prepared. Granting a measure of preparation to Duhem's 14th century heroes, however, was not the same as ascribing to them such a paramount position as to transform Galileo's pioneering discoveries into little more than an effort in mathematical elaboration. The history of medieval and Renaissance science, Koyré asserted, is to be divided into three stages, which correspond to three distinct modes of thought, even though these do not always obey strict chronological order. These stages are constituted by the Aristotelian, the impetus, and the Archimedean modes of thought, respectively, and they were traversed in condensed form yet in due order by Galileo in the period before his physical thought reached maturity.

Two consequences of this categorization *cum* periodization are important for our purpose. To begin with, we note that right from the start Koyré ties up his views on the upheaval in the science of the 17th century with the work of one of its main propounders — Galileo. Even though an unbridgeable chasm separates Koyré's conception of science from Mach's, he nevertheless shares, on a far more sophisticated and historically responsible level, not only the

15. Koyré, A., *Études Galiléennes*. Paris (Hermann), 1966 (2nd ed.); translated by J. Mepham under the title *Galileo Studies*. Atlantic Highlands (NJ; Humanities Press), 1978. The translations in the text are basically my own. The present quotation is from p. 12 (cp. Mepham, p. 1).

latter's discontinuity view, but also his focus on Galileo as the absolutely central figure. And secondly, it followed for Koyré that impetus physics was relevant to the advance of science mainly in its *failure* to bring about the revolution. Galileo's early thought, as embodied in *De Motu*, demonstrated the ultimate futility of the effort made there to treat impetus physics in the Archimedean way — impetus physics had to be given up before a genuine mathematization of nature was to become feasible. And this mathematization process was of the essence of the overall upheaval in science. It provided the framework for the ideal, abstract world of free fall in the void, of perfectly smooth planes, of completely frictionless motion, and so on, in which Galileo had already begun to delve in *De Motu*, and which was to remain his hallmark for the rest of his life.

Galileo's mature achievement, then, is to be found in his unprecedented awareness that, in Koyré's words,

> The geometrical form can be realized by matter. More than that: it always is. [...] The geometrical form is homogeneous with matter, and this is why geometrical laws have a real significance, and dominate physics.[16]

This is what enabled Galileo to go further than Plato in aiming for a true mathematization of nature. However, Koyré insisted, to say so was one thing; to do it, quite another. Galileo's greatest merit was his following up on his own insight that the only way to prove that physics can be treated in a mathematical fashion is to go ahead and actually *do* it. Koyré then concludes as follows:

> Motion obeys a mathematical law. Time and space are linked by the law of number. Galileo's discovery transforms the failure of Platonism into victory. His science is the vindication ["revanche"] of Plato.[17]

This vindication, he finally observes, was still incomplete: Descartes is to draw the final conclusions. Even though, unlike Galileo, Descartes did not regard himself as a Platonist, both share the idea of science as the reduction of physics to mathematics. But this new Platonism is far removed from the original, because space has now lost the cosmic value it still had for Plato. The new science has won. "But never has a victory been bought at a higher price."[18]

### *Dijksterhuis' picture – Galileo's mathematization at the edge of continuity*

Koyré and E.J. Dijksterhuis were almost exact contemporaries (both were born in 1892; Koyré died in 1964, Dijksterhuis one year later). Not only did

16. Op. cit., in note 15, pp. 282-3. (cp. Mepham, p. 204).
17. Op. cit., in note 15, p. 290 (cp. Mepham, p. 208).
18. Op. cit., in note 15, p. 291 (cp. Mepham, p. 209).

they share 72 years — they also shared the view that the key difference between early modern science and preceding natural philosophies is to be sought in its mathematical nature. I cannot discuss here in any measure of detail the fascinating set of similarities and differences between these non-identical twin brothers in the historiography of science. Suffice it to present as one conclusion of mine that, for Koyré, physics was actually the incarnation of mathematics, which is why the language of mathematics does indeed express the essence of reality, whereas Dijksterhuis held mathematics to be a language by means of which one can describe reality without actually penetrating into its essence.

Another striking difference is that, whereas Koyré proclaimed the revolutionary nature of the birth of early modern science, Dijksterhuis was to stick throughout his career far closer to his originally quite strong allegiance to the Duhem thesis in that he upheld that the entire history of science displayed fundamentally continuous, gradual progress. Throughout both his early masterpiece *Val en worp* (1924, "Free Fall and Projectile Motion') and his celebrated *Mechanization of the World Picture* (1950 in the original Dutch; 1961 in English translation), Dijksterhuis was concerned, in dealing, for example, with Galileo, to disentangle as carefully as possible the elements of the traditional and of the decisively new in his work. And yet a deep ambiguity can be discerned running through all his historiographical work. Despite his frequent protestations of continuity as the hallmark of scientific advance; despite even his persistent refusal ever to use the term 'Scientific Revolution', Dijksterhuis was also deeply convinced of the unprecedented character of the renewal of science that took place around 1600, and of which he regarded, not Galileo and Descartes, but rather Galileo and Kepler as the key figures.

However, there are also strong similarities between the views of Koyré and Dijksterhuis. Both regarded the mathematization of nature as the key characteristic of the process through which early modern science was born. The main function of experiments, they both believed, is to verify whether those phenomena whose existence has been predicted by mathematical deduction do actually take place in accordance with the prediction. As a consequence, neither Dijksterhuis nor Koyré was at all favorably disposed towards taking to the letter Galileo's accounts of his experiments insofar as these were presented as heuristic. Not only did they shrug off Mach's primitive empiricism and its projection backwards onto Galileo's achievement — for them the appeal to the empirical was an altogether second order phenomenon in doing science. One of the key elements of Galileo's greatness was precisely that, in Dijksterhuis' words,

> he did not experiment in order to find a law of nature, but in order to verify *a posteriori* a relation he had deduced by mathematical reasoning from suppositions that appeared more or less self-evident.[19]

In linking thus their appreciation for Galileo with their own views on what constitutes science, Koyré and Dijksterhuis obviously put at risk these views from the moment on when documentary or experimental evidence might show that Galileo had perhaps been far more of an heuristic experimenter than either was willing to allow. It is this issue which, as we all know, has come to dominate so much of present-day discussions of Galileo's actual performance.

*The connection all but severed*

When, in 1961, Thomas Settle let a reconstructed ball roll over a reconstructed inclined plane in allegedly precise accordance with the instructions given by Galileo in the *Discorsi*, and acquired an outcome very close to the one claimed by Galileo, he in that very act ushered in a debate over the true role played by experiment in Galileo's work that is still with us today, and which is united over just one thing, namely, that Koyré had been wrong. Following up on Settle's 'Experiment in the History of Science',[20] a renewed interest was taken in Galileo's scrap papers, and evidence was sought for·and against his having found at least some of his quantitative laws through actual, not thought experiment, employed heuristically, not for verification purposes only. The ensuing debate has over time become exceedingly technical, and is hard to follow for outsiders who do not possess instant recall of 'folio 116 verso' and like pet items of the aficionados. It is out of the question to deal with it here. For our purposes it suffices to note one important consequence of the debate in question.

This is that the very technicality of the debate has tended to rob the issue over which it is fought of the core importance it had rather uncritically been assigned from Mach down to Koyré and Dijksterhuis. Somehow, and strongly supported by Koyré's (and Dijksterhuis') own wording, the issue of the nature of early modern science had become closely linked up with the question of whether Galileo had actually carried out his experiments, and, if so, to what purpose. But, even if we fully accept the central position Galileo holds in the process of the Scientific Revolution, this simply puts more weight on the narrower issue than it can possibly bear. With this heavy load removed, the wider debate on the nature of the Scientific Revolution could now be continued in a less overcharged and perhaps healthier atmosphere.

19. Dijksterhuis, E.J., *The Mechanization of the World Picture*. Oxford (UP), 1961, IV:98.
20. Settle, Th.B., 'An Experiment in the History of Science'. *Science* 133, 1961, pp. 19-23.

However, the process of loosening up the overly intimate connection between the interpretation of Galileo's thought and the nature of the Scientific Revolution did not stop here. Increasingly we see that, since the onset of the sixties, the subspecialty of Galileology and attempts at defining what the Scientific Revolution was all about have gone their separate ways. Even though such a trend-setting scholar as Drake surely voiced opinions on the Scientific Revolution, and although scholars like Butterfield, the Halls, Kuhn, and Westfall produced variously tinged accounts of the Scientific Revolution, I am not aware of any attempt to incorporate, in the grand manner of a Koyré or a Dijksterhuis, a plausible, up-to-date overview of Galileo's achievement into a synthetic conception of the birth of early modern science.

What do we have instead? As mentioned, beside the products of specialized Galileo research we have a number of fairly divergent accounts of the Scientific Revolution which, however, tend to present fairly similar pictures of Galileo — essentially toned-down versions of Koyré's interpretation of his achievement, with a strong emphasis on the mathematical element but allowing somewhat more room for heuristic experimentation, and rather divergent views on lines of continuity — or lack of it — between Galileo and previous developments in science. Clearly, decades of ongoing research have made it immensely difficult to put forward a synthetic conception of the Scientific Revolution. That this is so, will now be illustrated by means of a very quick overview of some relatively recent attempts at *explaining* rather than at *defining* the Scientific Revolution, and of how these relate to Galileo.

### *The Randall/Wallace thesis – Galileo as the only scientist involved*

My first example is what, with due regard for the findings of Carugo and Crombie, may nevertheless be termed the Randall/Wallace thesis. In the original form given to it by Randall in 1940, it comes down to the idea that Galileo's science found its origin in methodological innovations at the University of Padua culminating just one generation before Galileo taught there. Lately, William Wallace has considerably shored up the thesis by showing in great detail how Galileo became familiar with this, essentially Aristotelian tradition through Jesuit professors of the Collegio Romano, and also how, through the same venue, Galileo may have acquainted himself with insights going back to the 14th century schools of Paris and Oxford — thus providing a missing link in Duhem's argument as well.[21]

21. Wallace, W.A., *Galileo and His Sources. The Heritage of the Collegio Romano in Galileo's Science*. Princeton (UP), 1984. Idem, 'Randall *Redivivus*: Galileo and the Paduan Aristotelians', *Journal of the History of Ideas* 49, 1, 1988, pp. 133-149.

Connections like those established by Randall and Wallace are not without their questionable sides, for example, the knotty issue of what methodological considerations may have to do at all with the transformation of science in the 17th century. Important for present purposes, however, is that, besides seeking to establish a specific form of continuity between Galileo and his predecessors, both Randall and Wallace claim something even grander as well. Randall was particularly emphatic on the issue, witness the title of his booklet in question, *The School of Padua and the Emergence of Modern Science.* In view of the Aristotelian convictions of his Padua methodologists, Randall went so far as to claim

> that the most daring departures from Aristotelian science were carried on within the Aristotelian framework, and by means of a critical reflection on the Aristotelian texts,[22]

which is rounded off with the daring flourish (in obvious allusion to Dante's characterization of Aristotle in the first circle of the Inferno):

> The 'father' of modern science, in fact, turns out to be none other than the Master of them that know.[23]

The chief problem with such a claim (and Wallace substantially subscribes to it, although he phrases it far more modestly) is that it is by its very nature confined to producing an Italian connection. Even if it could be proven beyond dispute that Aristotelians' reforms in Aristotelian methodology stood at the origin of Galileo's performance, this still means nothing for linking either Kepler, or Descartes, or Beeckman, or Gassendi, or Mersenne, or whomever you like around 1600, to reforms in Aristotelianism. Thus what aims to present itself as a contribution to our insight into the emergence of early modern science is in reality no more (though, of course, also no less) than a more or less plausible attempt at illumination of where Galileo came from.

### *The Yates thesis – Galileo as the protagonist left out*

Exactly the reverse seems true of Yates' thesis on the Hermetic influences on the Scientific Revolution which, in the stormy sixties and seventies, shook the profession almost to its very foundations. There are two sides to the Hermeticist coin. Partly, the thesis claims to supplement more conventional accounts of the Scientific Revolution by pointing to the persistence, through-

22. Randall, J.H., Jr., 'The Development of Scientific Method in the School of Padua', in: *The School of Padua and the Emergence of Modern Science.* Padova (Antenore), 1961, pp. 13-68. The quotation is from p. 63 in the book edition.

23. Op. cit., in note 22, same page.

out the 17th century, of 'Rosicrucian' elements which only through the application of anachronistic standards can be distinguished from science. But another, earlier side to the Yates thesis is the claim that early modern science owed its birth at least partly to the Hermetic body of thought that found such an outspoken articulation in Giordano Bruno. Whatever one may think of the validity of their specific claims, we can at least note that Yates, Rossi, Webster, Debus, and scores of other partial or wholesale adherents to the 'Hermeticist' approach managed indeed to link Copernicus, Paracelsus, Tycho, Kepler, Bacon, Descartes, Mersenne, Newton, and many other traditional protagonists of the Scientific Revolution with one typically Hermetic thought pattern or another. But what is missing almost completely from the Yates thesis is those perhaps rare, yet highly important, tough, 'no-nonsense' scientists, like Galileo, or Huygens, whose scientific biographies display no trace of coloring by whatever mystical or magical current of the time. For many pioneers of the Scientific Revolution the association with magical operations and with a mystical conception of the world is far from altogether far-fetched; for Galileo it would be totally out of place. Not that Yates even attempted to make him fit such a picture — she knew very well that such an exercise would be doomed to failure from the outset. And thus, whatever the Yates thesis may be said to have explained about the Scientific Revolution, the towering figure of Galileo remains outside it. Which, to say the least, does not serve to strengthen the force of the thesis to the extent that it claims to pertain to the Scientific Revolution in its entirety.

### *The Reformation and the rise of early modern science – Galileo as a stumbling block*

In the ongoing debates over whether and, if so, how precisely early modern science was shaped by religious forces, claims for the particularly favorable influence of Protestantism stand out. One such claim has become known as the Merton thesis, through a generalization of R.K. Merton's original argument (1938) that the more than proportional participation of Protestants in 17th century English science is due to a particular, Puritan ethic which fostered the cultivation of science. By many historians this thesis has been taken as a causal explanation of the Scientific Revolution. A different, though related view was put forward by R. Hooykaas. In 1972 he argued that the Biblical tradition, brought to the forefront of Western European thought through central tenets of Reformed belief, supplied natural science as hitherto cultivated in a predominantly Greek spirit with precisely the vital elements needed for early modern science to arise. In accordance with his own, radically contingent conception of science, Hooykaas defined these vital elements as a new respect for the factually given over *a priori* rationalizations of nature.

Although striking illustrations for their respective theses were adduced by both Merton and Hooykaas from among Protestants scientists of the 16th and 17th centuries, it is fairly evident that at least some of the features they find particularly characteristic are to be encountered as well in such Catholic protagonists of the Scientific Revolution as, obviously, Galileo. Was not Galileo's approach characterized by a similar awareness of the ultimately contingent nature of reality as Kepler is justly praised for by Hooykaas? Was not Galileo's devotion to science governed by motives quite similar to those plausibly derived from the Puritan ethos by Merton? Theodore Rabb, in particular, has argued on such grounds that the Merton thesis should not therefore be connected to the *birth* of early modern science, which was as much of a Catholic as a Protestant affair, but rather to the subsequent establishment and institutionalization phase of the entire process. Even though revisions along such lines are, in my view, far from rendering Merton's and Hooykaas' arguments worthless, it remains true that Galileo, for one, will continue to stand as a major stumbling block in the way of overly rigid historiographical constructions in this still hotly disputed area.

*The emergence of the social role of the scientist – Galileo fits the overall picture*

It might now almost seem as if Galileo stands in the way of whatever explanation of the birth of early modern science has so far been put forward, either as the only fitting case or as an insuperable exception. Fortunately, this is not quite the case. I shall now bring my exposé to a close by briefly discussing two from a number of approaches which at least do not suffer from this particular defect. The first one I select is Joseph Ben-David's sociological analysis of the Scientific Revolution — an analysis in which elements of the descriptive and the explanatory are almost undistinguishably intertwined.

Ben-David's key concept is that of the social role of the scientist, the idea, that is, that, in order for science to become an ongoing and persistently growing, relatively autonomous activity in its own right, scientists have been in need of a recognition in the wider society, borne by clearly distinct and sufficiently influential social strata, of the value of what scientists do. In Ben-David's view, the Scientific Revolution might well have remained an affair of one generation only, since the social basis upon which it got started in Italy provided insufficient sustenance to keep the movement going:

> Science and scientists remained dependent on the narrow circles within the upper classes that ruled both the country and the church and were interested in learning.

> These were the circles which had to be convinced that natural science was important and worthwhile enough for them to give to it the full blessing of public recognition and freedom of communication notwithstanding the weighty doctrinal difficulties that might arise from such recognition. The circles were not convinced however. Indeed, it could not have been otherwise at a time when the argument in favor of the Copernican theory was still inconclusive, and when science could offer no more than a few bits and pieces of intellectually interesting astronomical and mechanical theories and the unbounded confidence of Galileo's prophetic genius. In contrast to natural science was the vast body of learning, wisdom, and beauty represented by contemporary humanism and theology. As long as those who had to be convinced by the proponents of science were men who believed in the main depositories of traditional learning, the scientific movement was doomed to failure. For those in the upper class circles who mattered, and probably for the majority of those who did not, science was an intellectually and aesthetically second rate activity as well as a morally and religiously potentially dangerous one.[24]

Ben-David then goes on to show that, in the socially far more fluid situation in North-Western Europe, the scientific movement, when after the Galileo trial it definitively migrated to the North, met with a far more durably sympathetic reception. Crucial in his subsequent account, in which I shall not here follow him any further, is the emergence of an ideology for science which had failed to come into the making in Italy, and which had been lacking as well in ancient Greece.

Whatever the merits of Ben-David's frequently neglected or misunderstood account of the emergence of early modern science may be thought to be, it seems to me certain that, in addressing the Scientific Revolution in its full geographic and temporal scope he made it free of the type of shortcomings we have been addressing in the present context. Taking Ben-David's account as a starting point, one could go ahead and place Galileo's tireless effort to win over the aristocracy and the church for his views into its proper context. A first attempt to do precisely that has recently been undertaken by R.S. Westfall.

### *Patronage as a pervasive pattern in the Scientific Revolution – Galileo as a point of fresh historiographical departure*

In discussions of the Scientific Revolution, accounts of the financial means of scientists in the 17th century have rarely been raised to a higher level than the anecdotal and the biographical. Thus, we have known all along that certain

24. Ben-David, J., *The Scientist's Role in Society. A Comparative Study*. Englewood Cliffs (NJ; Prentice Hall), 1971. (The 2nd ed. came out with Chicago UP, 1984, and has a new introduction). The present quotation on p. 65.

individual protagonists of the Scientific Revolution were in a position to pursue their interests while living on independent means, or through close association with a court or with the church. However, no systematic survey has ever been made of what might somewhat roughly be called the funding of the Scientific Revolution. It seems very likely that the concepts of patronage and clientelism, as short-hand expressions for a pervasive social reality of the 17th century, may go a long way towards providing the framework for a unitary account of a vital aspect of the Scientific Revolution. Feingold, in a study of science at the Elisabethan universities, and Westfall, in a preliminary study of how Galileo's relations with the Medici court affected his scientific work, have begun to show the way.[25] Westfall, in particular, has demonstrated through a case study obviously applicable to a host of cases more or less similar to Galileo's, how deeply matters of priority, of publication habits, and the like, were embedded in a social reality best described as a patron-and-clients system. From this point of view, Westfall rightly claims at the end of his pioneering article, a far more promising approach to the social analysis of the Scientific Revolution may be gained than if one insists on squeezing this unique event onto the Procrustean bed of an obsolete, 19th century *a priori* framework of social interpretation.

### *Conclusion*

We have come a long way now. We have seen Galileo appear in the guise of an intuitive empiricist; of finishing what Buridan had begun; of absorbing and transcending the Alberti/Leonardo/Tartaglia tradition; of the acute observer who stumbled into the wholesale overthrow of the Aristotelian bulwark; of the revolutionary who, together with Descartes, shifted physics from the Aristotelian cosmos to the infinity of Euclidean space; of the man who, together with Kepler, led the way towards the mathematization of the world picture; of the inheritor of Aristotelian methodology; of the failed Hermeticist; of the Roman Catholic scientist; of the man whose attempt backfired to win over to the sake of science leading circles in 17th century Italy; of the client of the Medici, eager for patronage. So the question may not be altogether amiss, Will the real Galileo please step forward?

Note, I may beg you, that the striking variety of Galileo pictures we have worked our way through is still confined to no more than a selection of

25. Feingold, M., *The mathematicians' Apprenticeship. Science, Universities and Society in England, 1560-1640.* Cambridge (UP), 1984. (See in particular the last chapter).
Westfall, R.S., 'Science and Patronage. Galileo and the Telescope', *Isis* 76, 281, 1985, pp. 11-30.

accounts gleaned from far wider-cast analyses of the Scientific Revolution as a whole, in the formation of which Galileo's work either provided the prime stimulus or had somehow to be assigned a proper place.

It is clear that, but for the remarkable malleability of Galileo's work, such a wide variety of views would have been inconceivable. For, while admitting that my question is of course naive and that the 'real' Galileo is to elude us forever, there is a bit of him in virtually all these pictures. More in some than in others, to be sure, yet none of all these conceptions of what Galileo signified for the emergence of early modern science is totally silly or rejectable out of hand. One can indeed read Galileo as a Platonist, as an empiricist, as a theoretical technologist, and so on and so forth, without being fully unjust to him. More than any other pioneer of the Scientific Revolution, Galileo embodies the full scope, depth, and complexity of the event. In that sense it is indeed true to say that, three hundred and fifty years after the *Discorsi* appeared and some hundred years after Galileo was put at the center of the first definite interpretation of the Scientific Revolution, Galileo has stood, and will continue to stand, as the major touchstone for whatever new conception has been, and is going to be, put forward. His very versatility will never, I may safely predict, fail to encourage historians of science in their continuing search for a better understanding of this singular event in the history of Western Europe.

# ITALIAN AND DUTCH UNIVERSITIES IN THE SIXTEENTH AND SEVENTEENTH CENTURIES*

*Hilde de Ridder-Symoens*

I consider this paper as an introduction to the topic of this conference, namely the scientific relationship between Italian and Dutch scientists in the early modern period.

I do not intend to go into the historical debate over to what extent the Scientific Revolution took place in or outside the universities. Nevertheless it seems to me useful to present to this audience the universities as institutions of teaching and research during the Scientific Revolution. I will concentrate on the structure and population of the universities in the Netherlands and in Italy; I will not deal with the content of teaching and research.

The main centres for scientific and medical education and research in Europe in the 17th century were without doubt the Dutch Republic and Northern Italy, and more specially Leyden and Padua. Montpellier was losing its outstanding reputation and international audience, and we have to wait until the 18th century to see Edinburgh, Vienna, and to a certain extent Paris, becoming internationally reputed centres for medical learning.

The quality of the Dutch and Italian universities was not a consequence of similar structural factors or teaching methods which they had in common. On the contrary, both universities had a very different background and organisation, as we shall see. What Leyden and Padua had in common was their political environment. Both cities belonged to a republic whose government had different conceptions about academic and scientific freedom and about church influence. This stands in marked contrast to the more centralised and

*The term "Netherlanders" is not currently used in English, even though it appears in dictionaries. Nevertheless I intend to use it in this paper in order to avoid confusion. I will deal with three different groups of students. 1) The inhabitants of the Low Countries, the so-called Seventeen Provinces: in order to avoid long paraphrases I will call them *Netherlanders*, thus differentiating them from 2) the *Dutchmen*, being the inhabitants of the Northern Netherlands, the Dutch Republic. 3) For the inhabitants of the Southern Netherlands, who do not occur often in this paper, I will use an appropriate paraphrase.

absolutist monarchies and principalities existing at the time. However, it does not mean that the authorities of the Leyden and Paduan universities were completely independent from their state governments. It is a question of a like-minded policy of the government and the university even when the Dutch universities had a Calvinistic stamp and the Italian a Catholic one. Moreover, Catholic students visited the Protestant Dutch universities and Protestants the Catholic Italian studia, albeit that in Italy Protestants were largely confined to Padua, and in a lesser degree to Siena, while in the Netherlands Catholics are found mainly in Leyden.

## *Structures*

It is well-known that the mediaeval universities were divided into two types: the student university concentrated around the Mediterranean and the masters' university dominating North-Western Europe. Some universities, as e.g. Orléans and Prague, can be considered as mixed forms.[1]

The *universitas scolarium* or student university was born in Bologna, the university which claims to be the oldest *studium generale* of Europe and which celebrates this year its ninth centennial with great lustre. The other Italian universities followed the Bolognese pattern, albeit in adapted ways; one can discern some clusters of subtypes (see table 1). The most elaborate form of the student university is limited to Bologna and Padua. The *studium generale* consisted of two universities, the *universitas legistarum* (law faculty) and the *universitas artistarum et medicinorum* (faculty of arts and medicine), *universitas* meaning here a community of students. The faculty of theology was in the hands of the mendicant orders and existed at the margin of the *studium generale*. Each of these faculties was again divided into two "universities", the *universitas citramontanorum*, grouping the students from the Italian Peninsula, and the *universitas ultramontanorum*, housing the students from beyond the Alps. Both were subdivided in nations, according to smaller geographic areas. The number and the names of these nations altered regularly in the course of time. These variations were the expression of a changing geographical recruitment of the students. What is important in the present context is that the

1. General information on university history can be found in Rashdall (1936), d'Irsay (1935), Bannenberg (1953), Bellomo (1979), Kibre (1948), Kibre (1961); on the history of science and medicine in Corsi and Weindling (1983). For a survey of recent publications on Italian universities see Denley (1981). Data concerning the organisation of the Italian *studia generalia* can also be found in the introductions of Weigle's publications and in monographies of the different universities such as in Ermini (1971), Giard (1983-1984), Schmitt (1972), Simeoni (1940), Zdekauer (1894).

German nations which also housed the students from the Low Countries were, till the end of the 17th century, the most populated nations for not-Italians. Both the law university and the arts university or faculty as we may call it nowadays to avoid confusion, had a rector, a *commilito* elected by the *consiliarii* or presidents of the nations. The professors did not belong to the *universitates*; they were organised into *collegia doctorum*, with strong links to the City. In the 16th century the students lost most of their power. The structures and the forms remained but the *studium generale* came more and more under the immediate control of the city government.

The other Italian *studia generalia* did not have sufficient students to possess this very complex organisation.

Siena and Ferrara can be considered as a second cluster. In Siena the *universitas scholarium* had a less complicated structure. The legal faculty was divided into *citramontani* and *ultramontani*, each with several nations, among which a German nation. The *universitas artium et medicinorum* had no nations. In Ferrara, on the contrary, both universities were divided into nations but not into citramontane and ultramontane universities. Besides, many of these nations disappeared in the course of the 16th or 17th century because of a lack of students.

A third group of universities had a still more elementary structure, as e.g. Pisa and Perugia. Students in law, arts and medicine were concentrated in one *universitas scholarium*. Probably only the law students had organised themselves into a few nations. In Perugia there existed 7 nations, 4 for Italians and only 3 for *ultramontani* (Germans, Frenchmen, and Catalans). In Pisa it is not even sure whether the one ultramontane nation had sub-nations.

The fourth cluster of *studia* approach more closely to the Northern pattern, with three faculties (law, arts and medicine, and theology), e.g. Rome, Fermo, Parma. Their recruitment was too regional to permit the creation of nations.

This variety of patterns is the expression of the popularity of these *studia generalia* in and outside Italy. In the late 17th and 18th century, when the *peregrinatio academica* came to an end, many nations fused together or ceased to exist. As we noticed for Padua, Bologna, and Siena, the few law and medical students originating from the Germanic speaking countries or from regions under the political influence of the Holy Empire had the tendency to merge together in one association, looking more like a confraternity than like a mediaeval nation with wide powers in university government.

All the nations had almost the same organisation. *Consiliarii* or *procuratores* (presidents), sometimes vice-presidents, *assessores, syndici*, a treasurer, a librarian, a *bedellus*, etc. ran the nation, naturally with the democratic approval of the general assembly of all the members who payed their entrance fees. Till the end of the nation's existence the *consiliarii* chose the rector of the university. Fortunately for us the German nations kept their records very carefully (*matricula* and minute books) both in Italy and elsewere in Europe where they existed. The German nations are therefore the best-known student associations of the ancien regime.

By contrast, in Northern Europe the university structures were based on the Paris model, more corresponding to our present type of universities. *Magistri* or teachers played the dominant role; they ruled the universities and they chose the rector and the other officers of the university board from their midst. A *studium generale* consisted, besides any exceptions, of four faculties, the lower propedeutic faculty of arts and the three higher faculties of theology, law, and medicine, the last one being the poor relation of the university till the late 16th century. In Italy the law faculties were also the most populated of the *studium* (in a proportion of 4/1 or 3/1 of law against arts and medical students). About 1500 the arts faculties in the North dominated the universities with a average of 75% of the students, the medical faculties, at best, had a maximum of 5%; in some German universities they did not even reach 2%.[2] From the late 16th century things began to change. Thanks, *inter alia*, to the better quality of the grammar schools, less students attended the arts faculty, and when they did, it was for a very short period before going on to a higher faculty. Moreover, medical studies gained more and more prestige and became therefore more attractive for the social elites. They studied natural sciences and medicine not with a view to carrying on a profession, but as an interesting intellectual activity. Therefore medical studies were often combined with law or even theology in the late 16th and 17th century.[3]

A more striking difference between the North-European and Italian medical faculties lies in the number of chairs and ordinary professors. Already in the 15th century Padua had ten chairs for medicine (four for theoretical medicine, four for practical medicine, two for surgery and anatomy). Paris, and even Leyden around 1600 contented themselves with two or three chairs. Most of the smaller universities only had one or two full professors for medicine.[4]

Another important difference between the Northern and the Southern university system consists of the number and importance of the nations.[5] They

2. Schwinges (1986) 468-470; Brockliss (1978) 515; Nauck (1954) 175-186. See also Eulenberg (1904) and Frijhoff (1979), Chartier and Revel (1978).
3. Wansink (1981) 36; Frijhoff (1981) 40-41; Frijhoff (1983).
4. Lindeboom (1972) 35-36.
5. General information in Kibre (1948) and De Ridder-Symoens (1987).

played a less important rôle North of the Alps except in the mixed types, such as Orleans and Prague. In Paris the faculty of arts was divided into four nations (French, Anglo-German, Picard and Norman) grouping masters and students of similar origin. In the 16th century the Paris alma mater was no more the international centre of philosophical and theological learning. The number of foreigners decreased constantly and the nations lost their significance in the academic government. They evolved more into confraternities or *Burschenschaften*, as did the few nations on other universities (e.g. Louvain) as was later also to happen in Italy.

The first university of the Dutch Republic, founded in Leyden in 1575, had no nations at all, even although it followed in its structure and organisation the traditional mediaeval pattern of Northern Europe (Paris, Cologne, Louvain). The university was composed of the four traditional faculties (arts, theology, law and medicine) and was governed by a rector, *assessores* and *curatores*. The rector was appointed annually by the Prince, or in his absence by the States of Holland; the four *assessores* were chosen from each faculty and the two or three *curatores*, properly qualified persons from outside the university, were appointed by the States of Holland. The academic matters were treated by the *senatus*, composed of the *professores ordinarii* lecturing at Leyden university, and presided over by the rector. In the matters of the *forum academicum* (university law-court), tax-exemption, etc., Leyden followed the general trend. In the spirit of the age, the university even started by asking an oath professing the Protestant religion, the counterpart of the Catholic *professio fidei* required by the Council of Trent in 1564. Since the States of Holland noticed that this oath kept Dutch and foreign students away from the young university, the oath was abolished in 1578, "in order that everybody may be spiritually free".[6] The success was immediate. The number of the immatriculati increased from 26 in 1577 to 84 in 1578 and to 110 in 1590.[7]

The other Dutch universities created in the next 75 years (Franeker 1585, Groningen 1614, Utrecht 1636 and Harderwijk 1648) took over the Leyden model, albeit with some local adaptations or variations.[8] Utrecht, for instance, abolished the tax exemption on beer and wine in 1657, for reasons of abuse.[9]

6. Jurriaanse (1965) 15.
7. Calculated on the basis of the matriculations in *Album Stud. Leyden* (1875).
8. Frijhoff (1981) 14-18.
9. Roelevink (1986) 18.

One of the essential differences between the academic structures in the Dutch Republic and Italy is the place of the *artes liberales* in the university curriculum. The reception of the *corpus Aristotelicum* in the 13th century in the West profoundly changed the content of the arts faculties. Depending on the emphasis universities in the Middle Ages put on theology (e.g. Paris, Oxford) or medicine (e.g. Bologna, Padua, Montpellier) the propedeutic arts faculties diverged into different traditions, the theological and the medico-scientific ones.[10]

The Dutch universities had indeed taken over the North-Western pattern, but they were not bound by the mediaeval traditions in teaching and research. Therefore the professors, curators, and other authorities dealing with higher education could introduce the medico-scientific oriented system into the faculties of arts, insofar as these faculties could fit new disciplines and scientific methods into the straitjacket of the mediaeval university structures. If they could not fit, there remained only the possibility of inventing alternative forms of education. This is what happened in the Northern Netherlands in the late 16th and 17th century. The burghers of the new Republic tried to harmonize traditional theoretical teaching and scientific research with practical application, that is to say with applied research and experiment. The result was that the separation between *artes liberales* and *artes mechanicae*, essential for mediaeval intellectuals, gradually disappeared and technology became as important as book knowledge. Those studies of sciences that were essential for warfare (the Dutch Revolt!), commerce and navigation received the special interest of scientists and teachers. It is not my task to go into the content of teaching and research, but one will immediately have in mind Simon Stevin (1548-1620) and his engineering school, or the less well-known Adriaan Metius (1571-1635), professor of mathematics, who taught navigation, geodesy, and fortification in Franeker around 1600. He followed Stevin's example and lectured in Dutch, but his courses were more integrated into the university curriculum.[11]

In fact, the more practically orientated Netherlanders at Leyden and other Dutch universities followed the existing tradition of Italy where already in the 15th century engineering studies had evolved on a more scientific basis (mathematics and geometry).[12] Scholars worked together with cartographers,

10. Schmitt (1975) 488; Bullough (1972) 62-69; Randall jr. (1961) 20-27.
11. Van Berkel (1985) 14-34; Snelders (1983) 213-221.
12. Keller (1972) 11-27; Van Berkel (1985) 17-18; Siraisi (1987) 106-107; Hall (1969).

surveyors, instrumentmakers, etc., and were influenced by the latter's empirical knowledge. An analogous evolution took place in the field of the Arts and Literature (printing!) as is pointed out by P. Burke.[13] From the late 15th century scholars and highly skilled craftsmen from the North went to Italy, to the source of humanism, and brought back home ideas and techniques. In and outside Louvain university there evolved a similar group of scholars and technicians, with Mercator, Gemma Frisius, Stevin and others. These *Fiamminghi* as the Netherlanders were called in Italy did not hesitate to stay in Italy if their own country did not give the opportunity to follow their proper scientific calling as they saw fit. Vesalius left Louvain for Padua and Jozef van Goethuyse alias Giuseppe Casabona (ca. 1515-1596) from Munsterbilzen (diocese of Liège) was the inventor and supervisor of the botanical garden in Pisa.[14] As a result of the Dutch Revolt, and the brain drain to the young Republic, Leyden took over the torch of learning from Louvain and became the European centre for experimental and applied sciences.[15]

Since arts and medicine were structurally linked in Italy, it is easy to understand that in medicine too a bridge was built between book-science and empirical handicraft. In the Middle Ages surgery was already integrated into the curriculum of the Bolognese and Paduan medical schools, and surgeons were not kept away from academic teaching as was the case in the faculties of medicine in Northern Europe.[16]

Despite the fact that the arts faculty in Leyden was independent of the medical faculty, mentality, teaching and research conceptions of both Leyden and Padua were similar. Natural sciences were taught both in the faculties of arts and medicine. Consequently, *doctores medicinae* were all-round scientists (mathematics, natural sciences, natural philosophy, medicine). Students often combined lectures in arts and in medicine and they took degrees in arts and medicine simultaneously. The quadrivium of the arts faculty was no longer considered as a preparation for medical studies but as a full study in its own right.[17]

As has already been mentioned, in Padua emphasis was laid on practice, technology and experience (*experientia sensualis est mihi auriga*)[18] in medical education. This practical and experimental approach called for a special

13. Burke (1987) 43-87.
14. Goethuys (1985) 96-99.
15. Van Berkel (1985) 14-15.
16. Buck (1987) 280-281; De Moulin (1987) 40-46, 97-98.
17. Wansink (1981) 26; Roelevink (1986) 27ff.; Frijhoff (1981) 42ff.; Palmer (1983) 32; Schmitt (1975) 36-37; Hackmann (1975) 96.
18. Fichtner and Siefert (1978) 39 (Jacopo Berengario da Carpi 1530).

infrastructure. The creation of the hortus botanicus, first in Padua and Pisa circa 1545 and in Bologna in 1567 stimulated botany to evolve from a materia medica to a full independent discipline.[19] Nevertheless, in spite of its long tradition and famous professors of anatomy (Vesalius, Colombo, Falloppio) the first *theatrum anatomicum* was only built in Padua in 1594 by Gerolamo Fabrizi da Acquapendente (1533-1619). And the same concern for a more practical education of future physicians stimulated Montanus in the 1540s to bring the students to the beds of the sick in the Ospedale di San Francesco. The clinical teaching was combined with *consilia*, practical advice in specific cases. Already in the 14th century, Italian legal and medical scholars were famous for giving *consilia*. It was a welcome supplement to their rather low salaries.[20]

*Alumni* from the Italian *studia generalia* brought this pedagogical system to the North and if possible introduced it into their universities.[21] The first professors in the sciences and medicine in Leyden were Italian, and often Paduan *alumni*: Gerard Bontius (1536-1599) was a MD of Padua; Petrus Forestius (1522-1597), studied at Bologna, Ferrara, and Padua, and got his medical degree at Bologna; Rembertus Dodonaeus (1517-1585) made an *iter italicum*. In 1597 Rudolf Snellius (1546-1613), a graduate from Italy began a course of lectures in mathematics at the request of the students. The botanical garden was created in 1587 in an open field behind the Academy building at the Rapenburg and is still worth walking around. Dissections were already carried out from 1587, while the anatomical theatre was built in 1593, a year before the Paduan one, at the request of Peter Pauw (1589-1617) who was acquainted with Fabricius ab Aquapendente.[22] The cabinet of curiosa in the theatrum became a touristic attraction. Jan van Heurne (1543-1601), MD of Pavia in 1571, already wanted to start with clinical teaching in 1591; he asked the curators, albeit without success, for an appropriate clinical infrastructure. The Leyden medical faculty had to wait till 1637 to set up a university hospital. Despite this delay the *collegium medico-practicum* remained the only academic hospital in Europe till 1745 and got international fame with Franciscus dele Boë Sylvius (1614-1672) and Herman Boerhaave (1668-1738). The Leyden curators were stimulated by the fact that Wilhelm Stratenus (1593-1681), MD of Padua in 1613, had started clinical practice in 1636 in the newly erected

19. Schmitt ([2]1975) 39-44.

20. Dawson (1968) and Fischer (1961); general information in De Ridder-Symoens (1987) 21.

21. Data on individuals can be found in Lindeboom (1984) and Van Berkel (1985). See further Wansink (1981) 35-36 and Kroon (1911) 45ff.

22. Gogelein (1975) 100-103 based on H.J. Witkam, *De bouw van de anatomieplaats in Leiden*, archief U.B. Leiden. (Information from A.M. Luyendijk-Elshout).

university of Utrecht, wholly in the spirit of his Paduan masters, as is clear from his inaugural lecture.[23] The other Dutch universities were slower in introducing new teaching methods and techniques; in general their teaching remained rather traditional and theoretical till the end of the ancien régime.[24] That is the reason why most of the medical doctors in the Netherlands had studied at Leyden university even when taking their degrees elsewhere. Leyden was the European centre *par excellence* for experimental research: experimental physics was introduced by Willebrord Snellius (1580-1626); the use of physiological experiments for testing a hypothesis started in Leyden[25]; a *laboratorium chymicum*[26] and a *theatrum physicum* were instituted in the 1670s.[27]

Looking at the matriculation and promotion lists of Leyden and Padua we must conclude that Leyden took over the role of Padua in the course of the 17th century as the international academic centre for scientific learning.

This does not mean that scientific research was banned from the Peninsula — Galileo belies it — even when travellers agreed that such formerly famous universities as Padua and Bologna were "at present much decay'd thro' the Bigottry of the Professors ... and all advancement of Arts and Sciences pass with them for useless novelties and meer whimsies" (dixit E. Veryard about Padua in the 1680s).[28]

The transfer of the centre of experimental research to the North was also induced by the fact that in the early modern period the main current of Italian thinking was more dominated by universalistic philosophy than by sceptical ideas and the natural sciences (e.g. Campanella, Bruno).[29] According to L. Belloni[30] the experimental mentality broke through in the Italian universities only in the 18th century in spite of scientists such as Malpighi (1628-1694) who set out in his scientific testament a complete system of rational medicine based on experimental philosophy.

Instead of the universities, the scientific academies — such as the *Accademia del Cimento* of Florence (1657-67) and others — became the centres for the cultivation of experiment and research.[31] In Holland the academies did not

23. Ten Doesschate (1963) 5-10.
24. Frijhoff (1981) 105-107.
25. Lindeboom (1975) 279-280.
26. Van Spronsen (1975) 339.
27. De Pater (1975) 315.
28. Schudt (1959) 209; gives other examples. Cf. also Ornstein (1938) 218-219.
29. Van Kessel (1987) 123; Schmitt (1975) 505-507; Siraisi (1987) 77ff.
30. Belloni (1970) 106-107.
31. General information: Ornstein (1938) and McClellan (1985).

have the same importance as in Italy till the late 18th century. By their modernity the Dutch universities, or "Academies" as they called themselves in the 17th century, fulfilled all the functions, namely teaching, basic research, and the spread of knowledge among a larger public. In the late 18th century, when the universities stagnated, scientific societies were felt as a need. The *Hollandsche Maatschappij der Wetenschappen* at Haarlem, established in 1752, was intended to stimulate not only the natural sciences but also belles lettres and the visual arts.[32] This was the general trend in the academies of the Enlightenment.

The academies in the 18th century were a meeting place for the many individual scientists who flourished both in Italy and in the Netherlands.[33] These amateurs, as they were called, had laboratories in their private houses and made their own instruments. The scientific societies in the late 17th and particularly in the 18th century helped experimental science in creating laboratory facilities and in stimulating the spread of scientific knowledge among scientists. One of the ways was the exchange of publications and the correspondence between the members of the many provincial and local societies which were founded in the second half of the 18th century.[34] It is interesting that professors who took conservative scientific positions in their university lectures were protagonists of the new experimental science in the learned societies.[35]

### *Students*

If there were similarities in teaching and research between the Dutch and the Italian universities there was a world of difference between the mobility of their students and scholars. Dutch students are notorious for their mobility. For the Middle Ages, this is easily explicable, since the first "national" university of the Low Countries only opened her doors in 1425, at Louvain in the Duchy of Brabant. As a consequence they had to go abroad for an academic education. Yet neither the erection of Louvain nor the foundation of Dutch universities at Leyden (1575), Franeker (1585), Groningen (1614), Utrecht (1636) or Harderwijk (1648) after the secession of the Netherlands could keep them in their native land.[36]

32. Bots (1976) 11-17.
33. Ornstein (1938) 55ff; Van Berkel (1985) 35ff.
34. McClellan (1985) 153ff.
35. Kagan (1986) 173-174.
36. General survey for the Dutch Republic: Frijhoff (1981); see further: De Ridder-Symoens (1983); for a general overview of the European universities see Jílek (1984).

Some of the European universities were very international and they remained so, independent of external factors such as politics or religion, till the late 17th century, when the European *peregrinatio academica* came to an end. The success of a *studium generale* was connected with its teaching speciality but also with structural and more material factors. On the threshold of early modern times most of the countries and even regions possessed a proper university, considered mainly as a training centre for civil servants and Church officers. Only a few of these many *studia generalia* had an international reputation and an international audience. These universities were generally speaking the oldest institutions. In France, Paris remained attractive for foreigners not directly for its instruction, but as an intellectual, cultural and political centre. The faculties of arts and theology had lost their mediaeval fame as centres of learning. The unique law faculty of Orleans, famous in the Middle Ages, attracted large numbers of Germanic speaking people till the end of the 17th century, albeit that tradition, special privileges and aristocrats opportunities such as fencing, dancing, riding, and conversation in modern languages competed more and more with the legal studies. The international recruitment of the German and Swiss universities diverged and fluctuated, being mainly dependent on religious questions.

The Northern Italian universities, and more specially Bologna and Padua, kept their international audience till the late 17th century. In the Middle Ages the large majority of the foreign students studied law. From the middle of the 16th century the study of medicine in Italy became popular with students from North of the Alps and it started to compete with legal studies. However, in the course of the 17th century the medical faculties of the Dutch academies, with Leyden at the head became serious rivals. They supplanted their Italian counterparts in student numbers. But Dutchmen continued to visit the Northern Italian universities.

To evaluate the presence of Dutch students in Italy we must take several aspects into consideration. We must keep in mind that some students spent only a very short time in a university town, while others followed a whole *curriculum studiorum*. Therefore we must make a distinction between students attending lectures and those who just matriculated in a studium to receive a doctor's bull. Indeed, the place of graduation did not necessarily reflect the place of study. Studying and graduating were two different acts, not automatically linked to each other. There were a number of relevant factors to be considered, such as the graduation costs, the requirements for examination, the social function of an Italian degree (social prestige, snobbery, career improvement, family tradition, etc.). Hence it is important to know what is the proportion between *immatriculati* and *laureati* at a given university in order to define the character of the studium: is it a university at which to attend lectures, or a university, at which to graduate? The only way to know exactly the *cursus studiorum* and the intention of the student is to follow him in his *peregrinatio* and to list all the data of his studies, both in his homeland and abroad. In the 16th and 17th centuries it is not seldom that devoted travellers visited six and even more universities. Naturally their activities in all these places follow an erratic course. As an act of snobbery travellers of all ages wrote their name down in the registers of the student nation of an "aristocratic" university even if they had to pay the fees that go with matriculation. Padua, Bologna and Siena were considered as such. Furthermore, other advanced educational institutions, such as the colleges of the Society of Jesus or private academies (schools) and a growing army of *lettori privati* competed with the universities in attracting the sons of the upper classes.[37]

The tables in the Appendix reveal that from the end of the 16th century the *studio* of Padua was the most popular university for Dutch students. Bologna was out of favour, Siena was an alternative, while the other studia attracted hardly any Dutchmen.

Till the mid 16th century Padua could not compete with Bologna in numbers, certainly not with regard to the *Ultramontani*.[38] In 1562 the *Ultramontani* left the studium of Bologna en masse because the Papal legate had mishandled some of the 'heretical' students. In 1506 the university of Bologna came under the direct control of the Cardinal Legate of the Pope who had his residence in the town. Moreover, by the bull *In sacrosancta* of 1564 Pope Pius IV obliged all candidates for the doctorate in all faculties to take an oath of loyalty to the Catholic faith. For medical students the requirement of the *professio fidei* was confirmed in 1566 by Pius V in the bull *Supra gregem*. However, this constraint did not change the attitude of tolerance which the civil and ecclesiastical authorities of the Republic of Venice and its *studio* had long supported.[39] They accepted the Protestants as long as they remained inconspicious and did not debate about religious matters.[40] The civil government of Venice and the university authorities went a step further.[41] On the death of Pius IV in 1566 the Venetian Senate had hoped to annul the compulsory oath, but they did not succeed. Therefore both the university and

37. Kagan (1986) 173-174.
38. De Ridder-Symoens (1989); Brizzi (1988) 225-229.
39. Franceschini (1975) 135-138; Brugi (1888) 22-24; De Benedictis (1988).
40. Poelhekke (1961) 276.
41. For what follows see Van Kessel (1963) 101-102.

the senate encouraged the students to follow the long established but unpopular means of graduating under the authority of a Count Palatine. Emperor Sigismund (1367-1437) had given the competence to create doctors in the name of the Emperor to *comites palatini* and to Counts and Knights of the Holy See. Thus the Protestant Germans had the possibility of avoiding the *professio fidei* by taking a less prestigious private degree awarded by a Venetian Count Palatine in his private residence. The palatine doctor's bull could create some difficulties at home. The degree was awarded in a Catholic university which was regarded as suspect in the Protestant home country. Moreover, a private examination and an "unofficial" bull might cast doubt upon the scientific quality of the newly created doctor. In view of such considerations the student Frederick Sebiz asked the German nation of the Arts university for a certificate stating that the doctor in question had not taken an oath of Catholicity and that the Palatine bull had the same scientific value as one awarded in a public promotion by the university itself, and this practice was followed by other religious dissidents after him. This procedure was facilitated by the fact that several palatine counts were appointed as professors, e.g. the very popular Alramer (d. 1597) who was the promotor of many Dutch medical doctors; further, around 1600 the private bull was provided with the seal of the university. The last *doctor palatinus* was William Stratenus from Utrecht who graduated MD on 8 May 1613. This nobleman became professor at the new Utrecht university in 1636 and physician in ordinary of the Stadholders Frederik Hendrik, Willem II and Willem III. After his private promotion the Senate no longer accepted that in the Republic of Venice a degree was awarded by a foreign sovereign, in casu the German Emperor.[42] From 1616, students who wanted to avoid the public promotion in the presence of the bishop could graduate "auctoritate veneta" instead of "auctoritate imperiale".[43] From the 673 *promoti auctoritate veneta* listed by Weigle[44] between 1616 and 1663, two thirds were originated from the Protestant North of Europe, including 198 (or 30%) Netherlanders (125 from the North, 58 from the South, 15 unspecified).[45] We must not confuse the graduation "auctoritate veneta" with the examination carried out by the Venetian College of Physicians (Collegio dei medici).[46] Padua was the most expensive Italian university for taking degrees. Medical students could escape

42. Van Kessel (1963) 101-102.
43. Van Kessel (1963) 109.
44. Weigle (1965)
45. These Dutch students are listed in Poelhekke (1961).
46. For what follows see Palmer (1983).

the high costs by graduating in Venice. Indeed since the 15th century the Venetian College of Physicians was allowed to examine the candidates of the studium of Padua, as did the *collegium doctorum medicinorum* (*Sacro Collegio dei Filosofi e Medici*) of Padua. The doctor's certificate issued by both the collegi testified that the degree was awarded on the basis of the Papal and imperial authority by the studio generale of Venice, located in Padua. The rivalry between the two colleges in the long run kept down the price of graduation. Over half the Venetian graduates came from within the Venetian Republic, and only a small group from outside Italy (35 of the 618 *laureati* between 1501 and 1602; of these 8 came from the Low Countries).[47] The number of ultramontane graduates was disproportionately high in Padua, one third of them taking a degree at this university. The same proportion holds for students of the Low Countries. The influx of Dutch medical students in Padua occurs in the first half of the 17th century and represents more than 20% of the total population of the German nation of that faculty. Later it decreased to 10%. According to Poelhekke[48] 527 students from the Low Countries were members of the *Inclyta Natio Germanica Artistarum et Medicinorum* of Padua between 1553 and 1700. 260 of them took their degrees in Padua (194 from the North, 66 from the South); this means that almost half of them graduated. For the faculty of law I only have figures for the Spanish Netherlands. Law students very seldom took degrees in Padua. Further as can be seen in table 3 a new development occurred in the course of the 17th century. In 1600 75% of the student population was enrolled in the law faculty; whereas a century later they only represented about half the students.[49]

Altogether it was a constant concern for the Republic of Venice and of its alma mater to attract and retain foreign students, and more specifically the members of both the German nations. When we compare the Paduan figures with those of the other Italian universities it is clear that they succeeded, mostly to the detriment of Padua's rival, i.e. the studio of Bologna.

Bologna, the largest university in the Middle Ages had lost its dominant position in the course of the 16th century, both as regards its status as a centre of learning and in regard to the number of *suppositi*.[50] As a Papal university since 1506, its doors were closed to Protestant students and scholars not only because of the *professio fidei* for graduates but mainly because the municipal

47. Calculations based on the Catalogue of Graduations published by Palmer (1983) 67-192.

48. Poelhekke (1961) 281, 283.

49. Calculations based on the unpublished *atti* and *matriculae* of the German nation: Archivio Antico Universitario di Padova, mss 6-8, 459-464.

50. For general figures see Trombetti Budriesi (1988) and Brizzi (1988).

authorities were unable to give protection and to guarantee security to the *ultramontani* of the studium.[51] For this reason only a few Dutch medical students went to Bologna to graduate (see table 3). Before the erection of Leyden and the conversion to Calvinism, Bologna awarded 31 medical diplomas to Dutchmen (4 between 1500-24, 17 between 1525-49 and 10 between 1550-74) and 41 to subjects of the Southern Netherlands (respectively 4, 12 and 25).[52]

These figures may seem insignificant. However, this is not the case if we take into account that even the most outstanding medical faculties did not award more than a handful of medical degrees in a year. By comparison, I have counted in the still unpublished matricula of the German nation of the much larger law faculty of Bologna 37 Dutchmen from the Republic between 1573 (the beginning of the list) and 1601; after this date no Dutchman enrolled in the nation. Most of these students came from Frisia and Guelderland, and no one seems to have taken a degree. Predictably the situation in the Catholic Netherlands is different: 73 matriculations between the years 1573 and 1625 and 30 graduations.[53]

Nevertheless, there was a general decline in the peregrination of the students from the Spanish Netherlands, not only to Italy. This is mostly due to the fact that in 1570 Philip II prohibited his Nederlandish subjects from attending foreign universities. The *collegium doctorum juristarum* of Bologna had complained about this measure, though it appears without result.[54]

Another indication of the 'Italianization' and even the 'provincialization' of the *studio* of Bologna is the appointment of professors. Bologna had a long tradition of reserving four *cattedre eminenti* for foreign scholars; by the middle of the 17th century all the teaching positions were monopolised by Bolognesi.[55]

The alma mater of Siena also experienced the decline of foreign attendance after the fall of Siena, the last city-republic, in 1555 and the promulgation of the bull *In sacrosancta* of 1564. At that time, Germans, Bohemians, Netherlanders and Poles dominated the group of the ultramontani. English-

51. Kagan (1986) 167; De Benedictis (1988).

52. For the Bolognese figures see: Simeoni (1940) 101-102 and table 2 drawn up on the basis of Bronzino (1962).

53. Calculation based on the unpublished *matriculae* of the German Nation: Archivio di Stato di Bologna, Università, ms 4 and ibid. Libri segreti, mss 126-131 (=coll. can.), 137-143 (=coll. civ.).

54. Archivio di Stato di Bologna, Università, Atti, ms 25, fo 286.

55. Kagan (1986) 169, 171; Siraisi (1987) 121; cf. the *rotuli* of the teachers published by Dallari (1888-1924).

men, Portugueses and Spaniards had already left the Sienese "imperial" studium in large numbers in the course of the 16th century. In the 1560s there existed an "asilo di luteraneria" in Siena. However, German students began to be prosecuted by the Inquisition, with the agreement of the grand duke of Tuscany. The result was an exodus of foreigners, and also of members of the large colony of German craftsmen. In the following years many Germans came back as the climate became more tolerant and the *professio fidei* was no longer strictly upheld. Even if the Inquisition continued to take action against the heretics from time to time, student life was more than bearable[56] to such an extent that the German nation underwent a reorganisation in 1572 because of the affluence of new *suppositi*. In the 17th century Siena became the second most popular university after Padua in the attraction of students from Northern Europe. This relates mainly to law students; the future physicians preferring Padua or even Bologna. Of the 553 members of the German nation of Siena who graduated between 1485 and 1786 only 41 took a degree in arts and medicine, as opposed to 477 in law and 34 in theology; only 83 came from the Low Countries (North: 32 law, 1 medicine; South: 43 law, 6 medicine).[57] Siena was clearly not a "graduation university", but rather a stop in the *iter italicum*. Students were indeed offered more than purely academic attractions in Siena. At the demand of the foreign students the Sienese university authorities created a chair in the Tuscan language in 1588. It was the first "catedra di lingua toscana" in Italy; Pisa and Florence created their chairs in 1632, and Padua and Bologna only at the beginning of the 18th century. Moreover, in 1596 an *albergo* opened its doors serving "national" meals for the Germans, Bohemians, Poles and *Fiamminghi*.[58] Such a guesthouse was a meeting place for countrymen where they could speak their own language and feel a bit at home.

The other universities of Northern Italy, Pavia, Pisa, Ferrara and Perugia were not attractive for Netherlanders in early modern times, and certainly not for medical students.[59] This is understandable since these institutions were typical law faculties with very small medical faculties. The fact that these institutions adhered closely to Catholic orthodoxy also kept many Northern students away. Of the 224 German *promoti* in Perugia between 1489 and 1779,

56. Prunai (1959) 122-127.

57. Weigle (1944) 251 and calculations made for the Netherlands on the basis of the *matriculae*.

58. Prunai (1959) 99.

59. For Ferrara, see: Franceschini (1975) 155; Righini (1932) 215; for Pavia I went through the graduation registers in the Archivio di Stato di Pavia, Antico Archivio dell'Università, Inventario Doctoratus, 1525-1653.

195 were lawyers and 29 physicians and philosophers.[60] In Pisa in the 16th century between two thirds and three quarters of the studiosi were jurists.[61] In spite of this, both faculties provided fairly equal numbers of doctors after 1570, which means that proportionately more medical degrees than legal were awarded. In any case, Pisa seems not to have been neglected as a "graduation university" for the Germans. 37% of the matriculated students took a degree there between 1543 and 1775.[62] This is in sharp contrast with Siena where only 2.4% of all the Germanic speaking students took a doctorate between 1573 and 1738.[63] For the Netherlanders the ratio at Siena for the same period is 7.9%. The "provincialization" of these universities was a reality at the end of the 16th century. Only Perugia saw a kind of German revival in the first half of the 17th century. The authorities gave the Germans considerable facilities of various kinds in order to retain them in their city, to such a degree that the Pope gave special food prerogatives to the members of the German nation of Perugia concerning the consumption of eggs and dairy products during Lent.[64]

The picture which emerges of Italians going abroad and especially to the Dutch Republic in the early modern period is totally different. In general we can say that Italian students were not great *peregrinantes*. In the Middle Ages and the 16th century we find them mainly in Southern France, Spain and Switzerland, but not in great numbers.[65] A certain kind of Italian internationalism occurred in the Renaissance when Italian scholars and teachers of all disciplines were invited all over Europe and received high salaries to teach humanistic topics in foreign universities. We find them in Krakau, Leipzig, Paris or Louvain, as well as in Oxford or Salamanca, to give only a few examples.[66]

Till the 16th century no educational considerations could induce the youth of the Peninsula to cross the Alps. At this time we find some Italian *studiosi* travelling in France, Italy, Germany or Holland in two different contexts: on the Grand Tour or in religious exile. This is not the place to enlarge on the topic of the Grand Tour,[67] but two points may be made. First, medical

60. Weigle (1942) 113, 144.
61. Schiaffino (1935) 56; Paganini (1886-87) 125-126.
62. Weigle (1959) 194.
63. Weigle (1944) 210-212.
64. Ermini (1971) vol. I, 383-387.
65. Verger (1983).
66. Weiss (1960) passim.
67. Good introduction for the Netherlands in Frank-Van Westrienen (1983) and for Italy in Brizzi (1976).

students or physicians who were already qualified were not the most enthusiastic "tourists". In the 18th century Italian scientists travelled a lot North of the Alps. However, this cannot be considered as a Grand Tour in the sense of a *Bildungsreise*, but rather as a *Gelehrtenreise*.[68] Secondly, the Grand Tour of the Italians is, as far as I know, very little researched.

The only way to discover the exact number of Italians who studied and graduated in the Netherlands is to read through the *alba studiosorum* and *promotorum* of the five Dutch universities.[69] This is not an easy task since most of these editions are lacking indices of place names. The following figures therefore have to be treated with caution. Since one third of all the Leyden *immatriculati* were foreigners and 12% of them *graduati*[70] it is natural that Leyden also attracted most of the itinerant Italians. According to Colenbrander and Frijhoff 180 Italians enrolled at Leyden between 1575 and 1813. Only 4 of them also took a degree at this university.[71] Considering the thousands of Leyden *alumni* this number is insignificant. In Groningen[72] 8 Italians matriculated and 2 graduated; in Franeker[73] I noticed only two matriculations and no graduations. Both in Utrecht[74] and Harderwijk[75] the

68. See the paper of Mazzolini and others in these proceedings.

69. I am grateful to W. Frijhoff for information concerning Italian *immatriculati* and *promoti* at the Dutch universities. For an overview see table 6.

70. Colenbrander (1925) 300.

71. Colenbrander (1925) 294-295, 298-299 and Frijhoff (1981) 380. The *promoti* are: Samuel Nebelthavius from Naples, mat. 2-7-1612; DM 2-9-1612 (*Album Stud. Leyden* 107; Molhuysen (RGP 29) 36-37; Kroon 124: gives uncorrectly DM 5-9-1612); Johannes Vincentius Riccius from Lucca, mat. 22-10-1673, M.Cand., 39 years; DM 26-10-1672 (*Album Stud. Leyden* 579; Molhuysen (RGP 38) 317*); Josephus Constancius Romanus, Italus, mat. 10-2-1673 ling. Ital. mag., 25 years; DM 12-7-1677 (*Album Stud. Leyden* 581; Molhuysen (RGP 38) 327*): during his studies J.C.R. taught the Italian language; Jacobus Spinoza from Livorno, mat. 15-7-1720, 26 years; DM 19-7-1720 (*Album Stud. Leyden* 868; Molhuysen (RGP 45) 275*).

72. Thomas Astorinus Consentinus Italus, mat. 14-9-1686; DM 1-11-1686 (*Album Stud. Groningen* 127; *Album Prom. Groningen* 453); Franciscus Antonius Mannius cives Romanus, mat. 7-7-1655, DUJ, 35 years; DM 30-7-1655 (*Album Stud. Groningen* 72; *Album Prom. Groningen* 448). Mannius was already *doctor utriusque juris* before taking his degree in medicine.

73. De Ridder-Symoens (1985) 76, 82.

74. The *promoti* are: Eduardus Colonna, eques Romanus, mat. 1690; DM 23-1-1691 (*Album Stud. Utrecht* 88; *Album Prom. Utrecht* 57); Johannes Castillionaeus Florentinus, Phil. et L.Art. et Med. 23-1-1754 (*Album Prom. Utrecht* 157; no immatriculation). Castiglione was already *doctor utriusque juris* before taking his degree in arts and medicine.

75. The *promoti* are: Franciscus Riccardi Neapolitanus, mat. 26-7-1738, MCand.; DM 28-7-1738 (*Album Stud. Harderwijk* 90; *Album Prom. Harderwijk* 149); Jacobus Fregosse Neapolitanus, mat. 3-5-1745, MCand.; DM 5-5-1745 (*Album Stud. Harderwijk* 98; *Album Prom. Harderwijk* 161).

two *immatriculati* also took the degree. Many of these students went to the medical faculties. (See table 6).

It would be interesting to know the reason for their emigration to the North and to follow their future careers. For this we need exhaustive biographical research. At least three Italian students remained in the Dutch Republic. Two physicians started a private practice in Amsterdam.[76] Giovan Francesco Salvemini da Castiglione, after his graduation in jurisprudence at Pisa, went to Switzerland and in 1751 he was called to the university of Utrecht where he became professor of mathematics, experimental physics and astronomy, and where in 1754 he acquired a doctorate in the faculty of arts.[77] Some of these Italians have left explanations of why they came to the North. Ferdinando Capponi, a nobleman of Florence, went to Franeker in 1644 "ob amorem religionis reformatorae (sic) ex paternis aedibus profugus"[78], while Tommaso Antonio Astorino from Cosenza in Calabria matriculated in Groningen in 1686 "gratis quia patria Italia extorris (sic) ob religionem". Six weeks later he received his MD on a thesis dealing with obstetrics.[79] From the middle of the 16th century the Italian Peninsula was dominated by Papal and Spanish authorities who tried to restrain revolutionary and seditious opinions. The activities of the Roman and Spanish Inquisitions forced many scholars to leave the country to escape persecution. England and Holland were favoured countries for those who wanted a climate of free expression of thought, including scientific research.[80] Most of the Italians studying in the Netherlands originated from the Mezzogiorno, from the Kingdom of Naples. The Counter reformation and the Inquisition probably forced them to flee the country, both in the 17th and 18th centuries. Since these Italians could not profit from the privileges of their Northern brethren in Padua they were obliged to cross the Alps. In the 18th century we notice a number of Waldenses in Groningen and Leyden but not in the medical faculty. Through a comparison of the attendance of Italian students at Louvain university it seems clear that the *amor religionis* was the driving motive for travelling to the North and not the *amor peregrinationis*. Or was it the *amor scientiae*? The flourishing Louvain university attracted 114 *citramontani* between 1425 and 1569 (40 in 1426-1527, 74 in 1527-1569). In the 17th century, specifically from 1616 to 1699, only 10 Italians matriculated. The last Italian student visited the Brabant *alma mater*

76. Frijhoff (1981) 158.
77. The Doesschate (1963) 30; *Album Prom. Utrecht* (1936) 157.
78. *Album Stud. Franeker* (1968) nr. 4266.
79. *Album Stud. Groningen* (1915) 127; *Album Prom. Groningen* (1915) 453.
80. Van Kessel (1987) 121-123.

in 1712. Insofar as a place name is stated the students of the 17th and 18th century originated from Rome, Genoa, Florence and Locarno.[81]

In conclusion, if Dutch scientists did have an influence on Italian sciences this knowledge was certainly not transmitted by large numbers of wandering Italian students and scholars. By contrast, the numerous Italian *alumni* from the Low Countries had a direct influence on the transmission of the Italian sciences to the North.[82]

81. Figures based on the *Matricules de l'Université de Louvain.*
82. I wish to thank Dr D. Osler (Frankfurt/Main) for revising my English manuscript.

**Table 1**

STRUCTURES

BOLOGNA & PADUA

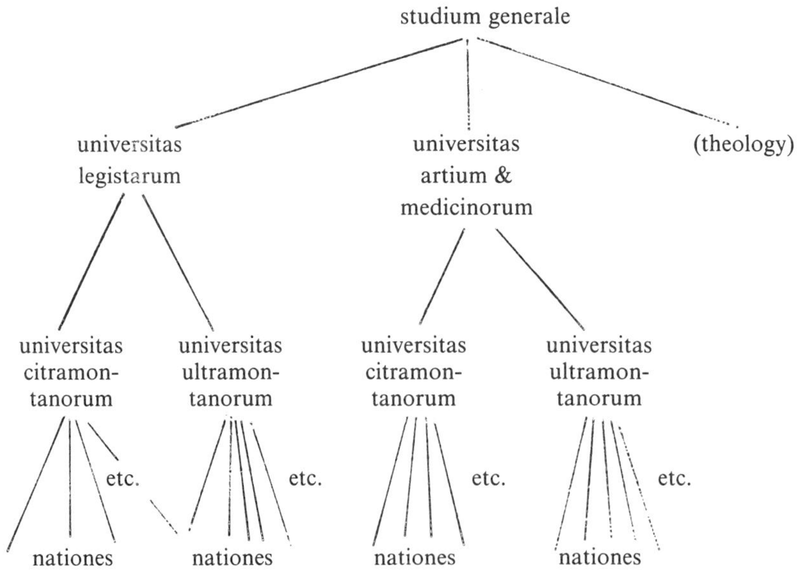

SIENA

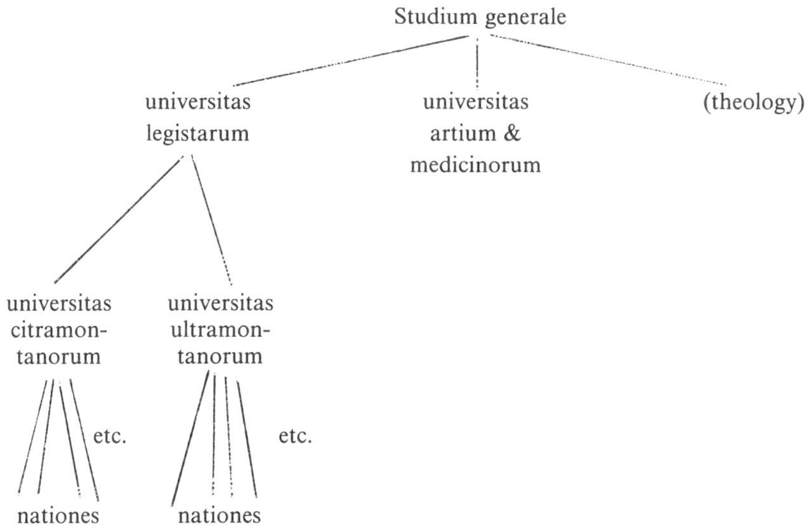


FERRARA

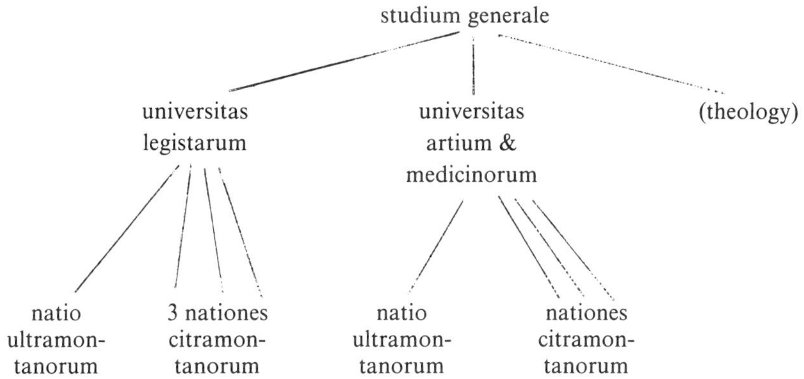

PAVIA

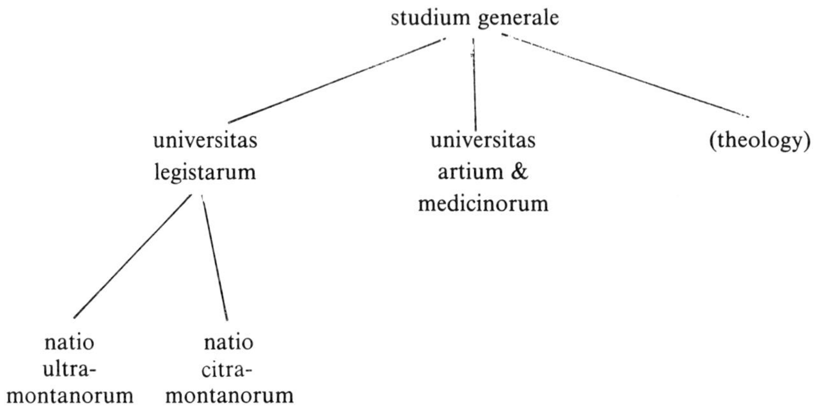


PERUGIA & PISA

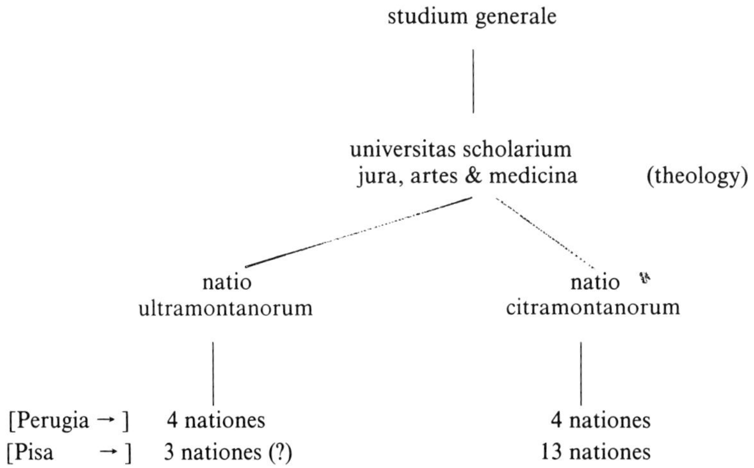

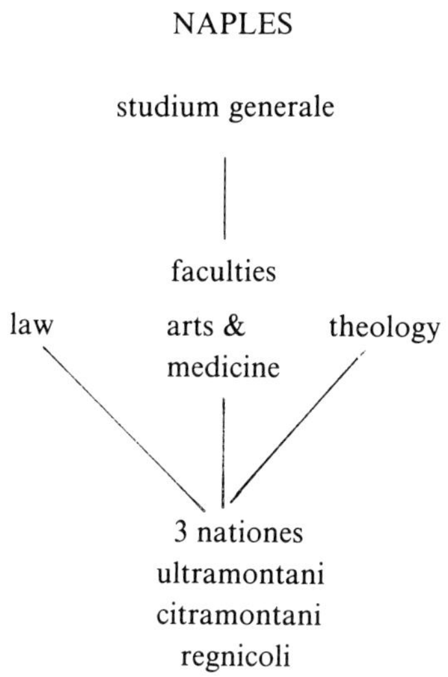
NAPLES
studium generale
faculties
law
arts & medicine
theology
3 nationes
ultramontani
citramontani
regnicoli

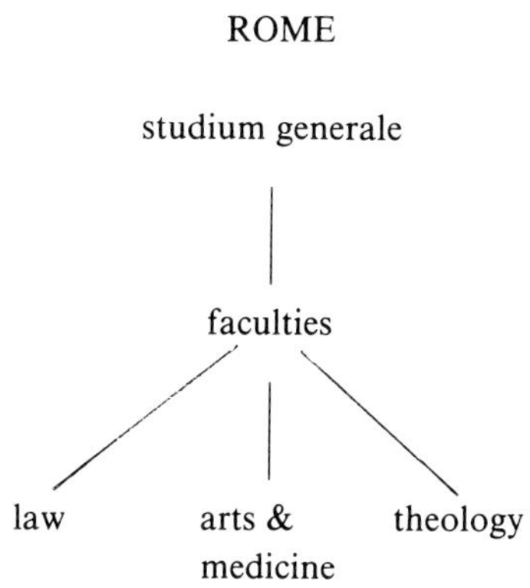
ROME
studium generale
faculties
law
arts & medicine
theology

**Table 2**

## GRADUATIONS IN MEDICINE AT THE UNIVERSITY OF BOLOGNA 1480-1799*

| | 1480-99 | 1500-24 | 1525-49 | 1550-74 | 1575-99 | 1600-24 | 1625-49 | 1650-74 | 1675-99 | 1700-24 | 1725-49 | 1750-74 | 1775-99 | 1480-1799 |
|---|---|---|---|---|---|---|---|---|---|---|---|---|---|---|
| *Citramontani* | | | | | | | | | | | | | | |
| Italy | 79 | 132 | 219 | 351 | 270 | 300 | 196 | 204 | 126 | 79 | 132 | 173 | 178 | 2439 |
| Bologna | 8 | 24 | 16 | 51 | 48 | 80 | 75 | 83 | 69 | 58 | 45 | 29 | 54 | 640 |
| Total Citram. | 87 | 156 | 235 | 402 | 318 | 380 | 271 | 287 | 195 | 137 | 177 | 202 | 232 | 3079 |
| *Ultramontani* | | | | | | | | | | | | | | |
| Portugal | 13 | 23 | 23 | 25 | 28 | 4 | 10 | — | 2 | — | — | 7 | 1 | 136 |
| Spain & Colonies | | | | | | | | | | | | | 2 | 2 |
| Holy Empire | 22 | 70 | 67 | 99 | 19 | 35 | 27 | 99 | 67 | 20 | 56 | 22 | 9 | 612 |
| Southern Neth. | 5 | 4 | 12 | 25 | 24 | 14 | 7 | 12 | 8 | — | — | 3 | 2 | 116 |
| Northern Neth. | 5 | 4 | 17 | 10 | 3 | 7 | — | 7 | — | — | 1 | — | — | 54 |
| Poland | 3 | 15 | 8 | 21 | 9 | 9 | 5 | 5 | 2 | — | 3 | 2 | 7 | 89 |
| France | 1 | 2 | 15 | 5 | 4 | — | 1 | 5 | — | 1 | — | — | — | 34 |
| Great Britain | 1 | 1 | 1 | — | 1 | 4 | 1 | 1 | — | — | — | — | — | 10 |
| Switzerland | — | — | 1 | 2 | 1 | 3 | 1 | 7 | 3 | 4 | 2 | 2 | 3 | 29 |
| Northern Countries | — | — | | 2 | 1 | — | — | — | — | — | — | — | — | 3 |
| Southern Europe | | | | 1 | 1 | 2 | 1 | 3 | 1 | | 3 | 1 | 1 | 14 |
| Bohemia, Moravia, Hungary | 1 | 1 | 5 | 3 | 6 | 2 | | 20 | 16 | 2 | 17 | 2 | — | 75 |
| Jews | 1 | | | 1 | | | | | | | | | | 2 |
| Total Ultram. | 52 | 120 | 150 | 193 | 95 | 79 | 52 | 159 | 99 | 27 | 82 | 39 | 25 | 1172 |
| | 36% | 40% | 38% | 32% | 22% | 17% | 16% | 35% | 35% | 33% | 30% | 16% | 10% | 28 |
| Unknown | 5 | 18 | 15 | 8 | 17 | 2 | 4 | 11 | 4 | 2 | 3 | 1 | 2 | 92 |

*Drawn up on the basis of Bronzino (1962).

**Table 3** GRADUATIONS OF STUDENTS FROM THE LOW COUNTRIES

| | MEDICINE | | | | LAW | | |
|---|---|---|---|---|---|---|---|
| | PADUA | | BOLOGNA | | PADUA | BOLOGNA | |
| | NNL | SNL | NNL | SNL | SNL | NNL | SNL |
| 1450-74 | | 2 | | | | 7 | 9 |
| 1475-99 | | | 5 | 5 | 1 | 15 | 6 |
| 1500-24 | | 1 | 4 | 4 | 5 | 8 | 5 |
| 1525-49 | | 6 | 16 | 12 | 5 | 2 | 8 |
| 1550-74 | | 11 | 8 | 27 | 11 | 11 | 42 |
| 1575-99 | 22 | 8 | 3 | 24 | 3 | | 26 |
| 1600-24 | 55 | 15 | 7 | 14 | | | 4 |
| 1625-49 | 85 | 35 | | 7 | | | 1 |
| 1650-74 | 29 | 18 | 9 | 12 | 1 | | |
| 1675-99 | 2 | 7 | | 8 | | | |
| 1700-24 | | | | | | | |
| 1725-49 | | | | | | | 2 |
| 1750-74 | | | 1 | 3 | | | |
| 1775-99 | | | | 2 | | | |
| 1450-1799 | 193 | 103 | 53 | 113 | 26 | 43 | 103 |
| 1550-1799 | 193 | 94 | 28 | 92 | 15 | 11 | 75 |

**Table 4** MATRICULATIONS AND GRADUATIONS (between brackets) OF STUDENTS FROM THE LOW COUNTRIES

| | PISA | | | | SIENA | | PERUGIA | |
|---|---|---|---|---|---|---|---|---|
| | NNL | | SNL | | NNL | SNL | NNL | SNL |
| | med. | law | med. | law | | | | |
| 1550-74 | 6(4) | 6(3) | 14(13) | 27(12) | (13L) | (2M,24L) | | |
| 1575-99 | 3(2) | | | 1 | 20(2L) | 40(2M,9L) | 3 | 11 |
| 1600-24 | | | 3(2) | 3(2) | 50(1M,3L) | 32(1M) | 11(2M,1L) | 13 |
| 1625-49 | | | 1 | | 42(1L) | 26(3L) | 6 | 33(4L,1T) |
| 1650-74 | | 1 | | | 18 | 23(1L) | 2 | 13(2L) |
| 1675-99 | | | | | 14 | 28 | 1 | 9 |
| 1700-24 | | | | | 2 | 5 | | 1 |
| 1725-49 | | | | | | | | |
| 1750-74 | | | | | | 1(1M) | | |
| 1550-1774 | 9(6) | 7(3) | 18(15) | 31(14) | 146(7) | 155(17) | 23(3) | 80(7) |

NNL = Northern Netherlands
SNL = Southern Netherlands + Liège

M = medicine
L = law
T = theology

**Table 5** GRADUATIONS IN ARTS AND MEDICINE AT ITALIAN UNIVERSITIES OF STUDENTS FROM THE LOW COUNTRIES

| | PADUA[1] | | BOLOGNA[2] | | SIENA[3] | | PISA[4] | | PERUGIA[5] | |
|---|---|---|---|---|---|---|---|---|---|---|
| | NNL | SNL | NNL | SNL | NNL | SNL | NNL | SNL | NNL | SNL |
| 1575-99 | 22 | 9 | 3 | 24 | | 2 | 2 | | | |
| 1600-24 | 55 | 11 | 7 | 14 | 1 | 1 | | 2 | 2 | |
| 1625-49 | 85 | 27 | — | 7 | | | | | | |
| 1650-74 | 29 | 11 | 7 | 12 | | | | | | |
| 1675-99 | 2 | 5 | — | 8 | | | | | | |
| 1700-24 | — | — | — | — | | | | | | |
| 1725-49 | — | — | — | — | | | | | | |
| 1750-74 | — | — | — | 3 | | 1 | | | | |
| 1775-99 | — | — | — | 2 | | | | | | |

1. Frijhoff (1981), p. 383.
2. Based on Bronzino (1962), see table 2.
3. Weigle (1944), pp. 216-240.
4. Weigle (1959), pp. 196-221.
5. Weigle (1956), p. 45.

**Table 6** ITALIANS AT DUTCH UNIVERSITIES

| | Total | | Leyden (1575) | | | | Franeker (1585) | | | |
|---|---|---|---|---|---|---|---|---|---|---|
| | | | total | | med. | | total | | med. | |
| | mat. | grad. | mat. | grad. | mat. | grad. | mat. | grad. | mat. | grad. |
| 1575-99 | 4 | — | 4 | | | | | | | |
| 1600-24 | 9 | 1 | 9 | 1 | 1 | 1 | | | | |
| 1625-49 | 20 | — | 19 | | | | 1 | | | |
| 1650-74 | 34 | 2 | 32 | 1 | 1 | 1 | | | | |
| 1675-99 | 21 | 3 | 19 | 1 | 1 | 1 | | | | |
| 1700-24 | 28 | 2 | 27 | 1 | 1 | 1 | 1 | | | |
| 1725-49 | 33 | 2 | 31 | | | | | | | |
| 1750-74 | 28 | 1 | 26 | | | | | | | |
| 1775-99 | 17 | — | 13 | | | | | | | |
| | 194 | 11 | 180 | 4 | 4 | 4 | 2 | | | |

| | Groningen (1614) | | | | Utrecht (1636) | | | | Harderwijk (1648) | | | |
|---|---|---|---|---|---|---|---|---|---|---|---|---|
| | total | | med. | | total | | med. | | total | | med. | |
| | mat. | grad. | mat. | grad. | mat. | grad. | mat. | grad. | mat. | grad. | mat. | grad. |
| 1575-99 | | | | | | | | | | | | |
| 1600-24 | | | | | | | | | | | | |
| 1625-49 | | | | | | | | | | | | |
| 1650-74 | 2 | 1 | 2 | 1 | | | | | | | | |
| 1675-99 | 1 | 1 | 1 | 1 | 1 | 1 | 1 | 1 | | | | |
| 1700-24 | | | | | | | | | | | | |
| 1725-49 | | | | | | | | | 2 | 2 | 2 | 2 |
| 1750-74 | 1 | | | | 1 | 1 | 1 | 1 | | | | |
| 1775-99 | 4 | | | | | | | | | | | |
| | 8 | 2 | 3 | 2 | 2 | 2 | 2 | 2 | 2 | 2 | 2 | 2 |

# BIBLIOGRAPHY

This bibliography contains all works to which reference is made in the notes.

*Album Prom. Groningen* = *Album Promotorum Academiae Groninganae*, ed. Historisch Genootschap te Groningen (Groningen 1915).

*Album Prom. Harderwijk* = *Het Album Promotorum van de Academie te Harderwijk*, ed. by O. Schutte (Zutphen 1980).

*Album Prom. Utrecht* = *Album Promotorum qui inde ab anno MDCXXXVI usque ad annum MDCCXV in academia Rheno-Traiectina quadam doctoratus adepti sunt*, ed. F. Ketner (Utrecht 1936).

*Album Stud. Franeker* = *Album Studiosorum Academiae Franekerensis (1585-1811, 1816-1844)*, ed. S.J. Fockema Andreae en Th.J. Meyer (Franeker 1968).

*Album Stud. Groningen* = *Album Studiosorum Academiae Groninganae*, ed. Historisch Genootschap te Groningen (Groningen 1915).

*Album Stud. Harderwijk* = *Album Studiosorum Academiae Gelro-Zutphanicae, MDCXLVIII-MDCCCXVIII*, ed. D.G. van Epen (The Hague 1904).

*Album Stud. Leyden* = *Album Studiosorum Academiae Lugduno-Batavae, MDLXXV-MDCCCLXXV* (The Hague 1875).

*Album Stud. Utrecht* = *Album Studiosorum Academiae Rheno-Traiectinae, MDCXXXVI-MDCCCLXXXVI* (Utrecht 1886).

Ascanelli, P. (1968) *I fascicoli personali dei lettori artisti della assunteria di studio all'Archivio di Stato di Bologna* (Archivio dell'Universitá/ Collana di monografie di 'Romagna medica' 25; Bologna).

Bannenberg, G.P.J. (1953) *Organisatie en Bestuur van de middeleeuwse universiteit* (Nijmegen).

Bellomo, M. (1979) *Saggio sull'Università nell'età del diritto comune* (Catania).

Belloni, L. (1970) Italian Medical Education after 1600, in: *The History of Medical Education*, ed. C.D.O. Malley (UCLA Forum in Medical Sciences nr. 12; Berkeley) 105-120.

Benedictis, A. De (1988) La fine dell'autonomia studentesca tra autorità e disciplinamento, in: *Studenti e Università* 193-223.

Berkel, K. van (1985) *In het voetspoor van Stevin. Geschiedenis van de natuurwetenschap in Nederland 1580-1940* (Amsterdam).

Bernardin, S. De (1974) La politica culturale della Republica di Venezia e l'Università di Padova nel XVII secolo, *Studi Veneziani* 16, 443-502.

Bots, H. (1976) *Van universitaire gemeenschap tot academische kring. Enige aspecten met betrekking tot de opkomst en ontwikkeling van geleerde academies en genootschappen in West-Europa* (Amsterdam).

Brizzi, G.P. (1976) La pratica del viaggio d'istruzione in Italia nel Sei-Settecento. *Annali dell'Istituto Storico Italo-Germanico in Trento* II, 203-291.

——(1988) Matricole ed effettivi. Aspetti della presenza studentesca a Bologna fra Cinque e Seicento, in: *Studenti e Università* 225-259.

Brockliss, L. (1978) Patterns of Attendance at the University of Paris 1400-1800, *The Historical Journal* 21, 513-544.

Bronzino, G. (1962) *Notitia doctorum, sive catalogus doctorum qui in collegiis philosophiae et medicinae Bononie laureati fuerunt ab anno 1480 usque ad annum 1800* (Universitatis Bononiensis monumenta IV; Milano).

Brugi, B. (1888) *La scuola Padovana di diritto romano nel secolo XVI* (Padua).

——(1893-94) Gli studenti tedeschi e la S. inquisizione a Padova nella seconda metà del secolo XVI, *Atti del R. Istituto Veneto di scienze, lettere ed arti*, S.7, t.5, 1015-1033.

Buck, A. (1987) Die Rezeption des Humanismus in den juristischen und medizinischen Fakultäten der italienischen Universitäten, in: *Der Humanismus und die Oberen Fakultäten*, Hg. von G. Keil, B. Moeller und W. Trusen (Mitteilung XIV der Kommission für Humanismusforschung (Weinheim) 267-284).

Bullough, V.L. (1972) Population and the Study and Practice of Medieval Medicine, in: *Science, Medicine and Society in the Renaissance. Essays to honor Walter Pagel*, ed. by A.G. Debus (New York) I, 62-69.

Burke, P. (1987) *The Italian Renaissance. Culture and Society in Italy* (Cambridge). Revised version of the first ed. of 1972.

Chartier, R. and Revel, J. (1978) Université et société dans l'Europe moderne: position des problèmes. *Revue d'histoire moderne et contemporaine* 25, 353-374.

Colenbrander, H.T. (1925) De herkomst der Leidsche Studenten, *Pallas Leidensis, MCMXXV* (Leiden) 275-303.

Corsi, P. and Weindling, P. (eds.) (1983) *Information Sources in the History of Science and Medicine* (London etc. 1983).

Dallari, V. (1888-1924) *I rotuli dei legisti e artisti dello studio bolognese dal 1384 al 1799*, 4. vol. (Bologna).

Dawson, J.P. (1968) *The Oracles of the Law* (Ann Arbor).

Denley, P. (1981) Recent studies on Italian Universities of the Middle Ages and the Renaissance, *History of Universities* 1, 193-206.

Ermini, G. (1971) *Storia dell'Università di Perugia*, 2 vol. (Florence).

Eulenburg, F. (1904) *Die Frequenz der deutschen Universitäten von ihrer Gründung bis zur Gegenwart* (Abhandlungen der Kön. Sächsischen Gesellschaft der Wissenschaften, XXIV, n. II; Leipzig).

Fichtner, G. and Siefert, H. (1978) *Padua* (Medizinhistorische Reihen 2; Stuttgart-New York).

Fischer, H. (1961) De adviezen van de Leidse professoren, in het bijzonder tijdens het professoraat van Everardus Bronchorst (1587-1627), *Tijdschrift voor Rechtsgeschiedenis* 29, 463-483.

Forni, G. (1948) *L'insegnamento della chirurgia nello studio di Bologna dalle origini a tutto il secolo XIX* (Bologna).

Franceschini, A. (1975) *Spigolature archivistiche prime* (Dep. prov. Ferrarese di storia patria. Atti e memorie, S.3, vol. 19, Ferrara).

Frank-van Westrienen, A. (1983) *De Groote Tour. Tekening van de educatiereis der Nederlanders in de zeventiende eeuw* (Diss. Leiden; Amsterdam).

Frijhoff, W. (1979) Surplus ou deficit? Hypothèses sur le nombre réel des étudiants en Allemagne à l'époque moderne (1576-1815). *Francia* 5, 173-218.

——(1981) *La société néerlandaise et ses gradués. 1575-1814* (Diss. Tilburg; Amsterdam).

——(1983) Non satis dignitatis ... Over de maatschappelijke status van de geneeskundigen tijdens de Republiek, *Tijdschrift voor Geschiedenis* 96, 401-405.

Giard, L. (1983-1984) Histoire de l'université et du savoir: Padoue (XIVe-XVIe siècle), *Revue de Synthèse* 104 (1983) 139-169; 105 (1984) 259-298.

Goethuys, J.P. (1985) G. Casabona alias Jozef van Goethuyse, *Nationaal Biografisch Woordenboek* 11, 96-99.

Gogelein, A.J.F. (1975) *Het "theatrum anatomicum" als gebouw. Leidse Universiteit 400 jaar* (Tentoonstelling Rijksmuseum Amsterdam 1975).

Greci, R. (1988) L'associazionismo degli studenti dalle origini alla fine del XIV secolo, in: *Studenti e Università*, 13-44.

Hackmann, W.D. (1975) The Growth of Science in the Netherlands in the Seventeenth and Early Eighteenth Centuries, in: *The Emergence of Science in Western Europe*, ed. by M. Crosland (London) 89-110.

Hall, R. (1959, reprint 1969) The Scholar and the Craftsman in the Scientific Revolution, in: *Critical Problems in the History of Science*, ed. by M. Clagett (Wisconsim UP, Madison) 3-32.

Irsay, S.d' (1935) *Histoire des Universités françaises et étrangères des origines à nos jours*, 2 vols (Paris)

Jílek, L. (ed.) (1984) *Historical Compendium of European Universities - Répertoire historique des universités européennes* (CRE; Geneva).

Jurriaanse, M.W. (1965) *The Foundation of Leyden University* (Leiden).

Kagan, R.L. (1986) Universities in Italy, 1500-1700, *Les universités européennes du XVIe au XVIIIe siècle. Histoire sociale des populations étudiantes*, Tome II (Paris) 153-186.

Keller, A. (1972) Mathematical Technologies and the Growth of the Idea of Technical Progress in the Sixteenth Century, in: *Science, Medicine and Society in the Renaissance. Essays to Honor Walter Pagel*, ed. by A.G. Debus, 2 vols (New York) 11-27.

Kessel, P.J. van (1963) *Duitse studenten te Padua. De controverse Rome-Venetië en het protestantisme in de tijd der contra-reformatie* (Van Gorcum's Historische Bibliotheek 71; Assen).

——(1987) Common Sense Between Bacon and Vico: Scepticism in England and Italy, in: *Common Sense. The Foundation for Social Science*, ed. by F. van Holthoorn and D.R. Olson (Boston) 115-131.

Kibre, P. (1948) *The Nations in the Medieval Universities* (Medieval Academy of America no. 49; Cambridge Mass.).

——(1961) *Scholarly Privileges in the Middle Ages* (Medieval Academy of America no. 72; London).

Kroon, J.E. (1911) *Bijdragen tot de geschiedenis van het geneeskundig onderwijs aan de Leidse universiteit 1575-1625* (Leiden).

*Leyden University* (1975) = *Leyden University in the Seventeenth Century. An Exchange of Learning*, ed. by Th. Lunsingh Scheurleer and G.H.M. Posthumus Meyjes (Leiden).

Lindeboom, G.A. (1972) *Geschiedenis van de medische wetenschap in Nederland* (Bussum).

——(1975) Dog and Frog. Physiological Experiments at Leiden during the Seventeenth Century, in: *Leyden University* 279-293.

——(1984) *Dutch Medical Biography. A Biographical Dictionary of Dutch Physicians and Surgeons 1475-1975* (Amsterdam).

——(1985) Johannes Antonides vander Linden (1609-1664), Medisch hoogleraar te Franeker en te Leiden. Zijn betekenis voor de Friese hogeschool, in: *Universiteit Franeker*, 356-370.

Luschin von Ebengreuth, A. (1884) Nuovi documenti riguardanti la nazione alemanna nello studio di Bologna, *Atti e Memorie della R. Deputazione di Storia Patria per la Provincie di Romagna* III S., vol. 2, fasc. 2-3, 3-20.

Luyendijk-Elshout, A.M. (1975) The Rise and Fall of the Mechanical School of Theodoor Craanen, in: *Leyden University* 295-308.

——(1985) De medische faculteit. Inleiding, in: *Universiteit te Franeker* 351-355.

McClellan, J.E. (1985) *Science Reorganized: Scientific Societies in the Eighteenth Century* (New York, Columbia UP).

*Matricules de l'Université de Louvain (1903-1967)* = *Matricules de l'Université de Louvain*, ed. by E. Reusens, J. Wils, A. Schillings, 10 tomes, 17 vol. (Brussels).

Minnucci, G. (1988) *I Tedeschi nella storia dell'Università di Siena*, a cura di S. Minnucci (Siena).

Molhuysen, P.C. (RGP 00) Catalogus Promotorum, in: *Bronnen tot de geschiedenis van de Leidsche Universiteit*, I 1574-1610 (RGP 20; The Hague 1913) 461*-473*; II 1610-1647 (RGP 29; The Hague 1916) 1-309; III 1647-1682 (RGP 38; The Hague 1918) 283*-334*; IV 1682-1725 (RGP 45; The Hague 1920) 191*-290*; V 1725-1765 (RGP 48; The Hague 1921) 214*-312*; VI 1765-1795 (RGP 53; The Hague 1923) 71*-144*; VII 1795-1811 (RGP 56; The Hague 1924) 107*-143*.

Moulin, D. de (1988) *A History of Surgery with Emphasis on the Netherlands* (Dordrecht).

Nauck, E.Th. (1954) Die Zahl der Medizinstudenten deutscher Hochschulen im 14.-18. Jahrhundert, *Archiv für die Geschichte der Medizin und der Naturwissenschaften* 38, 175-186.

Ohl, R.E. (1980) *The University of Padua: an International Community of Students and Professors* (Ph.D. University of Pennsylvania).

Ornstein, M. (1938) *The Role of Scientific Societies in the Seventeenth Century* (Chicago).

Paganini, P. (1886-7) Statistica degli studenti pisani del secolo XVI (1544-1609), *Rivista critica delle letteratura italiana*, 125-126.

Palmer, R. (1983) *The Studio of Venice and its Graduates in the Sixteenth Century* (Contributi alla storia dell'universitá di Padova 12; Padua).

De Pater, C. (1975) Experimental Physics, in: *Leyden University* 309-328.

Pinghini, C. (1927) La popolazione studentesca dell'Università di Ferrara dalle origini ai nostri tempi, *Metron*, 7.1, 120-168.

Poelhekke, J.J. (1961) Nederlandse leden van de Inclyta Natio Germanica Artistarum te Padua 1553-1700, *Mededelingen van het Nederlands Historisch Instituut te Rome* 31, 263-373.

Prunai, G. (1959) Lo studio senese nel primo quarantennio del principato mediceo (1557-1597), *Bullettino senese* 66, 79-160.

Randall jr., J.H. (1961) *The School of Padua and the Emergence of Modern Science* (Padua).

Rashdall, H. (1936) *The Universities of Europe in the Middle Ages*, new ed. by F.M. Powicke and A.B. Emden, 3 vol. (Oxford).

Ridder-Symoens, H. de (1983) La migration académique des hommes et des idées en Europe, XIIIe-XVIIIe siècles. *CRE Information*, 62, 69-79.

——(1985) Buitenlandse studenten aan de Franeker Universiteit 1585-1811, in: *Universiteit te Franeker*, 73-89.

——(1987) Eigenheid in den Vreemde. De studentennaties aan de Europese universiteiten, in: *Eigen en Vreemd. Identiteit en ontlening in taal, literatuur en beeldende kunst* (Amsterdam) 137-146.

——([2]1987) *Tot nut of onnut van 't algemeen. De functie van de universiteit in de middeleeuwen* (Inaugurale rede VU Amsterdam; Amsterdam 1987).

——(1989) La Place de l'université de Bologne dans la mobilité des étudiants européens, in: *Universitates e Università. Actes du colloque historique, 16-25 nov. 1987* (Bologna) in press.

Righini, G. (1932) *La laurea di Copernico allo studio di Ferrara* (Ferrara).

Roelevink, J. (1986) *Gedicteerd verleden. Het onderwijs in de algemene geschiedenis aan de Universiteit te Utrecht*, 1735-1839 (Diss. VU Amsterdam; Amsterdam).

Saibante, M., Vivarini, C., Voghera, G. (1924) Gli studenti dell'Università di Padova dalla fine del 500 ai nostri giorni, *Metron*, 4.1, 163-223.

Schiaffino, E. (1932-1935) Elenco dei dottorati di stranieri e di non Toscani nell'Università di Pisa, *Bollettino Storico Pisano* 1 (1932) 49-54; 2 (1933) 51-53; 3 (1934) 69-74; 4 (1935) 51-58.

Schmitt, C.B. (1972) The Faculty of Arts at Pisa at the time of Galileo, *Physis* 14, 243-272.

——(1975) Philosophy and Science in Sixteenth Century Universities: some preliminary comments, in: *The Cultural Context of Medieval Learning*, ed. by J.E. Murdoch and E.O. Sylla (Dordrecht) 485-537.

——([2]1975) Science in the Italian Universities in the Sixteenth and Early Seventeenth

Centuries, in: *The Emergence of Science in Western Europe*, ed. by M. Crosland (London) 35-56.

Schudt, L. (1959) *Italienreisen im 17. und 18. Jahrhundert* (Wien-München).

Schwinges, R.C. (1986) *Deutsche Universitätsbesucher im 14. und 15. Jahrhundert. Studien zur Sozialgeschichte des alten Reiches* (Veröffentlichungen des Instituts für europäische Geschichte Mainz, Abt. Universalgeschichte Bd. 123; Stuttgart).

Simeoni, L. (1940) *Storia della Università di Bologna II. L'età moderna (1500-1888)* (Bologna).

Snelders, H.A.M. (1983) Science in the Low Countries during the Sixteenth Century. A Survey, *Janus* 70, 213-227.

Spronsen, J.W. van (1975) The Beginning of Chemistry, in: *Leyden University* 329-343.

*Studenti e Università = Studenti e Università degli Studenti dal XII al XIX secolo*, a cura di G.P. Brizzi e A.I. Pini (Studi e memorie per la storia dell'Università di Bologna, NS vol. VII; Bologna 1988).

Ten Doesschate, G. (1963) *De Utrechtse Universiteit en de Geneeskunde 1636-1900* (Nieuwkoop).

Trombetti Budriesi, A.L. (1988) L'esame di laurea presso lo studio bolognese. Laureati in diritto civile nel secolo XV, in: *Studenti e Università* 137-191.

*Universiteit te Franeker = Universiteit te Franeker 1585-1811. Bijdragen tot de geschiedenis van de Friese hogeschool*, ed. G.Th. Jensma, F.R.H. Smit, F. Westra (Fryske Akademy, nr. 648; Leeuwarden 1985).

Verger, J. (1983) Les rapports entre les universités italiennes et les universités françaises méridionales (XII-XV siècles), in: *Università e società nei secoli XII-XVI* (Pistoia) 142-175.

Wansink, H. (1981) *Politieke Wetenschappen aan de Leidse Universiteit 1575-ca. 1650* (Utrecht).

Weigle, F. (1942) Die Deutsche Nation in Perugia, *Quellen und Forschungen aus italienischen Archiven und Bibliotheken* 32, 110-188.

——(1944) Die deutschen Doktorpromotionen in Siena von 1485-1804, *Quellen und Forschungen aus italienischen Archiven und Bibliotheken* 33, 199-251. Translated by R. Marcucci 'Le lauree Tedesche a Siena dal 1485-1804. Studenti Tedeschi in Italia, parte II, in: *I Tedeschi nella storia dell'Università di Siena*, a cura di G. Minnucci (Siena 1988) 33-92.

——(1956) *Die Matrikel der deutschen Nation in Perugia (1597-1727)* (Bibliothek des Deutschen Historischen Instituts in Rome 21; Tübingen).

——(1959) Deutsche Studenten in Pisa, *Quellen und Forschungen aus italienischen Archiven und Bibliotheken* 39, 173-221.

——(1962) *Die Matrikel der deutschen Nation in Siena (1573-1738)* 2 vol. (Bibliothek des Deutschen Historischen Instituts in Rom 22; Tübingen).

——(1965) Die Deutschen Doktorspromotionen in Philosophie und Medizin an der Universität Padua 1616-1663, *Quellen und Forschungen aus italienischen Archiven und Bibliotheken* 45, 325-385.

Weiss, R. (1960) Italian Renaissance in Western Europe: 1400-1520, in: *Italian Renaissance Studies*, ed. by E.G. Jacob (London) 69-93.

Zdekauer, L. (1894) *Lo Studio di Siena nel Rinascimento* (Milano, reprint Bologna 1977).

# SCIENCE AND RELIGION IN THE SEVENTEENTH CENTURY: THE CASE OF THE NORTHERN NETHERLANDS

*H.A.M. Snelders*

"In Holland the Calvinist Church was at first strenuous against the whole new system, but we possess a comical proof that Calvinism even in its strongholds was powerless against it; for in 1642 Blaer [sic!] published at Amsterdam his book on the use of globes, and, in order to be on the safe side, devoted one part of his work to the Ptolemaic and the other to the Copernican scheme, leaving the benevolent reader to take his choice". Thus Andrew Dickson White in his most influential study: *A History of the Warfare of Science with Theology in Christendom* (1896) which focused on the classical statement of the warfare thesis.[1]

White referred to the French translation of Willem Janszoon Blaeu's (1571-1638) vernacular work: *Tweevoudigh Onderwijs van de Hemelsche en Aerdsche Globen* (Twofold Instruction of the Heavenly and Terrestrial Globes; 1633), in which the Amsterdam cartographer first explained the globe according to "the opinion of Ptolemy with a fixed terrestrial sphere" and subsequently according to the "natural thesis of N. Copernicus with a rotating terrestrial sphere".[2] Blaeu had worked with Tycho Brahe at the latter's observatorium on the island of Hven in Denmark (1595-1596) and then settled in Amsterdam where he established himself as a merchant of maps and globes. His *Twofold Instruction* contains no refutation of Ptolemy's geocentric theory. "That it is believed that the earth rotates is the reason why globes are made like this and not with a fixed earth; I have contributed little or nothing to this; it was not my purpose to do so and others have done enough in this respect". Blaeu referred those who wished to know more about those theories to "two illustrious mathematicians, Simon Stevin and Philippus Lansbergius; the first short and colourful in his ideas, the other

1. A.D. White, *A History of the Warfare of Science with Theology in Christendom* (New York, 1896) Vol. I, 149-150.

2. W.J. Blaeu, *Tweevoudigh Onderwijs van de Hemelsche en Aerdsche Globen; Het een Na de meyning van Ptolemeus met vasten Aerdkloot; Het ander Na de Natuerlijcke stelling van N. Copernicus met een loopenden Aerdkloot* (Amsterdam, 1633).

broad and far-reaching in his opinions on the daily and yearly course of the earth".[3] Although Blaeu did not make a choice, he was a great protagonist of Copernicanism. Simon Stevin (1548-1620), who played an important role in the political, economical, and cultural life of the young Republic of the Northern Netherlands and who was the mathematics and science tutor to Maurice of Nassau, prince of Orange, was the first adherent of the heliocentric system in the Northern Netherlands, several years before Galileo became a Copernican. Stevin treated the system in his *Van den Hemelloop* (On the Heavenly Motions; published in 1608) as a theory which is physically true.[4] Philips van Lansbergen (1561-1632), a reformed minister and astronomer at Middelburg, was the first defender of the heliocentric system in the Northern Netherlands. In 1619 he taught the probability of the earth's motion in his book: *Progymnasmatum astronomiae restitutae de motu solis.* Ten years later van Lansbergen published a more direct defence of the Copernican system, now in the Dutch language: *Bedenckingen Op den Dagelijckschen, ende Iaerlijckschen loop van den Aerdt-kloot* (Reflections on the daily and yearly course of the terrestrial globe; 1629).[5] Like Simon Stevin van Lansbergen too believed that Copernicus's theory was physically true and that the cosmological texts in Holy Scripture were adaptations of the common way of speech. The book was published with the financial support of the strongly Calvinist states of the province of Zealand. It contains laudatory poems by the poet and pensionary Jacob Cats (1577-1660), the poet Johan de Brune (1589-1658), and the secretary of the Synod of Dordrecht (1618-1619) Daniel Heinsius (1580-1655). Opposition to van Lansbergen came mainly from the Southern Netherlands. It was the Roman-Catholic priest Libertus Fromondus who attacked the contents of van Lansbergen's book: *Progymnasmatum* in his work: *Ant-Aristarchus sive Orbis*

3. W.J. Blaeu, note 2, preface. Cf.: J. Keuning, *Willem Jansz. Blaeu. A biography and history of his work as a cartographer and publisher.* Revised and edited by M. Donkersloot-de Vrij (Amsterdam, 1973) 62-63.

4. S. Stevin, *Van den hemelloop.* Part 3 of volume I of: *Wisconstighe Ghedachtenissen* (Leyden, 1608). Although published in 1608, it was written at a much earlier date. Cf.: E.J. Dijksterhuis, *Simon Stevin* (The Hague, 1943) 146-167. On Copernicanism in the Netherlands, see: R. Hooykaas, "The Reception of Copernicanism in England and the Netherlands" in: *The Anglo-Dutch Contribution of Early Modern Society* (London, 1976) 33-44, and *G.J. Rheticus' Treatise on Holy Scripture and the Motion of the Earth* (Amsterdam, Oxford, New York, 1984) 178-183.

5. Ph. van Lansbergen, *Bedenckingen Op den Dagelijckschen, ende Iaerlijkschen loop van den Aerdt-kloot. Mitsgaders Op de ware Af-beeldinge des sienelijcken Hemels; daer in den wonderbare Wercken Godts worden ontdeckt* (Middelburg, 1629).

*Terrae Immobilis Liber Unus* (1631), in which he spoke of the "ridicula theologia Copernicana Lansbergii".

In a number of publications professor R. Hooykaas has sufficiently demonstrated contrary to White's thesis that when Calvinism in the Netherlands was in its heyday, nothing was done to prevent Copernicanism.[6] In the Northern Netherlands there was no 'warfare' between science and religion and the biblical arguments leading to Galileo's condemnation, were not used.

On the last day of 1619 the Dutch and the Venetian Republics signed a treaty of alliance. As a result both Republics exchanged ambassadors. From 1622-1627 Johan Berck (1565-1627), pensionary of the town of Dordrecht, was ambassador at Venice. In his train was the reformed minister Andreas Colvius (1594-1671), a poet and natural scientist,[7] who at Venice copied a number of manuscripts, among them Galileo's *Del flusso e reflusso del mare*. It was in this manuscript, which was completed in about 1616, that Galileo tried to prove the rotation of the earth by means of the tidal motion. In 1630 Colvius lent the manuscript to Isaac Beeckman (1588-1637), headmaster of the Latin School at Dordrecht.[8] Four years later, on 1 August 1634, Beeckman was given a copy of Galileo's *Dialogo sopra i due massimi sistemi del mondo tolemaico e copernicano* (1632) by a former pupil Martinus Hortensius (1605-1639).[9] Hortensius lectured, as professor of mathematics at the Athenaeum Illustre at Amsterdam, according to the Copernican system. About this time (1635) Franciscus de Weert (or: de Waert) entered the service of the Venetian Republic as an engineer.[10] He translated the *Dialogo* into the Dutch language and negociated with Lodewijk Elzevier (1604-1670) at Amsterdam about the printing.[11] Although a Dutch translation of Galileo's Dialogues of the two chief systems of the world never appeared, Galileo's

6. R. Hooykaas, "Hervorming en Natuurwetenschap", *Wending* 7 (1952) 640-647; "Science and Reformation", *Journal of World History* 3 (1956) 109-139; *Religion and the Rise of Modern Science* (Edinburgh, 1972; second edition 1973).

7. C.L. Thijssen-Schoute, "Andreas Colvius. Een correspondent van Descartes", *Nederlands Archief voor Kerkgeschiedenis* 38 (1953) 224-248. Also in: *Uit de Republiek der Letteren. Elf Studiën op het gebied der ideeëngeschiedenis van de gouden eeuw* (The Hague, 1967) 67-89.

8. *Journal tenu par Isaac Beeckman de 1604 à 1634 publié avec une introduction et des notes par C. de Waard* (The Hague, 1939-1954) Vol. III, 171. Cf.: K. van Berkel, *Isaac Beeckman (1588-1637) en de mechanisering van het wereldbeeld* (Amsterdam, 1983) 112-114.

9. *Journal* (o.c., note 8) Vol. III, 356.

10. C. de Waard, *Nieuw Nederlandsch Biografisch Woordenboek* (Leyden, 1914) Vol. III, col. 1398.

11. L. Elzevier to G. Galileo, 4 January 1638 (G. Galileo, *Le Opere* [Florence, 1890-1909] Vol. XVII, 251).

ideas and condemnation were generally known in the Northern Netherlands. This condemnation was — as is well-known — the result of a complex interplay of untoward political circumstances, personal ambitions, and wounded pride.[12] There was, however, no counterpart of Galileo's fate in the Northern Netherlands, although it stimulated the discussion of exegetical matters in the Dutch scientific and theological community. The Republic of the Northern Netherlands had no established church powerful enough to prevent the development and utterance of new scientific ideas which could threaten the intellectual foundations of religious orthodoxy. Galileo experienced the consequences of the fear of the Roman Catholic Church of new ideas which put to the test the ruling orthodoxy in the field of science and philosophy. Of course there were voices in the Dutch Reformed Church which proclaimed the same fear. But the reformed orthodoxy lacked the power to enforce their views on the whole society. It is evident that the universities as public institutions were exposed to more pressure than individual scientists. This situation may contribute to the explanation of the fact that more original scientific work in the seventeenth century was done outside the university milieu,[13] including the debate on Copernicanism.

A number of Dutch Copernicans appealed to John Calvin's influential theory of accomodation. Hooykaas has eloquently opposed the myth that Calvin (1509-1564) mentioned and rejected Copernicus's heliocentrism in his works.[14] Only in his commentary on the book of Genesis does Calvin point out that the story of creation does not compete with the "great art of astronomy", but accomodates and speaks in terms of the unlettered 'idiota', the common man.[15] "Moses wrote in a popular style things which, without instruction, all ordinary persons endued with common sense, are able to understand; but astronomers investigate with great labour whatever the

12. W.R. Shea, "Galileo and the Church", in: D.C. Lindberg and R.L. Numbers, eds., *God and Nature. Historical Essays on the Encounter between Christianity and Science* (Berkeley, Los Angeles, London, 1986) 114-135 (in particular 132). On the attempts to hold Galileo to the Netherlands, see: Van Berkel, *Beeckman* (o.c., note 8) 145-151.

13. K. van Berkel, "Universiteit en natuurwetenschap in de 17e eeuw, in het bijzonder in de Republiek", in: H.A.M. Snelders and K. van Berkel, eds., *Natuurwetenschappen van Renaissance to Darwin. Thema's uit de wetenschapsgeschiedenis* (The Hague, 1981) 107-130.

14. R. Hooykaas, "Calvin and Copernicus", *Organon* 10 (1974) 139-148; idem, *Religion* (o.c., note 6) 117-122.

15. H.O. Oberman, "Reformation and revolution: Copernicus' discovery in an era of change", in: O. Gingerich, ed., *The Nature of Scientific Discovery* (Washington, 1975) 134-169; G.B. Deason, "Reformation Theology and the Mechanistic Conception of Nature" in: Lindberg and Numbers, *God and Nature*, o.c., note 12) 167-191.

sagacity of the human mind can comprehend. Nevertheless, this study is not to be reprobated, nor this science to be condemned, because some frantic persons are wont boldly to reject whatever is unknown to them. For astronomy is not only pleasant, but also very useful to be known: it cannot be denied that this art unfolds the admirable wisdom of God. [...] Nor did Moses truly wish to withdraw us from the pursuit in omitting such things as are peculiar to the art; but because he was ordained a teacher as well of the unlearned and rude as of the learned, he could not otherwise fulfil his office than by descending to his grosser method of instruction. [...] Moses, therefore, rather adapts his discourse to common usage".[16]

Calvin's view, which in a certain sense we can also find in Galileo's *Letter to the Grand Duchess Christina* (composed in 1615 but eventually published in 1636) could not lead to verdicts about Copernicus's doctrine by the Reformed Church. As a Roman-Catholic Galileo was bound to obey the decree of the Council of Trent, which had decided that the interpretation of Scripture should conform to the common feeling of the Church Fathers. In a decree of 8 April 1546, the Council of Trent had declared: "Nobody has the right, basing himself on his own prudence, in matters of faith and morals that concern the edification of Christian doctrine, to violently divert the Holy Scripture towards his own opinion and to interpret it contrary to the sense that Holy Mother the Church holds and has held, to whom it belongs to judge the true meaning and the interpretation of the Holy Scripture, or even contrary to the unanimous agreement of the Fathers".[17]

In the Northern Netherlands Calvin's accomodation theory led to discussions between adherents and opponents of the heliocentric system who both had the opportunity to defend their views without the danger of a condemnation. It is interesting to note that the general impression abroad was that all inhabitants of the provinces of Holland and Zealand were Copernicans. Pierre Gassendi, who visited these provinces in 1629, believed this,[18] and Galileo thought that "all the most distinguished heretics", namely Protestants and especially Calvinists, in the North accepted this theory, as he wrote in his *Reply* (1624) to Francesco Ingoli (1578-1649), secretary of the

16. J. Calvin, *Commentaries on the First Book of Moses Called Genesis*. Translated by J. King (Edinburgh, 1847-1850) Vol. I, 86-87. Quoted from Deason, "Reformation Theology" (o.c., note 15) 171.

17. Paul Cardinal Poupard, ed., *Galileo Galilei. Toward a Resolution of 350 Years of Debate -1633-1983* (Pittsburgh, 1987) 120.

18. F. Sassen, "De reis van Pierre Gassendi in de Nederlanden (1628-1629)", *Mededelingen der Koninklijke Nederlandse Akademie van Wetenschappen. Afdeling Letterkunde*, N.S. 23 (1960) 263-307.

Propaganda Fidei.[19] Hooykaas pointed to the fact that the centres of oceanic navigation (that was in the provinces Holland and Zealand) brought forth most adherents of Copernicus, and further that the new system met wide approval from non-academic people, in particular Mennonites from the northern part of Holland.[20] Their arguments were mainly based on the interpretation of biblical texts. Philips van Lansbergen and his son Jacob (1590-1657) explicitly appealed to Calvin in their defence of Copernicanism. Scripture does not speak about astronomical matters. It is a book for laymen and written so as to be understood by both scientists and ignorants.[21] The mennonite shoemaker Dirk Rembrandtszoon van Nierop (1610-1682) tried to prove that the movement of the earth and the standstill of the sun were not contrary to Christian religion, although his co-religionist, the Amsterdam mennonite teacher of mathematics Sybrandt Hanszoon Cardinael (1578-1647) was an adherent of Ptolemy.[22]

In the northern provinces of the Republic, Groningen and Friesland, a number of astronomers rejected the earth's movement on scientific and religious grounds. Although Johannes Phocylides Holwarda (1618-1651) taught Copernicanism in his lectures at Franeker University,[23] his colleague Adriaan Metius (1570-1635) had followed Ptolemy's theory of the solar system.[24] Nicolaus Mulerius (1564-1630), headmaster of the Latin School at Leeuwarden (1608) and afterwards the first professor of medicine and mathematics at the new university of Groningen (1614), accepted at most the daily rotation of the earth.[25] The yearly movement of the earth he considered contrary to the Scriptures. In 1617 Mulerius edited the third edition of Copernicus's *De Revolutionibus Orbium Coelestium* (1543) as: *Astronomia instaurata*. He added short critical notes to the book, revealing that he

19. "Lettera a Francesco Ingoli in risposta alla *Disputatio de situ et quiete terrae*" (1624) in: Galileo, *Le Opere*, (o.c., note 11) Vol. VI, 509-561 (in particular 511).

20. Hooykaas, "Reception" (o.c., note 4) 39.

21. Van Lansbergen, *Bedenckingen*, (o.c., note 5; edition 1650) 18.

22. D. Rembrandtsz. van Nierop, *Des aertrycks beweging en de sonne stilstant; bewijsende dat dit geensins met de Christelyke religie is strijdende* ... (Amsterdam, 1661); S.H. Cardinael, *Mathematische ofte Wisconstighe bewysredenen, waar mede bewesen wort, dat de Aerdcloot stil staat en de Sonne daghelijcx en jaerlijcx sijnen loop doet* (Amsterdam, 1635).

23. J.P. Holwarda, *Dissertatio Astronomica* ... (Franeker, 1640).

24. A. Metius, *Institutiones Astronomicae et Geographicae* (Amsterdam, 1621) 10-14; *Primum mobile, astronomice, sciographice, geometrice, et hydrographice;* .... (Amsterdam, 1633) Vol. 1, 7-9; Cf.: H.A.M. Snelders, "Alkmaarse natuurwetenschappers uit de 16de en 17de eeuw", in: *Van Spaans Beleg tot Bataafse Tijd. Alkmaars stedelijk leven in de 17de en de 18de eeuw* (Zutphen, 1980) 101-122 (in particular 105-108).

25. N. Mulerius, *Tabulae Frisicae Lunae-Solares quadruplices; è fontibus Cl. Ptolemaei, Regis Alfonsi, Nic. Copernicus, et Tychonis Brahe* (Alkmaar, 1611).

considered Copernicus a restorer rather than an innovator of astronomy. There was also theological opposition to the Copernican system. On 29 September 1608 the theologian Ubbo Emmius (1547-1625) , who was to be the first Rector of Groningen University, wrote to his friend Sibrandus Lubbertus at Franeker a letter following the publication of Stevin's book: *Wisconstighe Ghedachtenissen* (Mathematical Memoirs). Emmius described Stevin's astronomical opinions as worse than absurd and preposterous. If they were true, he contended, Moses would have been a liar and Holy Writ would lie, too.[26]

But among theologians there were also those who did not reject the Copernican system on biblical grounds. The most influential theologian of the first half of the seventeenth century, André Rivet (1572-1651), did not deem the Copernican system to be against the Scriptures.[27] Although there were a number of theologians who agreed with Calvin's method of exegesis, others, however, adhered to scholastic philosophy, and for that reason were prompted to a traditional exegesis.[28] In particular Gisbertus Voetius (1589-1676), professor of theology at the Athenaeum Illustre (1634) and University (1636) of Utrecht was a fully fledged scholastic theologian. He deemed scholastic philosophy the only philosophy which conformed to Scripture. Biblical passages were — according to him — to be taken as literally as possible. Therefore he opposed new theories such as atomism and the movement of the earth supporting his reasoning by many references to scholastic authors like the Jesuits Suarez and Pereira, at the same time completely ignoring Calvin's ideas of accommodation. In particular Voetius was influenced by the Spanish Jesuit Benedict Pereira (1535-1610) who defended the literal interpretation of biblical texts in his chief work, a commentary on Genesis entitled *In Genesis commentarium et disputationes* (1591). Voetius was of the opinion that "Holy Scriptures teaches not only what is necessary for salvation, but also lays down the principles of all good sciences and arts, mathematics, astronomy, physics, logic, and so on".[29] He

26. *Briefwechsel des Ubbo Emmius.* Herausgegeben von H. Brugmans (The Hague, 1923) Vol. II, 51-52. Cf.: Dijksterhuis, *Stevin,* (o.c., note 4) 331.

27. A.G. van Opstal, *André Rivet. Een invloedrijk Hugenoot aan het Hof van Frederik Hendrik* (Harderwijk, 1937); H.J. Honders, *Andreas Rivetus als invloedrijk gereformeerd theoloog in Holland's bloeitijd* (The Hague, 1930).

28. J. Dillenberger, *Protestant Thought and Natural Science. A Historical Interpretation* (New York, 1960); Hooykaas, *Rheticus'* (o.c., note 4) 176-178 and Idem, *Religion* (o.c., note 4) 130-131.

29. G. Voetius, *Sermoen van de Nuttigheydt der Academien ende scholen, mitsgaders der wetenschappen ende consten die in de selve gheleert werden* (Utrecht, 1636) 35-36 and *Thersites Heautontimorumenos ....* (Utrecht, 1635) 266, 281, 283. Cf.: F. Sassen, *Geschiedenis van de*

also defended Ptolemaic astronomy on biblical grounds. In 1634 he posited the thesis: "That the heaven rests and the earth is rotating in a diurnal motion, can not at all be proved from Scripture and, duly considered, it is even clearly rejected by it".[30] The Copernican system was in flat contradiction to the text and the intention of the Bible.[31] According to Voetius the geocentric world view was inextricably entwined with the reformed doctrines so that there was little or no room for scientific research in the modern sense of the word.

Voetius not only combatted Copernicanism, but also Cartesianism. As is well-known, seventeenth-century Protestant Holland served as a refuge for threatened science. Descartes preferred to live and study there undisturbed. Most of his works were written in the Northern Netherlands. Already during his life René Descartes had found many adherents among orthodox theologians. It was, however, Voetius in particular who condemned Descartes's theory of methodic doubt and his demand that philosophy should be independent from theology as sharply as he condemned his physical theories.

Descartes made his impact on theology primarily in the Northern Netherlands.[32] "There his thought was frequently amalgamated with Aristotle's, with the result that the Cartesian opposition could equally well be held along with or in opposition to the Copernican system. Moreover, Cartesianism was theologically no less intellectual or rationalist than Aristotelism".[33]

From the 1640s until the end of the seventeenth century the struggle between the adherents and opponents of Cartesianism dominated the universities of the Republic. After the synod of Dordrecht (1618-1619), where Arminianism was condemned, orthodoxy was on the alert for a new false doctrine. The Voetian party tried to discredit Cartesianism and Copernicanism by connecting them with Armenianism and libertinism. The openly professed Cartesian physics of the Utrecht professor of theoretical medicine and botany Henricus Regius (1598-1679) was immediately attacked by

---

*wijsbegeerte in Nederland tot het eind der negentiende eeuw* (Amsterdam, Brussels, 1959) 143-148.

30. G. Voetius, *Assertiones theologicae de praejudiciis verae religionis ....* (Utrecht, 1634). Cf.: R. Hooykaas, *Rheticus*' (o.c., note 4) 169. Voetius referred to Psalm 19: 6-7 and 104: 5 and Ecclesiastes 1: 4-5. His rejection of the motion of the earth was more elaborated in 1635 (*Selectarum disputationum theologicarum pars I-V* [Utrecht, 1648-1669] Vol. I, 869-870). Cf.: A.C. Duker, *Gisbertus Voetius* (Leyden, 1897-1915) Vol. II, 152.

31. Duker, *Voetius*, (o.c., note 30) Vol. II, 32-33, 45.

32. J. Bohatec, *Die cartesianische Scholastik in der Philosophie und reformierten Dogmatik des 17. Jahrhunderts* (Leipzig, 1912).

33. Dillenberger, (o.c., note 28) 101.

Voetius because the former publicly upheld the thesis of a twofold motion of the earth.[34] This attack led to the condemnation of Regius in 1642 by the City Fathers of Utrecht. As a consequence Aristotelianism at Utrecht University remained the prevailing philosophy until the eighteenth century thanks to the social and institutional conditions of that town which made it possible for orthodox Protestants to have a direct influence on the university.[35]

At Leyden University Cartesianists saw to it that they were not connected with possible atheistic interpretations of Descartes's ideas. They refused an open confrontation with the Aristotelians. Though there was a lot of tumult there too, teaching became more and more influenced by Cartesianism.[36]

During the seventeenth century the conflict over Cartesianism between the theology and the philosophy faculties of the Dutch universities was steadily increasing. The general feeling of the Dutch universities about Descartes is to be illustrated by an advice from the University of Harderwijk to Count Lodewijk Hendrik of Nassau of 28 September 1651.[37] In that advice there was much praise for Descartes's mathematics and physics, but not for his medical theories. His metaphysics was thought to be pernicious to theology. On 30 September 1656 the States of Holland published a resolution, entitled: "Regulation concerning the mixing of Theology with Philosophy, and the abuse of the freedom of Philosophising to the detriment of Holy Scripture".[38] In the resolution an oath was formulated which stated that the philosophy faculty could not teach Cartesianism, and that neither faculty could encroach on each other's subject. The resolution added: "Although much has been said about the generous attitude of the States towards the liberty which is necessary for philosophizing, nobody must imagine that because of it the public and accepted manner of teaching may be dispensed with".[39]

34. M.J.A. de Vrijer, *Henricus Regius, en "Cartesiaansch" hoogleeraar aan de Utrechtsche Hoogeschool* (The Hague, 1917).

35. K. van Berkel, "Descartes in debat met Voetius. De mislukte introductie van het cartesianisme aan de Utrechtse Universiteit (1639-1645)", *Tijdschrift voor de Geschiedenis der Geneeskunde, Natuurwetenschappen, Wiskunde en Techniek* 7 (1984) 4-18.

36. E. Ruestow, *Physics at seventeenth and eighteenth-century Leiden: philosophy and the new science in the University* (The Hague, 1973).

37. A.A.M. de Haan, *Het wijsgerig onderwijs aan het Gymnasium Illustre en de Hogeschool te Harderwijk, 1599-1811* (Harderwijk, 1960) 65-67.

38. "Ordre jeghens vermenginge van de Theologie met de Philosophie, ende het misbruyck van de vreyheyt in het Philosopheren tot nadeel van de Heylige Schriftuyre", in: P.C. Molhuysen, *Bronnen tot de geschiedenis der Leidsche Universiteit.* Rijksgeschiedkundige Publicatiën (The Hague, 1918) Vol. 38, 47*-58*.

39. W.D. Hackmann, "The Growth of Science in the Netherlands in the Seventeenth and

This oath was only applicable at Leyden University and did not prevent the ideas of Descartes from being taught in the Republic. The first Cartesian theologians were Abraham Heidanus (1597-1678), from 1648 professor of theology at Leyden University, and Christoph Wittichius (1625-1687), from 1655 professor at Nijmegen and from 1671 at Leyden University.[40] Wittichius distinguished between biblical passages which needed to be understood in their vulgar sense and those which testified to doctrine. On the basis of the concept of accommodation, he defended Copernicanism.[41] But the Leyden theology faculty, standing in the tradition of Voetius, judged the heliocentric system to be against Scripture. Wittich further insisted that theology could not dictate the truth of philosophy. According to him philosophy did not exist to serve theology.

The most positive use of Descartes was made by the minister Balthasar Bekker (1634-1698). Descartes' strict separation between substance and mind gave him the possibility of explaining all natural phenomena as purely material. Bekker attacked the fear for comets. In his book: *De Betooverde Wereld* (Bewitched World; 1691-1693) he fought against superstition and in particular against sorcery. Bekker did not deny the existence of angels and devils, but he tought it unthinkable that these beings should act on the material world, unless of course it was the will of God. He tried to explain the biblical information on this as metaphorical. Bekker held a coordinate view of reason and Scripture. The last dealt with salvation, reason with nature.[42]

The discussion on the interpretation of biblical texts can among others be illustrated with a number of booklets written by the Utrecht physician Lambertus van Velthuysen (1622-1685) and the Leyden minister Jacobus du Bois (died 1661). In 1655 van Velthuysen, a fierce adversary of Voetius, wrote an anonymous: "Proof, that the feeling of those, who teach the stillstand of the sun and the movement of the earth, is not contrary to God's word". Du Bois immediately reacted in a number of refutations, whose contents are reflected by their titles: "Nakedness of the cartesian philosophy, bared in an

---

Early Eighteenth Centuries", in: M. Crosland, ed., *The Emergence of Science in Western Europe* (London, Basingstoke, 1975) 89-109 (in particular 90-91).

40. J.A. Cramer, *Abraham Heidanus en zijn Cartesianisme* (Utrecht, 1889); Dillenberger (o.c., note 28) 101-102; C.L. Thijssen-Schoute, *Nederlands Cartesianisme*. Verhandelingen der Koninklijke Nederlandse Akademie van Wetenschappen. Afdeling Letterkunde (Amsterdam, 1954) 58-59.

41. Chr. Wittich, *Consensus veritatis in Scriptura divina et infallibili revelatae cum veritate philosophica a Renato des Cartes detecta* (Nijmegen, 1659).

42. W.P.C. Knuttel, *Balthasar Bekker, de bestrijder van het bijgeloof* (The Hague, 1906); Thijssen-Schoute, *Nederlands Cartesianisme*, (o.c., note 40) 505-517.

answer on a Cartesian libel, called Proof that the feeling of those, who teach the stillstand of the sun and the movement of the earth, is not contrary to God's word" and "Harmfulness of the Cartesian philosophy, or clear proof of how harmful that philosophy is, both in the detaching of God's Holy Word, and in the introduction of new Harmful teachings". Apparently du Bois realised that the discussion was more than just the question of the movement of the earth. Cartesian philosophy was also at stake. Therefore, van Velthuysen extended the second edition of his booklet with a defence of Descartes: "Proof that neither the doctrine of the sun's stillstand and the earth's movement, nor the principle of Renatus Descartes are contrary to God's word" (1656).[43]

The history of relations between science and religion in the Northern Netherlands was a remarkable one with violent oscillations in feelings of ambiguity (before the seventeenth century), via cool distance (in the seventeenth century) and amorousness (in the eighteenth century) to an irreconcilable attitude (in the nineteenth century) and finally indifference (in the twentieth century).[44]

The seventeenth century tension between science and religion changed in the eighteenth century into a physicotheological natural science. The popularisation of the natural sciences in the eighteenth century had a strong religious connection. It was believed that investigations into the working of nature should lead to a better understanding of its Creator. In his well-known book: *Het regt gebruik der wereltbeschouwingen, ter overtuiginge van ongodisten en*

43. [L. van Velthuysen], *Bewys, Dat het gevoelen van die genen, die leeren der Sonne Stilstandt, En des Aertrijcks Beweging niet strydigh is met Godts-woort* (Utrecht, 1655); J. du Bois, *Naecktheyt Van de Cartesiaensche Philosophie, Ontbloot in een Antwoort Op een Cartesiaensch Libel, Genaemt Bewys, dat het gevoelen van die gene, die leeren der Sonne-stilstandt, en des Aerdtrijcks beweging niet strydig is met Gods Woort* (Utrecht, 1655); L. van Velthuysen, *Bewys Dat noch de Leere van der Sonne Stilstand, En des Aertryx Bewegingh, Noch de gronden vande Philosophie van Renatus Des Cartes strijdig sijn met Godts woort. Gestelt tegen een Tractaet van J. du Bois, Predikant tot Leyden: Genaemt Naecktheyt vande Cartesiaensche Philosophie ontbloot, etc.* (Utrecht, 1656); J. du Bois, *Schadelickheyt Van de Cartesiaensche Philosophie, Ofte Klaer Bewijs, hoe schadelick die Philosophie is, soo in het los maecken van Godes H. Woordt, als in het invoeren van nieuwe schadelicke Leeringen. Tot antwoort Op de tweede en vermeerderde druck van Doct. Velthuysens Bewys* (Utrecht, 1656); L. van Velthuysen, *Nader Bewys. Dat noch de Leere van der Sonne Stilstant, En des Aertryx Beweging, Noch de gronden van de Philosophie van Renatus Des Cartes strydig sijn met Godts Woort. Gestelt tegen een Tractaet van J. du Bois, Predikant tot Leyde; Genaemt Schadelickheyt van de Cartesiaensche Philosophie etc.* (Utrecht, 1657).

44. J. Bots, *Tussen Descartes en Darwin. Geloof en natuurwetenschap in de achttiende eeuw in Nederland* (Assen, 1972) 1.

*ongeloovigen* (The right use of world views for the conviction of atheists and unbelievers; 1715), Bernard Nieuwentijt (1654-1718) intended to prove the existence of God on the basis of teleological arguments. In the course of the eighteenth century a whole number of physicotheological books were published in which the existence of God was justified by the admirable structure of the world, including theologies based on snow, lightning, insects, and grass-hoppers. These books not only furnish ample evidence of God's existence, but also convinced men of the Creator's wisdom and omnipotence. Also Newtonian scientists such as Herman Boerhaave (1668-1738), Willem Jacob 's Gravesande (1688-1742) en Petrus van Musschenbroek (1692-1761) based their scientific empiricism on religious presuppositions.[45] Van Musschenbroek devoted a whole *Oratio de sapientia divina* (1744) to God's wisdom and care as they appear from all things created and from their preservation, and also from the efficient order of nature. 's Gravesande had revealed his religious belief by undertaking a "mathematical demonstration of the care God takes to direct that which takes place in this world, drawn from the number of boys and girls born daily".[46]

Returning to the seventeenth century we may conclude that in the Northern Netherlands there was much discussion between science and religion, but that discussion was not followed by censure by the Church or the State. Compared with other countries, there was a considerable amount of religious tolerance in the Republic, thanks to practical and ideological considerations.

The seventeenth century was a period in which social stability was preferred to religious orthodoxy. There was a large amount of intellectual freedom in the study of the natural sciences as long as there was no danger to religion or to the state. That was obviously not the case with Cartesian philosophy or with different dissenting religious views such as the writings of, for instance, Adriaan Koerbach (c. 1632-1669), a friend of Spinoza, who fought against every dogmatic religion, in particular against Calvinism and Roman Catholicism. Koerbach's notions caused so much scandal that the city government of Amsterdam, at the request of the church council, confiscated all unsold copies of Koerbach's explanatory dictionary of theology, philosophy, law, and medicine, which he had published in 1668.[47]

45. C. de Pater, *Petrus van Musschenbroek (1692-1761), een Newtoniaans natuuronderzoeker* (Utrecht, 1979).

46. W.J. 's Gravesande, *Oeuvres philosophiques et mathématiques* (Amsterdam, 1774) Vol. II, 221-248.

47. A. Koerbach, *Een Bloemhof van allerley lieflijkheyd sonder verdriet, geplant door Vreederijk Waarmond* (Amsterdam, 1668). Cf.: Thijssen-Schoute, *Nederlands Cartesianisme* (o.c., note 40) 362-366.

In official papers Calvinism was indicated as the only true Christian religion. But attempts to raise the Calvinist church to the status of the official established church had no results. In Galileo's time the climate for scientific research in the Northern Netherlands was much more favourable than in Italy, thanks to the political conditions in the Republic.

# MECHANICS AND HYDROSTATICS IN THE LATE RENAISSANCE: RELATIONS BETWEEN ITALY AND THE LOW COUNTRIES

*Carlo Maccagni*

One of the subjects that most interests me in my research into the history of science and techniques during the Renaissance is the way in which knowledge was passed among scientists and among technicians. In fact it is only possible to find a few cases, at least in Italy, of close links between the two sectors or people who, in themselves, had both specializations. The two fields are almost autonomous and each has its own characteristics: school or apprenticeship for education, demonstrations or analogous statements as ways of reasoning, theoretical processes or gaining experience as ways of developing knowledge, Latin and writing or the vernicular and drawings as means of expression, and writings and printed material or personal contacts of travelling workers for the circulation of knowledge.

In this Congress, devoted to the relations between Italy and the Low Countries in the XVIIth and XVIIIth centuries, it would be interesting to offer some examples of the circulation of knowledge between these lands in the two fields I have delineated. In particular, for the technicians, some engineers came from the Low Countries, for example Willem de Raet and Cornelis Meijer, and introduced techniques and methods which were new to us and which remained definitively in the field of applied hydraulics for the management of rivers, the maintenance of ports and canals and the draining of marshes for land reclamation. But I will limit myself to science along that complicated initerary, parts of which are little known, followed by the ideas of Giovanni Battista Benedetti up to those of Simon Stevin, while some hint about the presence of the ideas of the latter with reference to Galileo and his associates will close this paper.

* * *

The Benedetti affair arose at the beginning of his scientific activity.

Giovanni Battista Benedetti was born in Venice on 14th August 1530 and died in Turin on 20th January 1590. His life can be divided into three parts:

from his birth until 1558-1559 when he left Venice for Parma where he was appointed lecturer in philosophy and mathematics in the court of Ottavio Farnese; he remained there until 1567 when he went to Turin as mathematician to the court of the Savoys. He remained in Turin, under Emanuele Filiberto and Carlo Emanuele I until he died. The first and the last period are the most important for his scientific production.

In Venice he was educated by his father, of whom we know nothing, and was a pupil of Nicolò Tartaglia who read him the first four books of Euclid's *Elements*. He was thus educated outside the traditional school and university and was a typical example of that group of scientists, above all mathematicians, especially numerous in Italy, who played such an important role in the scientific revival in the XVth and XVIth centuries while remaining outside traditional educational institutions. Most probably the non-involvement in academic culture, dominated by the Aristotelian tradition in its different forms, left them more mental liberty which favoured their innovations.

Benedetti's first work had already been published in Venice in 1553 by the printer Bartolomeo Cesano: *Resolutio omnium Euclidis problematum aliorumque ad hoc necessario inventorum una tantummodo circini data apertura*, on the solving of problems of geometry with ruler and fixed compasses (see fig. 2).

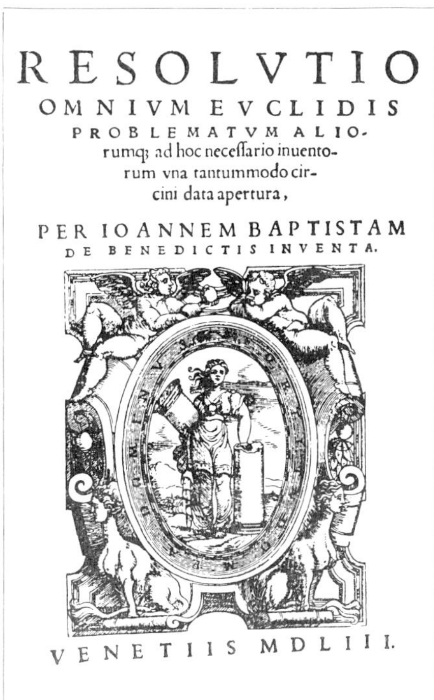

RESOLVTIO
OMNIVM EVCLIDIS
PROBLEMATVM ALIO-
rumq; ad hoc necessario inuento-
rum vna tantummodo cir-
cini data apertura,

PER IOANNEM BAPTISTAM
DE BENEDICTIS INVENTA.

VENETIIS MDLIII.

**Fig. 2 Title page of Benedetti's *Resolutio* (1553)**

While the mathematical part is not very important for us, the lengthy dedication of the book to the Abbot of Guzman is very interesting because it already shows a decidedly polemical attitude towards the Aristotelian tradition, which was to characterize all his activities, and also because it presaged in synthesis his theory about free falling bodies, to which, at least in part, he remained faithful until his last work *Diversarum speculationum mathematicarum et physicarum liber* (Taurini, apud haeredes N. Bevilaquae 1590). In the *Resolutio* Benedetti summarized his theory which he promised to publish again later in its complete form and said that he had already passed it on personally to other people, explaining that he had done this out of fear of other people taking his ideas by intercepting the letters in which he had described them:

> "Haec libentius privatim ad praeces scripsissem quam publice evulgassem, nisi (quod saepe accidere videmus) interceptionem litterarum timuissem ne ille, quisquis tandem fuisset, praeripuisset et nomine suo aut alicuius suorum in lucem hanc demonstrationem aedidisset, et ego oleum et operam perdidissem, prout multis aliis nostroque seculo accidit. Est siquidem genus quoddam hominum, fucos vocant, qui predicari magnique fieri gaudent, sed labores fugiunt, gloriam maximo et labore et sudore alieno partam per fas nefasque in sese transmoventes: eam ob causam coactus sum, eam, ita uti cernis, aedere, nimirum cum volumen de quo dixi evulgandi nunc animus non sit, in quo multa alia maiora his continentur" (*Resolutio* (1553), c.[10]v).

Benedetti's criticism aims to demonstrate the groundlessness of Aristotle's well-known theory that the velocities of falling bodies in a medium are proportional to the gross weights of the bodies themselves, and to the *ratio* of the gross weight of the falling bodies to that of the medium through which they fall. This opposition to the Aristotelian theory is based on Archimedes' hydrostatics (*De insidentibus aquae*). Benedetti drew two important conclusions from the discussion: bodies which are similar in form, or uniform (to avoid different reactions to the penetration into the medium), and of the same composition, or homogeneous, equal or unequal in weight, fall at the same velocity in the same medium:

> "dico quod si fuerint duo corpora eiusdem formae eiusdemque speciei, aequalia invicem vel inaequalia, per aequale spacium in eodem medium in aequali tempore feruntur" (*Resolutio* (1553), c.[9]r);

the velocity of fall is proportional to the *excess* of the gross weight of each of the falling bodies in relation to the gross weight of the medium through which they fall, which means to the *difference* of the respective specific gravities; and the author emphasizes explicitly the originality of this conclusion:

> "Ex his liquet motum magis velocem non causari ab excessu vel gravitatis aut levitatis corporis velocioris collatione tardioris (datis corporibus similis figurae), verum ex differentia speciei alterius corporis ad alterum, gravitatis levitatisve respectu, quae res non est ex mente Aristotelis aut alicuius suorum commentatorum quos mihi quidem videre et legere contigit, aut etiam contulisse cum eiusdem professoribus" (*Resolutio* (1553), c.[10]r).

Finally when the medium and the body immersed are homogeneous and therefore neither has an excess of specific gravity in relation to the other, the body is in indifferent equilibrium.

Substantially Benedetti's demonstration procedure consists in establishing the characteristic of natural motion of a body in a given medium, starting from Archimedes' hydrostatics, and therefore in the substitution of that body by a certain number of bodies, homogeneous and uniform to the same body, on the whole equal to it in weight, ideally connected rigidly by a weightless line and which meet no resistance to movement in the medium. In this case the smaller bodies will fall at the same velocity as the greater ones. By suitably varying the sizes and the types of the bodies and assorting them conveniently, the author manages to demonstrate his hypothesis.

The conclusive part *de motu* in the dedication of the *Resolutio* continues the controversy with the Aristotelian doctrine because Benedetti, attempting to extend the conclusions about falling bodies in a medium to violent motions, maintains not only the simple possibility but the *necessity* that any movement should be a mixture of natural and violent motion. This was opposed to the Aristotelian thesis by which one distinguishes the two types of motion qualitatively, considering them absolutely different in terms of their nature so that the two types of motion cannot be mixed.

Without any doubt Benedetti is right to claim the originality of his conclusions because they have little in common with those that went before him, but the adoption of Archimedes' hydrostatics and the anti-Aristotle controversy are not enough to separate his mechanics from tradition. Although he also makes efforts to discuss Aristotle's theories in relation to questions which are wider and more general than purely technical ones (an attitude that one does not see in Tartaglia for example, from many points of view close to Benedetti) he is still very attached to the old dynamics. In fact he conserves its terms and in part the concepts of *levitas* and *gravitas*, originally connected with the theory of the *loci naturales*; he sustains weight as *virtus movens* of falling bodies and velocity proportional to weight, albeit relative and not absolute weight; he continues, as many others in his time, to persist in wanting to deduce dynamics from statics. However, we can observe in Benedetti many attempts, although not always clear and conclusive, to go in new directions. He tends towards a simplified abstraction of reality as a compromise between its empirical complexity and the conceptual simplicity of theory, even in the sense of mental experimentation. He refuses bald empirical data and tends to relate these data not to metaphysical essence but to a structured and quantitative vision of physical reality. The rational instrument for articulating and organizing his address and inferring the observable consequences is mathematical logic hinged upon the space-time relationship rather than attributive logic hinged upon the substance-accident, essence-property relationships. As a result he interprets in the quantitative sense (according to Archimedes) the qualitative properties (according to Aristotle) of bodies by which the *gravitas* and *levitas* as previously absolute properties become variable and definable only in the consideration of the excess of weight of the body with respect to that of the medium. Finally he makes explicit the inevitable mixture of natural and violent motion that Tartaglia had set out with great hesitation.

The pre-announced work in the dedication of the *Resolutio*, was printed "Venetiis MDLIIII" by the same printer. It is a quarto booklet of 12 unnumbered leaves, with 9 woodcuts some of which had already been printed in the previous publication, entitled, with youthful boldness, *Demonstratio proportionum motuum localium contra Aristotilem et omnes philosophos* (see fig. 3).

The controversial intention of the author is already clear from the title, an intention which is developed in a more polemical way than in the preceding work. The conflict is played out above all on the contraposition between the statements of Aristotle and his Commentators and the conclusions of the mathematicians, to the point of asserting that his own reasoning as well as that of the latter is managed by mathematical demonstration and not by Aristotelian loquacity: "mathematica demonstratione et non aristotelica garrulatione (*Demonstratio* (1554), c.[8]v). To show that the time was ripe for science to claim its autonomy from philosophy, he calls for the first time upon *auctoritates* the great geometers of the past, Euclid, Archimedes and Apollonius, to contradict the philosophical tradition, sometimes with little respect for the disciplinary boundary-lines that that tradition had firmly established.

The booklet is divided into three parts: the citation of the specific points of the Aristotelian philosophy that the author intends to discuss; the repetition of the theses already presented in the dedication of the *Resolutio* with some additions to the discourse or more precise formulations, including an important brief addition concerning a question which he had not discussed

DEMONSTRA
TIO PROPORTIONVM
MOTVVM LOCALIVM CON-
TRA ARISTOTILEM ET
OMNES PHILOSOPHOS.

Per Ioannem Baptiſtam de Benedictis inuenta.

*Cum Gratia & Priuilegio.*

VENETIIS M D LIIII.

**Fig. 3 Title-page of Benedetti's *Demonstratio* (1554)**

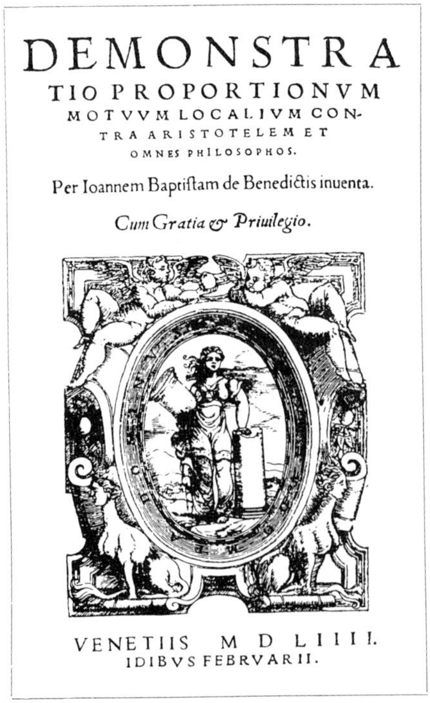

DEMONSTRA
TIO PROPORTIONVM
MOTVVM LOCALIVM CON-
TRA ARISTOTELEM ET
OMNES PHILOSOPHOS.

Per Ioannem Baptiſtam de Benedictis inuenta.

*Cum Gratia & Priuilegio.*

VENETIIS M D LIIII.
IDIBVS FEBRVARII.

**Fig. 4 Title-page of Benedetti's *Demonstratio* (1555)**

before, namely the movement of the bodies in vacuum; and the confutation by geometry of other Aristotelian "errors".

Concerning the motion of bodies in vacuum, which was one of the fundamental points of Aristotelian natural philosophy, Benedetti not only concedes it as a possibility, but maintains that bodies of any species, form and dimension fall at the same velocity:

> "si daretur motus in vacuo, ... corpora moventia ... moventur in tempore aequali, licet diversarum fuissent specierum formarumque corpora illa atque magnitudinum" (*Demonstratio* (1554), c.[8]r-v).

As for the last part, which is less interesting and often becomes a real forcing of logic with dubious formal justification, I will recall only one example, namely the recourse to the rectification of the circumference and then of curves in general. This was in accordance with the methods of Euclid, Archimedes and Apollonius with the purpose of contradicting Aristotle's assertion (*Physica*, VI.2.232*a*) "quod linea recta non est comparabilis lineae curvae", which, however, he had made in natural philosophy in the discussion *de motu naturali et violento.*

Again in Venice we have a rehashed version of Benedetti's theory *de motu* published "MDLIIII idibus Februarii", reduced from 12 to 10 leaves, with the same title except for *Aristotelem* instead of *Aristotilem* (see fig. 4). The dating is expressed *more veneto* with the year going from the first of March to the last day of February and the year number referred to as the first ten months with respect to the common manner: therefore the publication date in our language is 13th February 1555.

The new edition was set up using, in part, the typographic composition and the woodcuts of the previous edition. Benedetti cools his controversial ardour, especially reshaping the central part while the final part remains unaltered.

The substance of the contents does not change with reference to the free fall motion in vacuum and the mixture of natural and violent movement. As for the fall in a medium, beyond what he had already demonstrated about the proportionality of velocity in terms of the excess of the weight of the body over the weight of the medium, Benedetti emphasizes the friction between the body in motion and the medium, asserting that this friction bears proportionally on the maximum section of the moving body, normal to the direction of motion.

The treatise in its central part is reduced by about a third in comparison with the previous edition and the question is fragmented into a large number of objections to specific statements of Aristotle. We do not know why Benedetti modified his text in this way.

This small work does not seem to have been diffused much at the time in either edition. There are very few copies of the first edition in the public libraries today and only one copy of the second; the latter was absolutely unknown to historians until 1967 when I published the critical edition of the dedication of the *Resolutio*, of the two versions of the *Demonstratio* and of the variants of Taisnier's plagiarism which we deal with presently.

* * *

Jean Taisnier or Taisner of Ath in Hainout — in Latin Ioannes Taisnierius Hannonius — is a typical Renaissance figure with heterogeneous interests, great curiosity, a lively imagination and few inhibitions. He had a marked propensity for what we now call pseudosciences: astrology, chiromancy and physiognomy for which he became so famous as to be remembered as a chiromancer in the *Illustrium virorum elogia* by Tomasini, printed in Padova in 1630 (see fig. 5). He was fascinated by astronomy-astrology and by mathematics, curious about natural things in the widest sense that could then be given to those terms, and alert to the novelties of his time: discoveries, inventions and pseudo-inventions, magic wonders, astronomical and musical instruments,... In his works he relates that he travelled all the world, read mathematics and disputed of philosophy before cardinals and a hundred enthusiastic listeners, that he commanded troops and fought in battle, ... In reality, while he did not have by any means a quiet life, his account of it is only partly true.

He was born in 1508, had a good education and graduated in law; about 1535 he was appointed cantor to Charles V whom he followed on his journey to Spain, Italy, the Low Countries and in the expedition against Algiers in 1541. This relationship lasted until the Emperor's abdication in 1555; however, there were two interruptions when, still in an ambience closely linked with Spain, he was a musician first to the Archbishop of Palermo, Pietro Tagliania, and then to the Cardinal of Burgos, Francisco de Mendoza y Bobadilla. He lived in Italy in this period from 1546 to 1553, in Palermo, Rome, Ferrara, Bologna, Padua and Venice. Finally he returned to the Low Countries with the Cardinal of Burgos; after the abdication of Charles V he spent three years in Malines as head of a secondary school; in 1558 he went to Cologne as maestro in the chapel of the Archbishop Johannes Gebhard I. He died in that city presumably shortly after 1562 because the programme of publications he had planned and initiated that year, did not continue after this date.

Besides the profession of cantor and musician, which he exercised all his

**Fig. 5** **"Symbolum Ioannis Taisnerii praeclariss. mathematici" (from I.P. Tomasini, *Illustrium virorum elogia,* p. 164)**

**Fig. 6** **Taisnier's portrait (from his *Opusculum,*** c.[1]v)

life, he followed his own interests — teaching mathematics in Rome and Ferrara between 1546 and 1548, practising astrology and chiromancy with such success that he was also consulted by means of letters containing drawings of the lines of the palm and publishing various works in Italian and in Latin.

Among these, dedicated to practical geometry (Ferrara, 1548), to an astronomical ring invented by him (Panhormi, 1550; Antverpiae 1560), on the use of the solid sphere (Ferrariae, 1548; Coloniae, 1558 and 1559), on astrology in general (Coloniae, 1559), on chiromancy and physiognomy (Coloniae, 1562 and 1563), he also printed an *Opusculum perpetua memoria dignissimum de natura magnetis et eius effectibus. Item De motu continuo. Demonstratio proportionum motuum localium contra Aristotelem et alios philosophos. De motu alio celerrimo, hactenus incognito.* Authore Ioanne Taisnierio Hannonio, utriusque iuris doctore, poëta laureato, musico et rectore sacellis musices reverend. Coloniensis Archiepiscopi, etc., Coloniae, apud Ioannem Birckmannum, anno MDLXII (see fig. 7). ... It is a quarto volume of pp. [4]-86-[2] with numerous woodcuts (very well made) among which is the portrait of the author with the legend "I.T.H. aeta[tis] suae aº 53 1562" stamped on the *verso* of the titlepage and repeated on the *recto* of the unnumbered last page (see fig. 6).

Taisnier was not in reality the author of the *Opusculum* as he had declared on the title-page, but merely the person who had gathered together the texts included, which are really the *Epistola de magnete* by Petrus Peregrinus of Maricourt of 1269 and the *Demonstratio proportionum motuum localium contra Aristotilem et omnes philosophos* by Benedetti in the edition of 1554.

The work on natural magnets had recently been published: Petri Peregrini Maricurtensis *De magnete seu rota perpetui motus libellus* ... per Achillem P. Gasserum nunc primum promulgatus, Augsburgi in Svevis 1558; the plagiarism was already noticed by William Gilbert in *De magnete magneticisque corporibus et de magno magnete tellure physiologia nova plurimis et argumentis et experimentis demonstrata* (Londini, Petrus Short 1600, p. 5): "Ex hoc Petro Peregrino Iohannes Taisner libellum extraxit et ut novum divulgavit".

Benedetti's treatise can immediately be recognized from the title itself which in relation to the original is only slightly changed by the use of *Aristotelem* instead of *Aristotilem* and *alios* instead of *omnes*. As for the contents, Taisnier removed Benedetti's dedication to the Abbot of Guzman and replaced it with a short preface "Ad lectores" and a longer one "Ad pium et non aemulum lectorem", in which he recalled his teaching of mathematics in Rome, Ferrara and other places: "salvo omni arrogantiae scrupolo, nostrae lectiones saepe decorabat praesentia trecentorum et am-

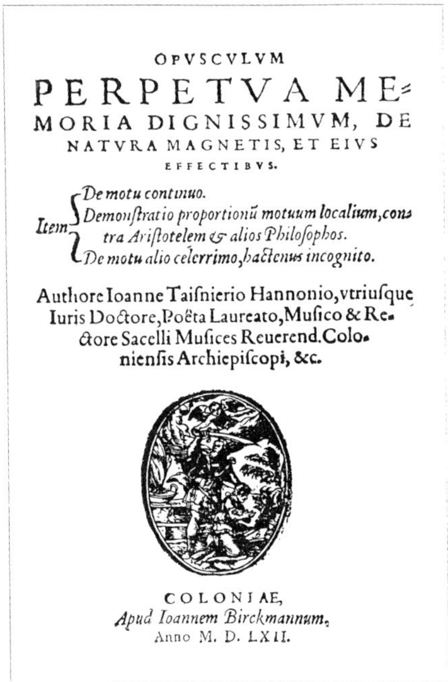

OPVSCVLVM
PERPETVA ME-
MORIA DIGNISSIMVM, DE
NATVRA MAGNETIS, ET EIVS
EFFECTIBVS.

Item { De motu continuo.
Demonstratio proportionũ motuum localium, contra Aristotelem & alios Philosophos.
De motu alio celerrimo, hactenus incognito. }

Authore Ioanne Taisnierio Hannonio, vtriusque Iuris Doctore, Poëta Laureato, Musico & Rectore Sacelli Musices Reuerend. Coloniensis Archiepiscopi, &c.

COLONIAE,
Apud Ioannem Birckmannum,
Anno M. D. LXII.

**Fig. 7 Title-page of Taisnier's *Opusculum* (1562)**

plius auditorium ... Aemuli et concurrentes aliquando medias disputationes iniuriis impediebant, ut aliquando (sicuti fit) viderentur finem disputationum verberibus et gladiis imponere"; he relates that his discussions of Aristotelian philosophy were maliciously revealed to the Pope and so he had to repeat them before cardinals, bishops and a large audience in the form in which they were published. As for the text, Taisnier removed every reference to Benedetti's life and friends and introduced about a hundred minimal variants of style or form, which do not affect the content. Taisnier planned as early as 1558 to publish Benedetti's *Demonstratio* which was pre-announced under the title "De motu locali", together with other future books in the preface to *De sphaerae materialis fabrica et usu* (Coloniae, Ioannes Bathenius 1558). Evidently the publication of the *Epistola de magnete* in that same year led Taisnier to combine, at a later stage, the two texts in the *Opusculum*.

We do not know when Benedetti became aware of the plagiarism, but in 1574, when publishing his treatise on gnomonics (*De gnomunum umbrarumque solarium usu liber*, Augustae Taurinorum, apud haeredes N. Bevilaquae 1574), he violently denounced Taisnier's impudence in daring to reprint, integrally, his treatise, changing only the name of the author and he said that Taisnier could not have done otherwise given his ignorance of mathematics;

he suspected that Taisnier had been courageous enough to print his, Benedetti's, writing because he thought he was dead; he laughed at the boast that his lessons in mathematics had three hundred listeners noting that in Italy even famous mathematicians could not muster a sixth of that — all this in a rising tide of abuse (c.[4]v).

> "Quid de manifestis furibus dicemus? qui vel ipsa integra aliorum volumina sibi imprudentes adscribunt, et quasi steriles ac scelesti plagiarii viventium filiorum (est enim haud dubie legitima proles quicquid fecundum ingenium lungo studio concepit, et peperit) miserandas infligunt piis parentibus orbitates, et se summa cum iactantia, eorum operum authores mentiuntur, quae magna cum infamia rapuerunt. Ut fecit impurissimus omnium Ioannes Taisnerus Hannonius. Qui opusculum nostrum, demonstrationis proportionum motuum localium contra Aristotelem, et alios philosophos, iamdiu antea a nobis editum, et iterum impressum Venetijs anno salutis 1554 ita integrum sibi desumpsit, ut nihil praeter authoris nomen immutaverit; quid enim mutavisset, qui nec percipere poterat quae in ea disputatione continerentur? homo vanus ab omni mathematica facultate alienus, qui merito propter crassissimam ignorantiam verebatur, ne vel aliqua syllaba sublata aut addita, totius tractationis inficeretur substantia. Credidit (ut opinor) me iam vita functum qui furti nunquam argui posse confidit. Et non intellexit suam temeritatem, qui seipsum mille argumentis qualis esset prodidit. Dum utre inflato inanior sese iuris doctorem, et simul etiam musici sacelli rectorem asseruit, quasi iura docere sit musici, aut iurisperiti sacellum regere, et dum de magnete, et motibus tractatus emisit, nusquam in titulis se mathematicum nominavit, sed poetam, eo quod crediderit poetae, aut musici, aut iurisperiti, esse de naturalibus motibus corporum diserere, debebat saltem et in hoc mentiri infamis impostor, ut se mathematicum in titulis praedicaret, ut in praefationem ad lectorem eiusdem usurpati opusculi fecit, dum se matheseos publice legisse Ferrariae, et alibi, trecentis, et pluribus auditoribus praedicat, cuius numeri auditorum ne sextam quidem partem quispiam vidit in Italia, in auditorio cuiusvis (etiam primi nominis) mathematici, quis inquam hos infames laudaverit in Flaviam legem committentes? ac non potius iuxta Constantini Caesaris sententiam, ad Celsum Aphricae Vicarium rescribentis, bestijs subijciendos censeat?"

Benedetti's theories had, however, been usurped notwithstanding the caution of publishing them repeatedly, and this happened not through the interception of private letters as he had feared initially, but through direct copying of the publication of 1554, and probably he never knew that while he was still alive the *Opusculum* of Taisnier was also published in an English translation between 1578 and 1579: *A very necessarie and profitable booke concerning navigation*, compiled in Latin by Joannes Taisnierius ... *named a treatise of continuall motions.* It was translated into English by Richarde Eden, London, R. Jugge, *sine anno.*

* * *

If, as has already been noted, the three versions of the theories *de motu* of Benedetti had little effect either on philosophers or mathematicians or on natural philosophers in the sense that there are no traces of it, Taisnier's plagiarism, printed in Continental Europe, nevertheless contributed to the spreading of his ideas to other countries and even reaching Simon Stevin who accurately referred to them as Taisnier's own work: and so they have been passed even by the Editors of the *Principal Works* of Simon Stevin, who did not recognize the true origin of Taisnier's pirate edition.

It does not seem desirable here to illustrate Stevin's work as amply as has been done for the little-known writings of the young Benedetti and the certainly less-known Taisnier: I will limit myself to recalling the occasion of the citation found in chapter II of the "Anhang van der Weeghconst" ("Appendix to the Art of Weighing"; "Appendix statices, ubi inter alia errores quidam Στατικῶν ἰδιωμάτων refelluntur") published at the end of *De Beghinselen des Waterwichts*, Leyden, Christoffel Plantijn 1586 (*The Elements of Hydrostatics*, in *The Principal Works ...*, I, pp. 65/509-68/515; *De hydrostatices experimentis*, in *Hypomnemata mathematica ...*, IV (1605), pp. 151-153).

In the first chapter of the "Appendix" Stevin rejects the so-called principle of virtual displacements which traditionally enables the foundation of statics on dynamical principles. In the second "dat de ghersorden met haer beletselen in gheen eueredenheyt en bestaen" ("that bodies in motion are not proportional to their impediments"; "res motas impedimentis suis non esse proportionales"), he combats various opinions about the motion of falling bodies in resistant media, explaining his theory as previously summarized in the preface "To the Reader" printed before *De Weeghdaet*, Leyden, Christoffel Plantijn 1586 (*The Practice of Weighing*, in *The Principal Works ...*, I, p. 5/297; *De statices praxi*, in *Hypomnemata mathematica ...*, IV, p. 81), where he had stated: "quanta vero moventis potentiae vis ... desideretur ..., quo datum pondus commoveatur vel impellatur, a Staticae doctrina alienum fuerit; cum via et ratione mathematica istae potentiae excessus inveniri neque explicari possit, etenim impedimenta et res motae constantem analogiam nullam habent" — that is to say: "as to ... the force which the moving body requires in order to set in motion the body to be moved ..., the Art of Weighing does not teach us to find that ... force mathematically; the cause of this is that the one moved body and its impediment are not proportional to the other moved body and its impediment" —.

Next he recalls the proper passage referring to the Aristotelian theory and summarizes what Taisnier (*i.e.* Benedetti) and subsequently Cardan had written against Aristotle; then the discussion follows, in which at first by empirical trials and afterwards by theory he disproves the two laws of falling bodies in a medium: namely, that velocity is proportional either to the *ratio* of the weight of the falling body to that of the medium, or to the *excess* of the weight of the falling body on that of the medium.

Against Aristotle he quotes an experiment (that is to say, an empirical proof) he carried out in collaboration with the Delft patrician Jan Cornets de Groot, "most industrious investigator of the secrets of Nature", which is of the same kind as that which Viviani says Galileo carried out some years later, dropping two different bodies from the leaning Tower of Pisa.

Against Taisnier (and Cardan) he describes two easily observable facts or experiences for falling weights and for ascending bodies respectively, both homogeneous and uniform and moving in a medium. In the first case, if from the same height and at the same time a small, single, short hair of cotton and a packet of the same, tightly tied together, are dropped, they will not fall in equal time through equal distances in the air; in the second, if a high, clear glass of water is stirred in such a way that many air bubbles are produced herein, the largest bubbles will ascend very rapidly, the smaller ones not so rapidly and the smallest ones still more slowly.

On the theoretical side, Stevin points out various remarks to state — as he has said before — that it is impossible to find any proportionality about the velocity of falling bodies between volumes, surfaces and specific gravities, because the medium through which the bodies fall is also a cause preventing such proportionality, even more so when concrete material conditions are concerned. In this way, he again reaches the general conclusion that "we should not rely on this appearance of proportionality, but consider the conclusions reached ... by other writers, to be errors, being satisfied with the mathematical knowledge of the equality of apparent weight of the moving body and the body to be moved as being sufficient for the purpose".

I wish to note the different treatments by Benedetti and by Stevin: the former has kept his discussion only on theorical ground, while the latter has examined the question both by experiment and theory, however, without setting up a unified approach in the experimental procedure.

I think it useful to quote the unabridged text of the second chapter of the *Appendix statices* in the Latin translation by W. Snellius:

"In Praxis Statices ad Lectorem praefatione diximus res motas suis impedimentis non esse proportionales, eiusque demonstrationi hunc locum destinavimus, ut argumenta aliter sententium refutemus. Principio *Aristoteles* eiusque sectatores 4 Physic. cap. de inani existimat corporibus duobus similibus, et materia aequipondiis per aërem delapsis eandem esse rationem ponderis ad pondus quae velocitatis illius ad velocitatem huius, idest quae sit impedimenti ad impedimentum. Quam sententiam variis locis clarius proponit ut 6 Physic. item 1, 2, 3, 4 de *Caelo*, aliisque compluribus. Sententiam hanc *Ioannes Taisnerus Hannonius* oppugnavit, proportionem quidem hactenus admittens ut corpora ista aequali temporis spatio aequalia permeent intervallo. Cui opinioni *Cardanus lib. 5 Proportion. propos. 110* consentit. Sed utrosque hallucinari ipsa experientia demonstrabimus, ac deinde eius causam declarabimus. Experientia vero contra *Aristotelem* istiusmodi est; sumito duos plumbeos globos (quod Cl. vir IOANNES GROTIUS sedulus naturae indagator, et ego quondam experti sumus) ponderis ratione decupla, eos altitudine 30 pedum pariter demittito in subiectum asserem, aliudve solidum unde sonus clare reddatur; manifeste cognosces leviorem non decuplo tardius graviore, sed pariter in asserem incidere ut sonitus utriusque illius redditus unus indemque videatur. Idem contingit in corporibus magnitudinis aequalis, gravitatis vero decuplae: quare dicta ista *Aristotelis* proportio a vero aliena est. Sed alterum experimentum huiusmodi contra *Taisnerum* facit: sumito e gossipio lanave tenue quoddam et exile filum, atque sarcinulam ex eadem materia pondere unius librae dense firmiterque colligatam, et forma filo simili, haec pariter quinque aut sex pedum altitudine demittito, re ipsa cognosces filum longe diutius in aëre morari, quam sarcinulam, etsi fili materia longe compactior densiorque sit sarcinula quae multum aëris admittit. Quare aequale spacium ab ipsi pari velocitate non transitur. Altera item experientia *Taisnerum* redarguit in pondere adscendente sive emergente, in phiala enim vitrea aquae plena agitata, ut multae excitentur bullulae simulac quievit videbis maiores bullas citissime atque unico momento, minores vero emergere tardius, minimas autem bullulas instar tenuissimarum arenularum lentissime, et tamquam testudineo gradu sursum prorepere, quarum omnium motus ab aequali velocitate vel tarditate longe abest. Atque hactenus de experientia. Superest ut dicamus cur hic nulla sit proportio, hoc modo. Quodlibet corpus movens habet quoddam motus sui impedimentum, quod in corpore per aërem delato est aëris et superficiei suae contactus; ideoque similium corporum maius, maiore quoque afficitur impedimento, sed quia similia solida superficiebus suis non sunt proportionalia (nam cubi in ratione octupla, habent superficies ratione quadrupla) nec impedimentis proportionalia esse possunt: atque hinc est quod minora corpora maius impedimentum patiantur, ratione proportionis, et propterea tardius descendant quam maiora.

Imo quamvis superficies corporibus suis essent proportionales, medium tamen per quod cadunt, quodammodo proportionem hanc evertit ut in duobus corporibus altero in aqua innatante, altero mergente animadverti facile potest, in quibus impedimenta superficierum quandam inter se habent rationem, tempora vero nullam, ideoque proportionalia non sunt. Sed dicat aliquis id intelligi solum caeteris paribus, videlicet quando utrumque corpus mergetur. Nego tamen in his ullam proportionem consistere. Sumptis enim duobus corporibus A, B quarum utrumque in aqua mergatur, sintque in dicta proportione. His positis, manifestum est infinita posse inveniri corpora inaequali gravitate minori quam B, et quae in aqua demergantur, paulatimque ita propius accedetur ad corpus immersabile, quius nulla cum corpore quod mergitur sit proportio. Sed illis eo continue accedentibus, atque A, B in data proportione consistentibus, certe nullum infinitorum illorum corporum ipsi A comparatum proportionem istam habebit; quia si in his esset, certe ad alterum non accederent quod thesi concessae repugnat. Quamobrem medium quoque per quod corpora permeant dictam proportionem prohibet.

Cumque in mediis ordinatissimis et ubique homogeneis nullam motus et impedimentorum proportionem inesse demonstraverimus, ubi simplex superficiei cum aëre vel aqua sit contactus. Longe firmiori ratione nulla proportio inerit exemplis magis inordinatis materiaeque non unius generis sed variae, ut in machinis partim ligneis partim ferreis caeterisque similibus, namque ibi hoc axungia illud oleo perungitur; aliud humido aëre turgescit, aliud erugine corripitur, quae omnia (ut multa alia omittam) machinarum motus modo expediunt, modo impediunt. Itaque ut in Statices Praxis praefactione dictum est, huiusmodi proportioni, quae probabilis videtur, nullo modo fidendum est: quin omnia ista quae *Cardanus 5 Proport. lib.* in variis *propositionibus*, aliique quam plurimi hinc deducunt, falsa et veri vana habenda sunt. Et solius moto et movendi aequilibrio contenti simus, quippe quae his abunde satisfaciat."

* * *

After considering how Benedetti's theories reached Stevin through Taisnier's plagiarism, I would like to refer at least to the question of whether the works of Stevin, especially after the Latin edition of 1605-1608 and the French one of the same years, were known by Galileo.

Several good objective reasons seem to point to his knowledge of Stevin's works, such as the relations between the Low Countries and Tuscany which were especially intense throughout the XVIIth century in the field of trade (the Dutch community in Livorno) as well as in the artistic, cultural and scientific fields up to the relations between Huygens and the Grand Duke at the time of the Accademia del Cimento, or the fact that Galileo was personally engaged in a series of controversies with Aristotelian philosophers about questions of hydrostatics (*Discorso intorno alle cose che stanno in su l'acqua o che in quella si muovono*, Firenze, Giunti 1612).

On this latter subject he had received some news in a letter — dated 11th

January 1611 from Linghen in the Low Countries — from his former pupil Daniele Antonini (1588-1616), who was here (1611-1612) serving as a soldier during the Flanders wars (G. Galilei, *Le opere* ..., XI, pp. 18-19), while in other letters he relates details to him about telescopes and observations of sunspots. Furthermore, it was Daniele's brother, Alfonso (1584-1657) who some years later in 1627 informed Galileo from The Hague about the prize to be offered by the States General of Holland for the solution of the longitude problem at sea (*ibidem*, XIII, pp. 376-377, 379-380), a problem which had well-known developments. We can add that the young Galileo, as early as 1588, had a first direct contact with the scientific culture of the Low Countries thanks to a letter on centres of gravity sent to him by Michel Coignet from Antwerp (*ibidem*, X, pp. 31-33).

Despite this, Stevin's name — like that of Benedetti — does not appear to be quoted in Galileo's works, neither in those he published, nor in the hand-written ones, nor in his letters; furthermore it does not even appear in his library, at least according to Favaro's careful reconstruction using various sources, which amounts to 521 titles.

However, some hydraulic experiments variously described and repeatedly discussed by Galileo both in his annotations to the *Considerazioni* of the Accademico Incognito (Arturo Pannochieschi) and also in the letter to Tolomeo Nozzolini of January 1613 (G. Galilei, *Le opere* ..., IV, pp. 172, 182-183; 297-310 respectively) are very close to examples I-III that in *De Beghinselen des Waterwichts* Stevin adds to the second proposition of the "Anvang der Waterwichtdaet" ("Preamble of the Practice of Hydrostatics", in *The Principal Works* ..., I, pp. 56/487-60/495; "De initiis praxis hydrostatices", in *Hypomnemata mathematica* ..., IV, pp. 143-147).

Furthermore, in the same period, more precisely on 23rd June 1614, in Rome, at the Collegio Romano before Prince Cesi and other members of the Accademia dei Lincei, Giovanni Bardi presented some experiments in hydrostatics considered by Galileo and by Stevin in their writings. Bardi explicitly referred to Stevin in the text in which he expounds and comments on what he showed in the assembly: *Eorum quae vehuntur in aquis experimenta a Joanne Bardio ad Archimedis trutinam examinata IX Kalend. Jul. anno Domini MDCXIV* (Romae, ex typ. B. Zanetti 1614). Unfortunately we know little about Bardi; however, from another brief note of Galileo at the *Considerazioni* of the Accademico Incognito (G. Galilei, *Le opere* ..., IV, p. 195) and from some letters in Galileo's correspondence (*ibidem*, XI, pp. 512-513; XII, pp. 76, 78, 79-80, 90-91, 95-96, 111-113), published in part also in the *Carteggio Linceo* (pp. 438-439, 440-441), and from his writing quoted above it seems that he was Florentine, that he was still young in those years, that he had been a pupil of Galileo remaining in correspondence with him, that at least from May 1613 he was in Rome where he had relations with the Jesuits of the Collegio Romano and the members of the Accademia dei Lincei, including Federico Cesi. We learn from the correspondence that Bardi sent his paper to Galileo and that none of the correspondents from Rome, in describing the assembly, dwells upon Stevin, behaving as though it were a question of something already known.

Finally, when Giovanni Battista Baliani, the Genoese patrician who had relations and controversies with Galileo, published his *De motu gravium solidorum* (Genuae, ex typographia I.M. Farroni, N. Pesagnii et P.F. Barberii soc. 1638), in which he explicitly recalls (at the beginning of the first book) Stevin's statics, in the Galilean milieu this work of the Dutch scientist was considered as something already known (G. Galilei, *Le opere* ..., XVIII, p. 43).

As was the practice at the time, Galileo was not in the habit of making many references to other writers except for the Classics and quotations with controversial intent. An additional reason explaining his silence on Stevin may also be a difference concerning both their approach to general theory and many final conclusions.

It is surprising, however, that not even pupils of Galileo speak of Stevin even in printed works: in particular not even Vincenzio Viviani who was close to Niels Stensen, the Danish scientist who was an admirer of Stevin. In fact, it is clear from some manuscript papers in Viviani's own hand (Firenze, Biblioteca Nazionale Centrale, *Collezione Galileiana*, "Discepoli" vol. 141, ff. 7-12), that Stensen dictated to him the translation from the Dutch of some propositions from the book of the Amsterdam burgomaster Nicolas Witsen, *Aeloude en hedendaegsche scheeps-bouw en bestier* (Amsterdam, C. Commelijn, B. en J. Appelaer, 1671). Furthermore, Stevin's works were not to be found even in the following century in the well-stocked library of a famous hydraulic expert such as Leonardo Ximenes.

Clearly this silence had very deep roots that I think have to be sought among those motives that slowed down the development of Italian science in the period after Galileo; among those motives we may also place an excessive faith in his teaching amongst his own scholars.

## BIBLIOGRAPHY

G.B. Baliani, *Opere diverse* (Genova: Franchinelli, 1792) (Pars secunda: *De motu gravium solidorum*).

D. Barsanti, *La biblioteca di Leonardo Ximenes. La cultura di uno scienziato italiano del XVIII secolo*. (Firenze: Osservatorio Ximeniano, 1988).

*Resolutio omnium Euclidis problematum aliorumque ad hoc necessario inventorum una tantummodo circini data apertura.* Per IOANNEM BAPTISTAM DE BENEDICTIS inventa (Venetiis: apud Bartholomaeum Caesanum, 1553). The critical edition of the Dedication to the Abbot of Guzman, cc.[7]r-[10]v, in C. Maccagni, *Le speculazioni ...*, pp. 1-15.

*Demonstratio proportionum motuum localium contra Aristotilem et omnes philosophos.* Per IOANNEM BAPTISTAM DE BENEDICTIS inventa (Venetiis, 1554). Critical edition, with Taisnier's variants, in C. Maccagni, *Le speculazioni ...*, pp. 17-61.

*Demonstratio proportionum motuum localium contra Aristotelem et omnes philosophos.* Per IOANNEM BAPTISTAM DE BENEDICTIS inventa (Venetiis, 1554 Idibus Februarii [1555]). Critical edition in C. Maccagni, *Le speculazioni ...*, pp. 63-84.

*Le due edizioni della "Demonstratio proportionum motuum localium contra Aristotelem et omnes philosophos" di GIOVANNI BATTISTA BENEDETTI*, Ristampa anastatica a cura di C. Maccagni, Venezia, Istituto Veneto di scienze, lettere ed arti 1985.

G. Bordiga, "Giovanni Battista Benedetti filosofo e matematico veneziano del secolo XVI", in: *Atti del R. Istituto Veneto di scienze, lettere ed arti* 85 (1925-1926), pp. 585-754 (new edition "con un aggiornamento bibliografico ragionato di P. Ventrice", Venezia, Istituto Veneto di scienze, lettere ed arti 1985).

R. Caverni, *Storia del metodo sperimentale in Italia*, Firenze, Civelli 1891-1900.

M. Clagett, *The science of mechanics in the Middle Ages*, Madison, University of Wisconsin Press 1959.

C. Costantini, *Baliani e i Gesuiti. Annotazioni in margine alla corrispondenza del Baliani con Gio. Luigi Confalonieri e Orazio Grassi*, Firenze, Giunti 1969.

*Cultura scienze e tecniche nella Venezia del Cinquecento.* Atti del Convegno internazionale di studio "Giovan Battista Benedetti e il suo tempo", Venezia, Istituto Veneto di scienze, lettere ed arti 1987.

J. Dewert, "Jean Taisnier, d'Ath", in *Annales du Cercle archéologique d'Ath et de la région* 1 (1912), pp. 3-36.

*Dictionary of scientific biography*, edited by C.C. Gillespie, New York, Scribner's 1970-1978.

E.J. Dijksterhuis, *Il meccanicismo e l'immagine del mondo dai Presocratici a Newton*, Milano, Feltrinelli 1971.

S. Drake — I.E. Drabkin, *Mechanics in sixteenth-century Italy*, Madison, University of Wisconsin Press 1969.

R. Dugas, *Histoire de la mécanique*, Neuchatel, Griffon 1950.

R. Dugas, *La mécanique au XVII^e siècle*, Neuchatel, Griffon 1954.

A. Favaro, "Amici e corrispondenti di Galileo Galilei, XXIX. Vincenzio Viviani", in *Atti del R. Istituto Veneto di scienze, lettere ed arti* 72 (1912-1913), pp. 1-147.

A. Favaro, "La libreria di Galileo Galilei descritta ed illustrata", in *Bullettino di bibliografia e di storia delle scienze matematiche e fisiche* 19 (1886), pp. 219-293.

*Il carteggio Linceo della vecchia Accademia di Federico Cesi (1603-1630)*, a cura di G. Gabrieli, Roma, Bardi 1938-1942 (*Memorie della R. Accademia Nazionale dei Lincei*, s. VI, Memorie della classe di scienze morali, storiche e filologiche, vol. VII).

G. Galilei, *Le opere*. Edizione nazionale, Firenze, Barbera 1929-1939.

A. Koyré, *Études d'histoire de la pensée scientifique*, Paris, P.U.F. 1966, pp. 122-146 ("Jean-Baptiste Benedetti critique d'Aristote", previously printed in *Mélanges offerts à Etienne Gilson*, Toronto-Paris, Pontifical Institute of Mediaeval Studies 1959, pp. 351-372).

C. Maccagni, "Contributi alla biobibliografia di Giovanni Battista Benedetti", in *Physis* 9 (1967), pp. 337-364.

C. Maccagni, "Galileo, Castelli, Torricelli and others. The Italian school of hydraulics in the 16th and 17th centuries", in *Hydraulics and hydraulic research. A historical review*, edited by C. Garbrecht, Rotterdam-Boston, Balkema 1987, pp. 81-88.

C. Maccagni, "Le scienze nello Studio di Padova e nel Veneto", in *Storia della cultura veneta*, $3^{III}$, Vicenza, Pozza 1981, pp. 135-171.

C. Maccagni, *Le speculazioni giovanili "de motu" di Giovanni Battista Benedetti*, Pisa, Domus Galilaeana 1967.

E. Mach, *I principii della meccanica esposti criticamente e storicamente nel loro sviluppo*, Roma-Milano, Albrighi Segati 1909.

S. Moscovici, *L'espérience du mouvement. Jean-Baptiste Baliani disciple et critique de Galilée*, Paris, Hermann 1967.

G.B.C. De' Nelli, *Vita e commercio letterario di Galileo Galilei*, Losanna 1793.

H. Rouse and S. Ince, *History of Hydraulics*, Iowa, Iowa Institute of Hydraulic Research 1957.

G. Sarton, "Simon Stevin of Bruges (1548-1620)", in *Isis* 21 (1934), pp. 241-303.

W.R. Shea, *La rivoluzione intellettuale di Galileo*, Firenze, Sansoni 1974.

*Hypomnemata mathematica* ... a Simone Stevino conscripta et e Belgico in Latinum a WIL[LEBRORD] SN[EL] conversa, Lugduni Batavorum, ex officina Ioannis Patii 1605-1608 (*Tomus quartus ... de statica*, 1605).

*The principal works* of Simon Stevin, edited by E. Crone, E.J. Dijksterhuis, R.J. Forbes, M.G.J. Minnaert, A. Pannekoek, Amsterdam, Swets and Zeitlinger 1955-1966.

*Opusculum perpetua memoria dignissimum de natura magnetis et eius effectibus* ... authore Ioanne Taisnierio Hannonio, Coloniae, apud I. Birckmannum 1562 (The item *Demonstratio proportionum motuum localium contra Aristotelem et alios philosophos*, pp. 16-40, is reprinted in the variants of the *Demonstratio* (1554), in C. Maccagni, *Le speculazioni* ..., pp. 17-61).

*A very necessarie and profitable Booke concerning Navigation*, compiled in Latin by Joannes Taisnierius ... *named a treatise of continuall Motions*. Translated into

English by Richarde Eden, London, R. Jugge 1578-1579 (reprinted: Boston, Massachusetts Historical Society, 1923: "American Series Photostat Reproductions" 86).

G. Targioni Tozzetti, *Notizie degli aggrandimenti della scienze fisiche accaduti in Toscana nel corso di anni LX del secolo XVII*, Firenze, Bouchard 1780 (II[1], pp. 2-10: *Eorum quae vehuntur in aquis experimenta, a Joanne Bardio Florentino ad Archimedis trutinam examinata IX Kalend. Jul. Anno Domini MDCXIV*, Romae, ex typ. B. Zanetti 1614).

L. Thorndike, *A history of magic and experimental science*, V, New York, Columbia University Press 1941.

I.Ph. Tomasini, *Illustrium virorum elogia*, Patavii, apud D. Pasquardum 1630.

H.Th. van Veen and P. McCormick, *Tuscany and the Low Countries. An introduction to the sources and an inventory to four Florentine Libraries*, Florence, Centro Di 1985.

# GALILEO IN HOLLAND BEFORE THE *DISCORSI*: ISAAC BEECKMAN'S REACTION TO GALILEO'S WORK

*K. van Berkel*

Galileo was already known as a distinguished scientist in Holland well before the Elzeviers published the *Discorsi* in Leyden in 1638. His *Dialogo*, with the ensuing trial and condemnation of its author, was discussed in intellectual circles, while a somewhat wider public was interested in his proposal for determining longitude by observation of the satellites of Jupiter. For a study of this early Dutch reception of Galileo, the scientific diary or 'Journael' of Isaac Beeckman (1588-1637) is a most important source. Beeckman, himself a mechanical philosopher of some distinction, was the Dutch scientist most congenial to Galileo in the period before Huygens. A study of his Journael not only gives us some information about the ways in which Galileo's work travelled to Holland, but it also informs us about the intellectual response to his ideas.

## 1. *Beeckman as a mechanical philosopher*

The city of Middelburg, where Isaac Beeckman was born in 1588, as the eldest son of a candlemaker whose parents had fled from the Southern Netherlands, occupies a special place in the study of Galileo and his work.[1] According to every story told about the invention of the telescope, lens-grinders in Middelburg play an important part in it. Albert van Helden, who has published the latest study of this subject, states that the lens-grinder Sacharias Janssen may have been the first to realize that a toy which he had seen in the possession of an unknown Italian in Middelburg could be used as

For convenience's sake, all references to letters of Descartes, Mersenne and others are to the *Journal tenu par Isaac Beeckman de 1604 à 1634*, edited by C. de Waard, and not to the standard editions of Adam-Tannery, De Waard, Favaro and others.

1. For literature on Beeckman: K. van Berkel, *Isaac Beeckman (1588-1637) en de mechanisering van het wereldbeeld* (Amsterdam, 1983). Additional information in: H.F. Cohen, *Quantifying music. The science of music at the first stage of the Scientific Revolution* (Dordrecht-Boston, 1984).

an instrument for different purposes.[2] Actually, one of the statements on which van Helden bases his reconstruction of the invention of the telescope is derived from the Journael of Beeckman, who in the 1630's tried to learn the craft of lens-grinding from Sacharias Janssen's son Johannes. But Beeckman left Middelburg to study theology and mathematics in Leyden before another lens-grinder from Middelburg, Hans Lipperhey, was able to show a telescope to an international audience in The Hague (1608). So the fact that Beeckman was born in the city from which Galileo ultimately derived the idea of his telescope, is a pure coincidence. Middelburg, however, had a small Italian community (mainly glass-blowers and soldiers) and Beeckman may have learned to read Italian during his early years there.[3]

After his study of theology and mathematics in Leyden, Beeckman returned to the province of Zealand, where he became a candlemaker like his father. He settled in the small town of Zierikzee. There, in almost total isolation, without friends interested in scientific matters, he developed his own natural philosophy, which can be called mechanistic. In Zierikzee, for instance, he formulated a modern principle of inertia and he also began to explain natural phenomena by postulating tiny particles in constant motion. There is no indication that he had sources of inspiration beyond what classical authors like Hero of Alexandria and sixteenth-century compilators like Cardanus could supply.[4]

In 1616, however, Beeckman disposed of his business and returned to Middelburg to work with his father and to study medicine. To obtain a degree, he travelled to Caen in 1618 and on September 6th 1618 he was formally promoted to the degree of doctor medicinae. During his stay in Caen, he visited some of the professors there and one of them, the mathematician François-Gilles Macé, showed him a sketch of Galileo's telescope. In that same year Macé, who was professor of mathematics and medicine from 1613 to 1618, was to publish a pamphlet on the comet, in which he would refer to the work of Galileo. Unfortunately, we do not know whether he discussed Galileo's work on astronomy with his Dutch visitor.[5]

After his departure for Zealand, Beeckman travelled to the city of Breda in Brabant, where an uncle was living at that time and where Beeckman also

2. A. van Helden, *The invention of the telescope* (Philadelphia, 1977).

3. C. de Waard, *De uitvinding der verrekijkers. Eene bijdrage tot de beschavingsgeschiedenis* (The Hague, 1906), 105-114.

4. K. van Berkel, "Vruchtbaar isolement. Isaac Beeckman in Zierikzee (1611-1616)", in: *Kroniek van het land van de zeemeermin (Schouwen-Duiveland)* 11 (1986), 59-74.

5. *Journal tenu par Isaac Beeckman de 1604 à 1634*, publié avec une introduction et des notes par C. de Waard (4 vols., The Hague, 1939-1953), I, 208-209.

hoped to make love to a lady whose identity is unknown. In that period, in Breda there also lived a young French soldier with a gift for mathematics, the young René Descartes. By coincidence, Descartes met Beeckman and the two became friends. In November and December 1618 they discussed quite a number of mathematical and physical problems, including the problem of free falling bodies. Alexandre Koyré has discussed their conversations — rendered to us both by Descartes and by Beeckman — and he was able to show how Beeckman and Descartes had a totally different opinion of the best way to solve the problem. Descartes, being first of all a mathematician, approached the problem from the mathematical side, while Beeckman's approach was more that of a natural philosopher. It was Beeckman who came forward with the physical principles they used for the solution (including the principle of inertia), while Descartes showed him the mathematical methods needed to arrive at the correct solution. This difference in approach would always characterize the relationship between Beeckman and Descartes.[6]

As we will see later on, Beeckman's mainly natural philosophical approach to scientific problems would also distinguish his work from Galileo's. Even in this case, the difference becomes clear as soon as we start comparing Beeckman's solution of the problem of free falling bodies to Galileo's solution. While Galileo's approach was purely kinematical, no use being made of supposed forces in actio, Beeckman's reasoning started from the assumption that unsupported bodies fall because of the pressure exercised upon them by tiny particles rushing in from the skies. Although Beeckman would not have disapproved of Galileo's solution, he himself preferred a solution that started with some assumptions concerning the physical lay-out of the universe.

## 2. *Beeckman's introduction to Galileo's work*

Beeckman and Descartes parted ways on the last day of 1618, when Descartes left Breda and travelled to Amsterdam. Beeckman stayed in Breda for a while and later that year became a schoolteacher in the city of Utrecht. But by 1620 he had moved to the more enterprising city of Rotterdam, where he stayed until 1627. During that period, most of the entries in the Journael are concerned with technical and practical matters, although through the reading of books he did not lose contact with the international world of science. However, only after he had left Rotterdam to become headmaster of

6. A. Koyré, *Etudes galiléennes* (Paris, 1966[2]).

the Latin school in the prestigious city of Dordrecht, was Beeckman introduced to the world of international science personally. One of the first visitors from abroad was his old friend René Descartes, and two other French scientists, Pierre Gassendi and Marin Mersenne, also paid him a visit on their trip through the country. Beeckman also came into contact with the French scientist Fabri de Peiresc, who later on would prove to be an important link between Galileo and the Dutch scientists at that time.[7]

But it would be wrong to suggest that Beeckman was introduced to the work of Galileo through these international contacts. More important was his acquaintance with the protestant minister in Dordrecht Andreas Colvius, one of his best friends. Colvius had been a member of the Dutch Embassy to Venice and had spent some years in Northern Italy. In those years (1622-1627), Colvius had collected a number of important books and manuscripts, including work by Paolo Sarpi and Galileo. For instance, Colvius possessed a copy of an unpublished treatise by Galileo on the tides. This manuscript, entitled "Discorso sopra il flusso e reflusso del mare", he showed to Beeckman in 1630, on which occasion Beeckman made some notes on it. Also Beeckman had a chance to study Galileo's *Istoria e Dimostrazioni intorno alle macchie solari*, his *Discorso intorno alle cose, che stanno in sù l'acqua* and *Il Saggiatore*.[8] Reading these books and manuscripts, Beeckman congratulated himself on the fact that his own thoughts accorded so well with those of a great man like Galileo.[9]

Colvius, however, could not oblige Beeckman with a copy of Galileo's *Dialogo* after its publication in 1632. Already in 1631, Beeckman had asked Mersenne for more information on what he called Galileo's *Cosmographia*, but after its publication he experienced how difficult it was to obtain a copy in Holland. Having heard some rumours about Galileo's trial, he wrote to Mersenne in May 1633 how uneasy he felt about Galileo's position (the first interrogations were held in April and Galileo was sentenced in June) and he once more asked for a copy of the book. Later that year, a former pupil of Beeckman, the astronomer Martinus Hortensius tried his luck with Gassendi, who seemed to be one of the very few in Paris to have a copy of the book.[10] Enclosed in this letter to Gassendi, another letter was sent by

7. Cf. F.J. Baumgartner, "Galileo's French correspondents", in: *Annals of science* 45 (1988), 169-182.

8. *Journal*, III, 216-217, 223, 224.

9. *Journal*, III, 223: "Hic, et procul dubio in multis aliis, ut vidi in ejus *Il Saggiatore*, eadem fere dicit quam ego ante in hoc libro sum meditatus mihique ab id gratulor quod vir tantus eadem quae ego, excogitaverit".

10. Cf. R.S. Westman, "The reception of Galileo's *Dialogue*. A partial world census of

Hortensius to Galileo himself. Gassendi passed on the request for a copy of the book to Peiresc in Aix-en-Provence and through his kind intervention, a copy of the *Dialogo* at last was sent from southern France to Holland in January 1634. This copy only reached Gassendi — who acted as an intermediary — in the summer and by the time he was able to pass it on to Hortensius, another copy had been sent by Galileo to Hortensius straight from Livorno. So in the end, in the summer of 1634, Hortensius possessed two copies of the much desired *Dialogo*.[11]

Being a pupil of Beeckman, Hortensius lent one of these copies to his former teacher when he visited him on August 1st, 1634. Immediately, Beeckman made an extensive study of its contents, noting down in his Journael all passages "laudanda vel corrigenda". Four points in particular seem to have interested Beeckman: he says something about the light that comes from the moon, he criticizes Galileo's theory of the tides, he proposes an experiment to demonstrate the relativity of movement and he makes a remark on the oscillation of the pendulum. The notes on the tides are very extensive, but although he was a Copernican himself, he did not say a word about Galileo's defence of the Copernican system.[12]

Two weeks after he had received the book from Hortensius, Beeckman took the *Dialogo* with him to Amsterdam, where he served his apprenticeship in lens-grinding with an English lens-grinder. In Amsterdam, Beeckman also paid a visit to Descartes. Although Descartes had quarrelled with Beeckman some years before, they had restored their relations and were on friendly terms once again. Beeckman was kind enough to leave his copy with Descartes while he was grinding lenses himself, and so, for two days, on the 12th and the 13th of August, Descartes had a chance to read Galileo's masterpiece. Later on, in a letter to Mersenne, Descartes tells his correspondent about this visit and comments on the *Dialogo*. He too mentions the theory of the tides, which he thinks untenable and at least inferior to his own

---

extant copies", in: P. Galluzzi, ed., *Novità celesti e crisi del sapere. Atti del convegno internazionale di studi Galileiani* (Firenze, 1984), 329-355.

11. *Journal*, III, 356; IV, 215-216 (a letter of Beeckman to Mersenne, 30th May 1633, asking him to send him a copy, or otherwise to lend it to him, or otherwise to send him some excerpts of the book); III, 220-222 (Peiresc to Hortensius, 23th January 1634, informing him about Galileo's situation and announcing the arrival of a copy of the *Dialogue* by separate post); IV, 224 (Descartes to Mersenne, 14th August 1634, reporting about Beeckman's visit to Descartes).

12. *Journal*, III, 356-357. It is not known whether Beeckman really owned the book; Hortensius may also just have lent it to his former teacher. This minor problem may be solved by the publication of the recently found catalogue of Beeckman's library. See: H.H. Kubbinga, "Isaac Beeckman Catalogus ... librorum", in: *De Zeventiende Eeuw* 4 (1988) 81-82.

explanations, developed in his unpublished treatise *Le Monde*. He also tells Mersenne that Beeckman did not leave the book with him very long; on the 14th of August, Beeckman took it with him to Dordrecht, where the pupils of his Latin school were waiting for him.[13] It is a pity that Beeckman's Journael breaks off around this time, for now we do not know what his opinion of Galileo's theories was after he had had some time to think them over and to discuss them with someone like Descartes.

### 3. *The problem of determining longitude*

The story of Beeckman's reaction to Galileo's work does not end with the last pages of his Journael. In a sense, the most interesting was still to come. During the last years of his life, Beeckman became involved with Galileo's proposal to the States General of the Dutch Republic for a solution of the problem of determining longitude in the open sea.

Shortly after his discovery of the satellites of Jupiter, Galileo had realized that these new 'planets' could be used as a heavenly clock with which the difference in local time between a certain place on earth and a known location could be determined. From this difference in time the difference in geographical longitude would follow immediately. The first nation Galileo presented his solution to was Spain, but the dealings with the Spanish crown did not lead to any result. In 1627, however, Galileo was informed, for the first time, that the States General had offered a prize to anyone who would be able, as his correspondent Alfonso Antonini put it, "insegnare il modo di trovare la longitudine per uso della navigatione".[14] At first, Galileo was not interested, but after his trial things had changed. The problem of determining longitude was still unsolved and as the Dutch diplomat in Swedish service, Hugo Grotius, wrote in 1634, there were even those who were beginning to think that the problem was insoluble. To a friend in Paris he wrote: "In our country, many have, without success, worked to find a solution, so much so that some idiot in Amsterdam has nailed to the wall four heresies, as he called them: a means to determine longitude, the stone of the philosopher, a perpetuum mobile and the quadrature of the circle. But the more desperate this search becomes, the greater will be the glory of the inventor and the delight of the readers".[15] That glory might become Galileo's.

A year later, in a letter of May 17th 1635, Grotius tells his Dutch friend Vossius about Galileo. He reports his trial and the way he is held in custody

13. *Journal*, IV, 224 (cf. note 11).
14. *Journal*, IV, 127 (Alfonso Antonini to Galileo, 25th October 1627).
15. *Journal*, IV, 223 (Grotius to Guillaume de Lauson, 26th June 1634).

in Tuscany. Then he informs his friend, who was a professor at the Athenaeum Illustre in Amsterdam, that in Paris a group of friends of Galileo is thinking of bringing him to Amsterdam, where he would find the necessary rest and quietness to continue his studies. Grotius asks his friend to go around in Amsterdam and find out about the possibilities of helping Galileo. Vossius did what he could and he discussed the matter with Laurens Reael, one of the regents of Amsterdam and an amateur scientist himself. He also went to see Hortensius, who had by then become his colleague in Amsterdam, and the map-maker Blaeu. But Vossius had to confess that those in power were not really interested. "It would be easier", he reported in a letter of July 1st, "if not so many of those who are in charge were more concerned with money than with the truth and the glory of the city".[16]

In reply, Grotius added another argument that might persuade the regents of Amsterdam, an argument perfectly designed to stimulate their self-interest. He told Vossius that Galileo had found something very important for a sea-faring nation like the Dutch, a way of determining longitude. But it was to no avail.[17] Already a month later, he had to report to Vossius that Galileo had refused to leave his country to go to Holland. For a man of his age, to go to a foreign country and to make new friends was simply too much.[18] Nevertheless, Galileo was thinking of presenting his method for determining longitude to the Dutch government and with the help of Galileo's friend Elia Diodati, a member of the Parliament of Paris, Grotius hoped to be able to persuade Galileo really to do it.

And so, in the end, Galileo presented his method to the States General. Already in 1635, a committee of experts was installed to test his proposal. For the Dutch, Galileo's proposal would only be of any value if it could be used in practice; theoretical correctness itself was not enough. Members of the committee were Reael (who acted as its president), Hortensius (who acted as its secretary), Blaeu, the Leyden professor of mathematics Golius, and, after a while, Beeckman. Beeckman was not among the first members of the committee, but he seems to have been elected a member by the others, all of whom were very familiar to him. Perhaps he replaced Golius, of whom not much is heard in this context.[19]

16. *Journal*, IV, 235 (Grotius to Vossius, 17th May 1635), 236 (Vossius to Grotius, 1st July 1635).

17. *Journal*, IV, 236 (Grotius to Vossius, 6th July 1635).

18. *Journal*, IV, 236 (Grotius to Vossius, 2nd August 1635).

19. In the Resolutions of the States General of 11th November 1636 only Reael, Hortensius and Blaeu (all of them living in Amsterdam) are mentioned as members of the commission to investigate Galileo's proposal, while it is said that they have the liberty to co-opt Golius if

There may have been one more reason why Beeckman was elected a member of the committee. At least Hortensius, who in 1634 had had the exceptional privilege of reading Beeckman's Journael, must have known that Beeckman had once suggested practising the same solution to the problem of determining longitude as Galileo had presented to the States General. In 1631, Beeckman had suggested, independently of Galileo, making ephemerides of the four moons of Jupiter and especially of their eclipses as a means of determining the exact time at some predetermined place on earth. Although Beeckman knew about the prize being offered by the States General, he did not consider it worth while presenting his idea to the States. But he may have discussed the matter with Hortensius and Hortensius may have suggested taking Beeckman onto the committee.[20]

As is well-known, however, the affair did not lead to any result. Hortensius did his very best to make a success of Galileo's proposal, but he did not succeed. Most members of the committee died before anything could be done. Beeckman passed away first, on May 19th 1637. That same year he was followed by Reael, while Blaeu died in 1638. Hortensius was the only survivor when he too died in August 1639, shortly before he was to travel to Italy to meet Galileo in person.[21]

## 5. *The theory of the tides*

The Dutch being a sea-faring nation, it is not surprising that so much attention was paid to Galileo's proposal for determining longitude. Perhaps the special relationship of the Dutch with the sea can also explain why Beeckman gave so much attention to one specific aspect of Galileo's work, the theory of the tides. But whatever the reason, a short analysis of Beeckman's explanations of the tides will reveal in what sense his approach to natural philosophy and science differed from Galileo's.

Beeckman first read about Galileo's theory of the tides in 1630, when he was able to read the manuscript "Discorso sopra il flusso e reflusso del mare". Galileo explained the alternation of ebb and flow not by any

---

necessary (*Journal*, IV, 253). In a letter to Diodati of 24th November 1634, Hortensius informs his correspondent about the commission, mentioning Beeckman as one of the experts involved (*Journal*, IV, 254). Much later, in another letter to Diodati of 27th April 1637, Hortensius tells his correspondent that Beeckman had written a letter to him, asking him for information about Galileo's proposal, "quia inter Commissarios delectus fuerat" (*Journal*, IV, 270).

20. *Journal*, III, 229-230.

21. Of all the participants, Hortensius had been the most ardent supporter of Galileo's case in Holland. He deserves a special study. For further information about Galileo's proposal, see the paper by G. Vanpaemel in this volume.

attraction between the moon and the earth or by a corpuscular mechanism, but solely by the combination of the diurnal and the annual movement of the earth. In his view, this was not only a good explanation of tidal movement, but also a very strong argument in support of the Copernican system of the world. Beeckman approved of this theory and claimed that it accorded well with his own principles.[22] In a sense, he was right. Both he and Galileo rejected attractive forces as the cause of the tidal movement, but on the other hand, Beeckman's own explanation, developed before 1630, was purely corpuscular. This explanation was based on the assumption that small particles of light rushing in from the moon and the sun made the sea swell and thereby caused the phenomenon of the flow.[23] After reading and approving of Galileo's theory Beeckman actually had two theories of the tides: one corpuscular (his own theory) and one based purely on the mechanical properties of the moving earth (Galileo's theory).

Although in 1630 Beeckman said he approved of Galileo's theory, later on he both elaborated on it and criticized it. He elaborated on it by extending the influence of the combination of the two movements of the earth to the air. Just as the tides were caused by the movements of the earth, so in Beeckman's view were the winds. The explanation of the winds had attracted his attention ever since he had started his natural philosophical speculations, and his first ideas about it tended, just as in the case of the tides, to circle around some invisible corpuscular mechanism. But after reading Galileo, a second approach presented itself, also mechanistic, but not corpuscular. But Beeckman did not stop there. In the end, he even speculated on the explanation of the movements of the moon. Might not the moon be considered as a body floating on top of a sea of air, influenced let us say, by the 'tidal movements' of the air? In this way, a connection between the tidal movement and the moon would have been re-established, but in a totally different way.[24]

But at the same time, Beeckman criticized Galileo's conceptions and he sometimes suggested a different approach. For instance, he suggested that the movement of the tides was not caused by the movements of the earth, but by

22. *Journal*, III, 171, 205.

23. *Journal*, III, 11, 38, 86. Before 1630, he had another theory; specially shaped particles coming from the moon cause low air pressure right under the moon and this in turn causes the water to gather there. This mechanism of 'under pressure' Beeckman also used for the explanation of magnetism. After 1630, nothing is heard of this theory again (*Journal*, II, 167, 229, 317-318, 387-388).

24. *Journal*, III, 206.

the movement of the centre of gravity of the moon-earth system.[25] From his scattered notes on this topic it is not possible to deduce one coherent theory. If Beeckman had really done what he always hoped to be able to do, that is to publish his theories in a clear and orderly way, he might have tried to systematize his ideas and choose one of the conflicting approaches to one and the same problem. But he never managed to do so and in a sense this is not to be deplored. In their present state, his notebooks confront us with a fundamental problem of the mechanical philosophy that was emerging in the early seventeenth century, the tension between a corpuscular and a merely mechanical way of treating physical phenomena. Galileo had, at least in the case of the theory of the tides, opted for an explanation derived purely from the mechanical properties of material bodies; Beeckman hesitated between both approaches, as is shown by his notes on the theory of the tides.[26]

25. *Journal*, III, 281.

26. R.S. Westfall, *The construction of modern science. Mechanisms and mechanics* (Cambridge, 1977).

# SCIENCE DISDAINED: GALILEO AND THE PROBLEM OF LONGITUDE *

*G. Vanpaemel*

While on a visit to Italy in 1635, the French mathematician Jean de Beaugrand asked Galileo, who was then 71 years old, to give his opinion on a certain method, recently proposed by Jean Baptiste Morin, to determine geographical longitude. Galileo, not wishing to get involved in an already heated debate, rather reluctantly complied, pointing out some of the practical flaws in Morin's method. But he also in a casual way mentioned his own solution of the problem of longitude: "I pretend to have found such an easy and certain method, that without the need of astronomical instruments or calculations and only using eyesight and an accurate clockwork [...] one can determine the longitude of any place on earth and at sea more exactly than if we had a lunar eclipse every night in every horizon."[1]

Galileo's formidable statement was in every respect an accurate description of the method he had worked out over the past 23 years, but at the time he wrote to de Beaugrand, he could only have had little hope left of ever being able to provide convincing proof of his invention, for which he needed the support of a great, sea-faring nation. On several occasions Galileo had tried to awaken the interest of the Spanish King, but his efforts had been in vain. At the very time he was writing to de Beaugrand, he initiated new negotiations, now offering his method to the States General of the United Provinces of the Netherlands. As before, the official bodies concerned were not very impressed.

How could this have happened? Why could neither the personal fame of Galileo as an astronomer, an inventor and philosopher, nor the established scientific soundness and value of his proposed method itself, create anything more than a few polite letters with vague promises that eventually came to

*With an appendix where the 'Resolution of the States General' on Galileo's first proposal on longitude (5 December 1635) is published.

1. G. Galilei, *Le Opere di Galileo Galilei*, Edizione Nazionale, Firenze (1892-1909); repr. 1968 (hereafter *Opere*), XVI, pp. 340-344, letter from Galileo to de Beaugrand, 11 November 1635.

nothing? What obstacles were raised against the pursuit of a scientific enterprise, that had such an obvious economic and military utility? The recurring failure of Galileo's proposals, with which we shall deal in this paper, may help us to clarify some of the issues that affected the relationship between science and its social use during the Scientific Revolution.

During the seventeenth century the problem of longitude was constantly at the forefront of scientific research. As a result of the maritime expansion of the Western World since the fifteenth century, the practical problem of determining the exact position of a ship at sea, and in particular its geographical longitude, acquired an important economical and military meaning. Spain, Holland, France and probably also Portugal and Venice[2] offered huge rewards for anyone who could find a reliable and practical method. The prize money was attractive: Spain offered a perpetual pension of 6,000 ducats together with a life pension of 2,000 ducats. The States General of the Dutch Republic promised in 1600 a reward of 5,000 florins, which was raised in 1611 to 15,000 florins — and sometimes they were prepared to pay even more.[3] As could be expected, a large number of proposals was submitted, some of them extremely bizarre, but others well-founded and perfectly feasible. In particular in the fields of astronomy, cartography and magnetism the wide-scale investigation of the problem of longitude can be said to have provided a constant stimulus for the advancement of scientific knowledge.

Yet the progress made in solving the problem before the end of the seventeenth century was very slight. The most popular method of finding longitude was to plot the variation or declination of the compass needle from true north. But since no simple relationship between variation and change of longitude existed and particularly after it had been established in 1635 that the magnetic variation at a given location was gradually changing with time, the use of magnetic variation as an aid to navigation declined.[4]

Another important method was based on the astronomical observation of the moon and the fixed stars. However, this method presupposed a perfect knowledge of the movement of the moon, an accurate measurement of the position of the moon as observed from a particular place on earth, and complicated calculations to account for such inevitable errors caused by

2. L.A. Brown, *The Story of Maps*, Boston (1949), repr. New York (1977), p. 209.

3. C.A. Davids, *Zeewezen en wetenschap. De wetenschap en de ontwikkeling van de navigatietechniek in Nederland tussen 1585 en 1815*, Amsterdam (1985), pp. 69, 73 and 129-131.

4. The use of magnetic variation to determine longitude ceased in the Dutch *Verenigde Oostindische Compagnie* (East India Company) around 1650, but it was revived during the eighteenth century through the work of Edmond Halley and others. Davids, *Zeewezen en wetenschap*, (op.cit., in note 3) pp. 79-80 and 192-196.

parallax and atmospherical refraction. For these reasons, this method of lunar distances was considered impractical.[5] Still, in the opinion of many scientists in the seventeenth century, astronomy continued to be regarded as a promising road to the solution of the problem of longitude.

Galileo's method was indeed based on astronomy, but it had none of the disadvantages associated with the method of lunar distances. As he wrote to de Beaugrand, longitude could be determined without the use of astronomical instruments or calculations. His method was at once easy and certain, "only using eyesight and an accurate clockwork." Why then was it so difficult to have the method accepted?

## 1. *Negotiations with Spain*

On 7 January 1610, when Galileo was observing the fixed stars with his telescope, he remarked that Jupiter was accompanied by three little stars, positioned in a certain arrangement almost in a straight line. Returning to his observations on the next night, he found, much to his surprise, a very different arrangement. He continued his observations and within the next week, he reached the conclusion that no less than four satellites, which he called the "Medicean Stars", were circulating Jupiter.[6]

Probably soon after this discovery, Galileo must have realised the possibilities offered by the regular movements of Jupiter's satellites in solving the problem of longitude. If one could predict the time when a certain arrangement of the satellites would be seen under a prime meridian, every other observer of the same arrangement of satellites could at once determine the difference between his local time and the time at the prime meridian. From this time difference the difference in longitude could be easily deduced.

Galileo set himself to work out the periods of revolution for every one of the satellites and published his first approximate estimates in March 1612.[7] With this numerical data he could accurately predict the future arrangements of the satellites a few months ahead. This was no easy job, since Galileo realised quite clearly that the observed positions of the satellites and the apparent radii of their orbits were sensibly affected by the annual movement of the Earth around the Sun, a problem he tried to deal with by introducing a

5. The practical obstacles to the method of lunar distances were explained in Galileo's letter to de Beaugrand (see note 1).

6. The discovery of the satellites of Jupiter was described by Galileo in his *Sidereus Nuncius* (Venice, 1610). See S. Drake, *Discoveries and Opinions of Galileo*, New York (1957), pp. 51-58.

7. G. Galilei, *Discorso intorno alle cose che stanno in su l'acqua*, Florence (1612). Galileo in this work declared to have found these values in April 1611.

correcting calculation, the *prostaphaeresis*.[8] To assist and simplify his calculations, Galileo had already designed in 1611 the first of a series of nomograms, which he called *giovilabio*, and which enabled him to determine all the necessary parameters of the future constellations only by means of a simple geometrical construction on paper.[9] Of course, the same simple nomogram, once it had been perfected, could also be used by navigators at sea to determine longitude.

In October 1610, Galileo had entered the service of Grand Duke Cosimo II de' Medici, partly as a result of his discovery of the satellites of Jupiter and partly through his friendship with the Grand Duke's secretary, Belisario Vinta. Galileo undoubtedly discussed his invention at the court of his employer. It was then decided that Galileo's invention would be offered to the Spanish King, as part of the negotiations the Grand Duke was conducting at that time with Spain concerning commercial trade with the Indies. The negotiations did not lead to an agreement, and as for Galileo's proposal, it was said that for the time being it could not be accepted since a similar proposal made by a Spanish mathematician was currently being investigated.[10]

Four years later, Galileo made a second proposal to the Spanish King, this time on his own behalf. This second proposal followed shortly after Galileo had been called to Rome, where he was personally informed of the decree against Copernicanism. It may well have been that as a result of the action taken by the Holy Office, Galileo turned his attention away from cosmology and, at least for a while, focused on practical astronomy and longitude.[11] In June 1616, Galileo sent a letter with his proposal to Spain. He was careful, however, not to be too specific about the actual content of his invention. Galileo explained in his proposal how longitude could be determined in the most exact way by means of lunar eclipses, but that this procedure suffered greatly from the rarity of the eclipses. "If a way were found," he then continued, "to make eclipses more frequent, [...] this affair could be turned to great advantage [...]". And he added: "And even if they were not truly lunar eclipses, but things in the sky and appearances similar and equivalent

8. G. Righini, "Contributo alla interpretazione scientifica dell'opera astronomica di Galileo", *Monografia 2. Supplemento agli Annali dell'Istituto e Museo di Storia della Scienza di Firenze*, 1978, fasc. 2, pp. 55-71.

9. The *Opere* contain five different versions of the *giovilabio,* of which the first one is dated posterior to 15 March 1611. See G. Righini, "Contributo", pp. 58-61; S.A. Bedini, "The Galilean Jovilabe", *Nuncius,* 1 (1986), 25-46.

10. *Opere*, XI, pp. 392-393, letter from Vinta to Orso d'Elci, 7 September 1612; p. 417, letter from d'Elci to Vinta, 16 October 1612.

11. S. Drake, *Galileo at Work*, p. 257.

to lunar eclipses, it is manifest that the gain would be great."[12]

Galileo's proposal was to be supported in Spain by the Tuscan ambassador, Orso d'Elci. Galileo assured d'Elci that the practical execution of his invention was no more difficult than the normal astronomical procedures used to determine latitude. However, the observations could not be made without the most perfect telescopes, which Galileo was willing to provide. Galileo furthermore proposed to come to Spain in order to demonstrate his invention before the King. Thereafter, he would reveal his whole method, and for the rest of his life, he would provide personally the ephemerides of the stars, calculated for every day and every hour.[13]

In June 1617, Galileo wrote again to Orso d'Elci, announcing a new invention of great utility.[14] Three months before he had been discussing with a few sea captains what use could be made of the telescope in navigation. It was agreed that enemy ships could be spotted from a long distance, so that there still would be enough time to prepare the attack or to retreat. But owing to the continual agitation of the ship, it was deemed impossible to keep the telescope fixed on any object, even for the shortest time. To overcome this difficulty, Galileo devised the *celatone*, a headgear fitted with one or two telescopes, which made it possible to hold the objects in view with the same ease as with the naked eye.[15] However, the *celatone* was not suited for the astronomical observation of Jupiter's satellites, because of its low magnifying power. Galileo then proceeded to construct another machine in order to detach the observer's position from the movements of the ship.[16] It obviously did not work to his entire satisfaction, since the problem of the movement of the ship kept creeping up in all future negotiations on the subject.

Through the efforts of Orso d'Elci, the invention was discussed among

12. *Opere*, V, pp. 419-425 (translation S. Drake, *Galileo at Work*, pp. 258-259). I equally follow S. Drake in placing this "Proposta della Longitudine" in 1616 rather than in 1613 as Favaro did.

13. *Opere*, XII, pp. 291-295, letter from Galileo to Orso d'Elci, 13 November 1616.

14. *Opere*, XII, pp. 321-328, letter from Galileo to Orso d'Elci, June 1617.

15. The actual fabrication of the *celatone* is not known. Galileo asserted that with the *celatone* it was possible to estimate the distance of the objects. This could be done by superposing the images from two telescopes, one of which was movable. See S. Drake, *Galileo at Work*, p. 261, n. 10. But later, in a letter to Reael (see note 16), Galileo described his *celatone* as consisting of only one telescope, the second image being produced by naked vision. See A. Favaro, *Documenti inediti per la storia dei negoziati con la Spagna per la determinazione delle longitudine in Mare*, reprinted in: *Nuovi Studi Galileiani* , Venice (1891), pp. 101-148.

16. *Opere*, XVII, pp. 96-105, letter from Galileo to Laurens Reael, 5 June 1637. Galileo agreed that it could only be used on a moderately agitated ship.

Spanish officials, which resulted in a decision of the Crown to investigate the matter further. In order to facilitate the negotiations, d'Elci agreed to translate Galileo's proposal and arranged for him to come to Spain and demonstrate his method. But, by the end of 1617, it had become clear that the reception of Galileo's proposal was not undividedly favourable. It was feared that telescopic observations were impossible aboard a moving ship and in the hands of ordinary sailors, and furthermore, that the satellites would not be visible during the day and part of the night.[17]

Galileo replied with the greatest care to these objections. It was not necessary, he assured his correspondent, to have a determination of longitude at every hour, or even day by day. It was enough if the approximate value of the longitude, found through the variation of the compass needle or through an estimate of the distance travelled, could from time to time be corrected. Galileo conceded that Jupiter would not be visible during two months every year, but, he added, "we do not depreciate the science of medicine because it can not cure all diseases."[18] Furthermore, the alleged difficulties in making good telescopic observations at sea should be put on the same level as the obstacles associated with determining a ship's latitude: a cloud sky, a stormy sea or an unqualified navigator were equally damaging to the determination of latitude as to Galileo's method of longitude.

In the mean time, Galileo was still improving his estimates of the periods of the Jovian satellites. Every night from 1610 until 1619, almost without interruption, he observed the satellites and added corrections to his tables. In January 1617, while in full negotiation with Spain, Galileo put together his so-called "Table of Bellosguardo", which contains his best and definitive values for the mean motions of the planets.[19] At this time, Galileo appears to have been very confident that his proposal would soon lead to the final solution of the problem of longitude.

But his hopes were not realised. The negotiations died out in 1618, probably after negative advice given by the Spanish State Council. The proposal was renewed in 1620, and again in 1629 and 1631, but always with the same result.[20] It was deemed impossible or of little use to make reliable telescopic observations on a moving ship at sea. But if the gain in navigation may indeed have seemed very slight to the officials, we may still wonder why

17. *Opere*, XII, p. 353, letter from Orso d'Elci to Curzio Picchena, 30 November 1617.

18. *Opere*, XII, pp. 358-363, letter from Galileo to Orso d'Elci, 25 December 1617.

19. A. Favaro, "Avvertimento", in: *Opere*, III (Pt. 2), pp. 403-424.

20. The complete story is told in Favaro, *Documenti inediti* (see note 15). In 1630 Galileo offered a telescope to the Spanish King, but to no avail.

none of the other possible advantages incorporated in Galileo's proposal were properly recognized and adopted? The improvement of geographical maps by observations made on land, the use of a low-power telescope to spot enemy ships, the corrective value of a sporadic exact determination of longitude — any of these was undoubtedly worth a prize.

The rejection of Galileo's proposal may have been caused by the fact that he had "to deal with diplomats and naval officers alone in this matter."[21] Yet, although the objections raised against Galileo were indeed of a purely practical nature, his proposal was also inspected by Spanish mathematicians, who totally agreed with the scientific claims of the method but expressed the same doubts as to its practical execution. Moreover, Galileo often discussed the matter with sea captains and admirals in Livorno, and showed himself in no way disturbed when hearing their comments. Probably, the most important obstacle was the sheer novelty of the telescope, perhaps also its high price. The failure of the Spanish Crown in rejecting Galileo's invention was not that it was unimpressed by the heavenly clockwork of Jupiter's satellites, but rather that it did not appreciate the enormous possibilities of the telescope.[22]

## 2. *Negotiations with the States General*

On 25 October 1627 Galileo received a letter from The Hague, written by Alfonso Antonini, informing him that the States General of the United Provinces of the Netherlands had deposited a prize of about 30,000 *scudi* for anyone who could establish a correct method of finding longitude, for the use of navigation.[23] Galileo asked for more information, especially about the qualifications of the people who were assigned to judge the proposals, but as far as we know, he did not at any time have the intention of stopping his lingering negotiations with Spain nor of sending his invention to the

21. S. Drake, *Galileo at Work*, p. 307. In his letter to d'Elci of 13 November 1616 (see note 13) Galileo had also expressed his fear of the possible ignorance and envy of the court officials, who were to inspect his proposal, and he added "I believe there is no greater hatred in the world than between ignorance and wisdom."

22. Another proposal, based on the use of a telescope, was offered in 1631 to the Spanish King by Michel Florent Van Langren. It was not accepted although it was supported by many scientists. In 1644, Van Langren received notice that in Spain "there was no interest nor understanding of such matters". See O. Van de Vyver S.I., "Lettres de J.-Ch. della Faille S.I., cosmographe du roi à Madrid, à M.-F. Van Langren, cosmographe du roi à Bruxelles, 1634-1645," in: *Archivum Historicum Societatis Iesu*, XLVI (1977), pp. 73-183, esp. p. 177.

23. *Opere*, XIII, pp. 376-377, letter from A. Antonini to Galileo, 25 October 1627. The alleged amount of 30,000 *scudi* should be compared with the 15,000 florins as it was stated by the States General in 1611 (see note 3).

Netherlands. Probably, he had less faith in the organizing capacity of the small Dutch Republic than in the enormous resources open to the King of Spain. For the completion of his invention Galileo needed a powerful associate, who could bring into action the necessary ships, crews, navigators, colonies, ports and maritime power. In this respect, the democratic organization of the Dutch Republic possibly appeared to Galileo inferior to the absolutist Spanish Kingdom.[24] It may have been, then, a sign of despair when in 1636 Galileo at last sent his proposal to the Dutch States General.

A year before, friends of Galileo had conceived of a plan to let the Italian philosopher leave his country, where his scientific activities were too hindered by the disastrous condemnation of 1633. Hugo de Groot, a Dutch exile in Paris, wrote a letter to Gerard Vossius in Amsterdam in May 1635 to inquire whether it could be arranged to find an appropriate living for Galileo in Amsterdam.[25] Vossius discussed the matter with Laurens Reael (1583-1637), regent of the city of Amsterdam, Martinus Hortensius (1605-1639), professor at the *Athenaeum Illustre*, and the cartographer Willem Jansz. Blaeu (1571-1638), who were all enthusiastic about the affair. To support the project, de Groot added in his next letter that Galileo would bring with him his invention of the determination of longitude, which would be of the greatest interest to the Dutch Republic.

It is not known at what time Galileo was actually informed of the project. Presumably, he had no part in the initiative. Before long, de Groot had to report to his Dutch correspondents that Galileo did not wish to leave Italy, because of his old age. However, de Groot still hoped to convince Galileo at least to offer his proposal to the States General. In Paris, he found an ally in Elia Diodati, who at the time was helping Galileo to have his *Discorsi* printed abroad.[26] Galileo may have agreed to start negotiations with the Dutch around the end of 1635. Indeed, on 5 December a request on behalf of Galileo was made to the States General to start the examination of his proposal. Since, as far as I know, this request has never been published

24. A. Favaro, *Documenti inediti per la storia dei negoziati con gli Stati Generali d'Olanda per la determinazione delle longitudini*, reprinted in: *Nuovi Studi Galileiani* (see note 15), pp. 289-338. Galileo's attitude may also be judged from his choice of employment between the Republic of Venice and the Grand Duke of Tuscany. On the other hand, Galileo in his proposal to the Dutch States General pointed out that in a democratic government it was less likely that the negative attitude of one person in power towards the project would inevitably lead to the total breakdown of the enterprise.

25. *Opere*, XVI, p. 266. See also note 24.

26. Diodati informed Galileo in July 1635 that Galileo's *Discorsi* could be printed by the Elzeviers in Holland. Galileo only consented to it in May 1636.

before and has not been mentioned by earlier historians, I have added the original text and an English translation at the end of this paper. Hortensius, Blaeu and the Dordrecht Rector Isaac Beeckman (1588-1637) were authorised to ask the Italian philosopher for a written exposition of his method.[27] But Galileo was in no hurry. It is perhaps questionable whether this first request made to the States General indeed originated from Galileo himself. If not, Galileo may not have welcomed the precipitate action of his Dutch supporters. Moreover, at the time, Galileo was very busy recopying and polishing up parts of his *Discorsi*, which was forwarded to Elzevier in August 1636. Consequently, the official proposal to the States General was only written on 15 August 1636. "So what?" Galileo apologized to Hortensius, "Things that have been occult during all the years of the world, can very well remain hidden for another three of four months!"[28]

The text of Galileo's new proposal differs in some respects from the one he had sent to Spain twenty years earlier. The secrecy and vagueness that were always present in his earlier writings on the subject, had now completely disappeared. He went into considerable detail as to how exactly longitude could be found from the appearances of Jupiter's satellites. First of all, the satellites could be observed when they entered the cone of shadow of Jupiter. Since the immersion only lasted for a very short time, the exact moment could be accurately determined, with a margin of error of less than one minute. Another possible procedure was to observe the conjunctions and separations of the satellites with Jupiter itself. In this case the error of observation would not exceed one half minute. A third way was to observe the conjunctions and separations among the satellites themselves, which could be determined with an error of at the most a few seconds. These appearances could be observed every night and from every place on earth, at least during the time of the year when Jupiter was visible.

From these observations the longitude of any place on Earth could be

27. For the resolution of the States General on Galileo's request, see the appendix. Presumably the proposal was made by Hortensius and/or Willem Borel (see *Opere*, XVII, pp. 46-49, letter from Elia Diodati to Constantijn Huygens, 20 March 1637). The inclusion of Beeckman among the commissioners at least partially clears Hortensius from the reproach made by Diodati, for having informed Beeckman (and through him Mersenne) on this matter, which ought to be kept secret (see *Opere*, XVII, pp. 44-45, letter from Elia Diodati to Hortensius, 16 March 1637).

28. "Ma che? quello che è stato occulto tutti gli anni del mondo, ben poteva, Sig. Ortenzio mio, celarsi tre o quattro mesi ancora." *Opere*, XVI, pp. 534-536, letter from Galileo to Hortensius, 1636. The text of Galileo's proposal is printed in *Opere*, XVI, pp. 463-468. Probably Galileo's original letter and a copy are to be found in Algemeen Rijksarchief (Den Haag), Staten-Generaal, 12561, nr. 72 (Loketkas Admiraliteit).

determined with the utmost accuracy, even without ephemerides of the satellites. It was enough that the observations made at one place were compared with those made in Amsterdam to find the difference in longitude between the two locations. But if the method was to be used for navigation, four more elements were required for its proper use. First, it was necessary to compute the ephemerides of all the appearances of the satellites and this ought to be done by qualified astronomers. Second, the observations could only be made with the most excellent telescopes, without which the satellites were not visible. Third, a solution was to be worked out to overcome the agitation of the ship, and finally, the observer needed a timepiece to measure the time in hours and minutes which had elapsed since noon or sunset.

Galileo was confident that every one of these difficulties could be overcome. He had worked out ephemerides, that had proved to be correct for several months ahead and it was to be expected that future astronomers would continually improve his tables. Galileo moreover claimed that his telescopes not only made the satellites visible at night, but even during sunset when no other stars could be seen. The agitation of the ship still posed a problem, but actually Galileo thought it less disturbing for telescopic observations than it was for astronomical measurements, e.g. with a quadrant.[29] Finally, Galileo claimed that he had designed a suitable timepiece, easy to construct and less subject to external alterations than any other similar instrument.

Galileo's proposal was first sent in August 1636 to Elia Diodati in Paris, together with letters to Reael, Grotius en Hortensius. Reael was asked to submit the proposal to the States General, which he did on 11 November 1636, after having made a Dutch translation of Galileo's proposal. Reael himself, Hortensius, Blaeu and, if need be, Jacobus Golius (1596-1667), professor of mathematics at Leyden, were appointed to examine the method proposed by Galileo.[30] The report of their discussions was presented to the States General on 7 April 1637. It was then resolved that a second commission was to be formed, consisting of three gentlemen: Rantwijck,

29. Astronomical measurements to determine the altitude of the sun or a star in order to find latitude were indeed equally affected by the movement of the ship. During the seventeenth century many technical improvements were proposed, but with most seamen astronomical measurements remained unpopular, often replaced by less accurate, but much easier procedures. See C.A. Davids, *Zeewezen en wetenschap*, (see note 3), pp. 124 and 172.

30. The choice of this second commission may have been suggested by the fact that they were all living in Amsterdam. Beeckman also appears to have been involved in the further discussions: K. van Berkel, *Isaac Beeckman (1566-1637) en de mechanisering van het wereldbeeld*, Amsterdam (1983), p. 149.

Weede and Schonenburch, who would deliberate with Reael what action was to be taken. Two weeks later it was resolved that a golden chain, worth 500 florins, would be donated to Galileo (who had offered his method for free) and that 1,000 florins would be handed over to Reael for the purchase of instruments needed for further research.[31]

The normal procedure when solutions to the problem of longitude were proposed to the States General, consisted of three stages. At first, the proposal was examined by theoretical scientists to establish its soundness. If it was found to be correct in principle, the judgment of men with practical experience was sought. Only at the third stage was the method put to the test. Few inventions reached this final stage.[32] The resolutions made about Galileo's invention all refer to the first stage, which was clearly not over yet. The method was in principle favourably judged as to its astronomical possibility, but the four difficulties, summed up in Galileo's proposal, still remained to be solved.

During the following months, the commissioners heard nothing from Galileo after his first letters in August 1636, but Galileo received no news from Holland either. All his correspondence had to pass through Paris, because Galileo, living in Arcetri under strict surveillance, wished to keep his contacts with the Dutch a secret. Diodati became impatient and asked Constantijn Huygens to use his influence with the States General to speed up the negotiations. Reael, on the other hand, could not find the time before March 1637 to inform Galileo of the progress made and the resolutions taken. Hortensius, an astronomer himself, had already written an informal letter to Galileo in January, but it reached Arcetri at a time when a severe inflammation of his right eye did not permit Galileo to read a syllable. Due to his bad health, Galileo could only finally respond to the objections made by the Dutch commissioners in June 1637.[33]

The main objection raised was, again, the difficulty of operating a telescope aboard a moving ship. Galileo suggested to Reael that the observer would be placed in a small vessel, floating on a sufficient quantity of water contained in a larger one. In Galileo's opinion this device should work very well when the ship was only moderately agitated, as could be easily demonstrated with a few experiments. Furthermore, once navigators were used to this kind of operation, Galileo claimed they would soon be able to

31. It is not known what was purchased with the 1,000 florins.

32. C.A. Davids, *Zeewezen en wetenschap*, p. 73. After the method was tried on a sea-voyage, the results still had to be interpreted before a reward could be assigned.

33. *Opere*, XVII, pp. 96-105, letter from Galileo to Reael, 5(?) June 1637.

perform the observations even without all this machinery. Also the use of Galileo's *celatone* could very much facilitate the whole procedure.

Essential to the whole project was the telescope. Hortensius had informed Galileo that there were no lens-grinders in Holland who could polish lenses with enough precision to suit the required observations of Jupiter's satellites. Above all, they did not know anything about dioptrics.[34] Galileo offered to have his own lens-maker prepare a lens for Hortensius.[35] Galileo also promised to work on the ephemerides of the satellites, once his eyes were better again. It appears that he wanted to recalculate all his previous observations in order to construct a most perfect and updated *giovilabio*.[36]

In the same letter to Reael, Galileo also largely expounded on a new invention, his time-measurer. The precision of this clockwork was not only that it could give the number of hours, minutes, seconds and even *terzi* but also that if two, four or six similar clocks were made, they would not in the space of an hour, a day or even a month differ more than one pulse (*pulsazione di polso*).

> "I base the construction of such a clock on an admirable proposition that I prove in my book on motion which is now being printed at Elzevirs in Leyden. The proposition is this: If in a vertical circle the perpendicular, which is the diameter of the circle, is erected, and from the point of contact, or the upper end of the diameter, as many chords as you wish are drawn along which bodies descend as along inclined planes, the times of their descents will be equal to the time of descent along the [vertical] diameter."[37]

From this theorem, Galileo deduced the supposed isochronous movement of the pendulum.[38] Since it followed that all vibrations of the pendulum had an equal duration, the period of the pendulum could be used as an accurate

34. *Opere*, XVII, pp. 18-21.

35. Galileo even proposed to send his own telescope to Holland. *Opere*, XVII, pp. 174-175, letter from Galileo to Reael, 22 August 1637.

36. *Opere*, XVII, pp. 136-137, letter from Galileo to Diodati.

37. "Ed il fondamento di tal fabbrica traggo io da una ammirabile proposizione, che io dimostro nel mio libro *de motu* che ora *est sub prelo* de i Sig. Elzeviri in Leida; e la proposizione è tale: Se in un cerchio eretto all'orizonte s'ecciterà dal toccamento la perpendicolare, che in conseguenza sarà diametro del cerchio, e dal punto del contatto, ovvero dal termine sublime del diametro, si tireranno quante si vogliono corde, sopra le quali s'intendano scendere mobili come sopra piani inclinati, i tempi de i loro passaggi sopra tali corde e sopra il diametro stesso saranno tutti eguali" *Opere*, XVII, p. 100 (see note 33). The theorem is proved in the Third Day of the *Discorsi e dimostrazioni matematiche intorno à due nuove scienze*, Leyden (1638), Theorem VI, Proposition VI.

38. P.E. Ariotti, "Aspects of the Conception and Development of the Pendulum in the 17th Century," in: *Archive for the History of Exact Sciences*, 8 (1971/72), pp. 329-410.

measure of time, even when the amplitude of the oscillation was continually diminishing through friction. Galileo consequently worked out a device through which the regular movement of the pendulum could be transferred to a counter, introducing another novelty, the pin-wheel escapement. Galileo, however, never actually constructed his time-measurer. In 1641 he discussed the matter with his son Vincenzio, but nothing came of it.[39] It has since been recognized that Galileo's clock was as important a contribution to the solution of the problem of longitude as his discovery of the satellites of Jupiter.[40] In fact, the construction of a perfect and transportable clock could entirely solve the problem, without the need for a heavenly clockwork. Still, it should be remarked that Galileo only meant his time-measurer to serve for very short intervals of time, starting at noon or at sundown.

Galileo's letter to Reael contained an entire programme of research, for which he invited the help of scientists, mechanics and astronomers. The challenge was readily taken up by Hortensius who, from the very start of the negotiations, had strongly supported Galileo's proposal and had shown himself prepared to make all the necessary astronomical observations and calculations. Galileo showed himself very pleased with Hortensius' enthusiasm and pleaded with Reael to have an observatory built and equipped in Amsterdam to be used by Hortensius.[41] He must also have realised that, without the support of Hortensius, the method would never be accepted by the States General. Indeed, Galileo was informed by Reael that the whole project was considered by the States General to be

> "... at first sight implausible, indeed impossible to investigate; and, once it has been investigated, in no way practical to our sailors, who are rude people, men only superficially acquainted with mathematics and astronomy, who content themselves with those few propositions which are useful to their needs. Men moreover, who still find insuperable the problem of using your discovery on a moving ship, continuously tossed about."[42]

Galileo's proposal was not rejected by the States General; it literally died out. Laurens Reael, to whom the responsibility of the investigations was entrusted, died on 10 October 1637, without even opening the latest letters

39. V. Viviani, "Sull'invenzione dell'orologio a pendolo," *Opere*, XIX, pp. 647-659. Vincenzio Galilei died in 1649.

40. D.W. Waters, "Galileo and Longitude: Fundamental Contributions to a Fundamental Problem," in: *Physis*, VI (1964), pp. 287-302.

41. Letter from Galileo to Reael, see note 33. On the other hand, the intellectual abilities of the ambitious Hortensius were severely downgraded by Descartes, probably as a result of a personal antagonism: K. van Berkel, *Beeckman*, (see note 30), p. 150, n. 95.

42. *Opere*, XVII, pp. 116-117, letter from Reael to Galileo, 22 June 1637.

from Galileo. Two other members of the first commission, Beeckman and Blaeu, died respectively on 19 May 1637 and 26 October 1638. The Leyden professor, Jacobus Golius, who lived until 1667, appears not to have entertained any connection with the further negotiations of the whole affair. Only Hortensius was left. He was assigned by the States General to visit Galileo in Italy, for which he received 2,070 florins in advance to cover his expenses. In Italy, however, the impending visit of Hortensius roused suspicion with the Inquisition and Hortensius was advised to postpone his voyage for a short while until the atmosphere was cleared. Also, Galileo's bad health made a meeting with Hortensius at this time not appropriate. The following ycar Hortensius became professor in Leyden. After he had taken up his new residence there, he died on 17 August 1639.

With the death of Hortensius, Galileo's invention lost its supporting ground in Holland. Diodati tried one more time, with the help of Constantijn Huygens, to have another commission appointed, but nothing seems to have come of it.[43] The problem of longitude continued to occupy Galileo's mind, until he died on 8 January 1642.

### 3. *From Italy to France*

From an historical point of view, Galileo's method to determine longitude has a double heritage. The original invention mainly focussed on the appearances observed in the satellites of Jupiter. Galileo had worked on the ephemerides, but much remained to be done. On the other hand, while working on his own invention, Galileo hit on the idea of a pendulum-clock, inspired by his work on motion. Galileo's time-piece remained largely unknown to his contemporaries, but his theoretical discussion of the pendulum had opened a vigorous debate on its alleged isochronism. As was long known, if a pendulum-clock could be made whose vibrations were isochronous and independent of external circumstances, the problem of longitude would be solved in a most elegant and simple way.[44] The search for a perfect pendulum-clock in the seventeenth century owes much to Galileo's mathematical investigations of the phenomenon.

The double heritage left by Galileo was taken up in a different way in Italy

43. The States General demanded the refunding of the money, given to Hortensius, from his heirs. This may have caused some resentment over the whole affair.

44. The first theoretical discussion of this method is usually attributed to Gemma Frisius. See A. Pogo, "Gemma Frisius, his method of determining differences of longitude by transporting timepieces (1530), and his treatise on triangulation (1533)," in: *Isis*, 22 (1934-1935), pp. 469-505.

and Holland. Although Galileo's doctrine of the pendulum was heavily discussed and criticized all over Europe, its special use, as far as the problem of longitude was concerned, was only brought to the fore by the Dutchman Christiaan Huygens, who had the idea of applying the pendulum to a clockwork, thus renewing Galileo's own design. Whether Huygens was aware of Galileo's time-measurer, or whether he was merely developing his own thoughts while reflecting on Galileo's achievements, is really a matter of taste.[45] The improvements Huygens introduced in the design of the clock are of such enormous weight as to obliterate any possibly underestimated indebtedness to Galileo. Moreover, even if Galileo had indicated how the pendulum could be used in principle as a time-measuring device, only Huygens effectively dealt with the many mechanical difficulties in making it work.

Huygens proposed his clocks several times as a method to determine longitude.[46] After his original invention in 1657 of the pendulum-driven clock, he constructed a portable clock that could be used at sea. In 1663 and 1664 his clocks were tested by the Royal Society with some success, and again in 1668-1669 by the *Académie des Sciences*. However, the next voyage, in 1670, was a disaster. Huygens then started a new line of inquiry, working on spring-regulated clocks. In 1684, however, he again returned to his older cycloidal pendulums to carry out a new trial at sea, financed by the Dutch East India Company (VOC). The bad results of this voyage convinced him of the variation of gravity with latitude, which caused a change in the period of the pendulum. A final trial, carried out in 1690, offered no solution. The determination of longitude by means of a marine chronometer had to wait another seventy years before it was to become a reliable and successful method.[47]

Beside Huygens, there were of course other Dutch inventors to propose their solutions to the States General. The majority of them based their method on astronomy, relying in particular on the observation of the moon. Since many of the proposals were kept secret, we can hardly judge the soundness of these inventions. None of them, however, except Huygens',

45. Huygens, in his *Horologium* (The Hague, 1658), only refers to Galileo's doctrine of the pendulum, published in the *Discorsi*. Another interpretation of Huygens' debt is given by P.E. Ariotti, "Development of Pendulum" (see note 38).

46. M.S. Mahoney, "Christiaan Huygens: The Measurement of Time and Longitude at Sea", in: H.J.M. Bos et al. (eds.), *Studies on Christiaan Huygens*, Lisse (1980), pp. 234-270.

47. However, because of their expensiveness and the low output of manufacturers, marine chronometers were not often used by Dutch navigators during the eighteenth century. See C.A. Davids, *Zeewezen en Wetenschap*, op.cit. in note 3, pp. 182-184.

appear ever to have been tried at sea.[48] From this may be inferred that the astronomical contributions of the Dutch were far inferior to Huygens' mechanical and mathematical research on the problem of longitude.

Indeed, observational astronomy had almost completely disappeared in Holland after the death of Hortensius. Good lens-makers were still scarce, so that when Christiaan Huygens in 1652 wanted to build a telescope according to his own ideas, he wrote to the Louvain professor Gerard van Gutschoven (1615-1668) to ask for information on the practical grinding of lenses in order to make them himself.[49] Although Huygens subsequently became famous as an astronomer with the discovery of a satellite of Saturn, he was not someone who would gladly give up his nights to collect observations. In the Spanish Netherlands, observational astronomy was primarily represented by Govaart Wendelin (1580-1660), who made the pertinent suggestion of applying Kepler's third law to the study of Jupiter's satellites.[50] This at least resulted in the disqualification of some new satellites, reported to have been discovered around Jupiter.

But the real centre of observational astronomy, in particular concerning the satellites of Jupiter, was Italy. Several explanations have been offered to account for this preference among Italian scientists. First of all, we can point to the presence of Galileo, who for a long time possessed the best telescopes in Europe and who stimulated his friends and students to make astronomical observations. In his old age, Galileo had especially taken care that his work on the satellites would be continued. He sollicited the cooperation of Vincenzio Renieri, who in 1639 published the result of his own calculations in the *Tabulae mediceae secundorum mobilium universales*. After Galileo's death he started to work over all of the observations made by Galileo since 1610. But Renieri died before this task was finished, and his papers were lost. Another admirer of Galileo, Giovanni Battista Hodierna, introduced some improvements in the theory of the satellites, which he published in 1656 under the title *Menologiae Jovis Compendium seu Ephemerides Medicaeorum*. In the same vein, Giovanni Alfonso Borelli, with his *Theoria Mediceorum planetarum* (1666) and Giovanni Domenico Cassini, who in 1668 published his *Ephemerides Medicaeorum syderum*, all belong to the same tradition which started with Galileo.

Apart from the intellectual heritage, Italy also enjoyed the advantage of having the best quality glass in Europe. The most important aspect in the

48. C.A. Davids, *Zeewezen en Wetenschap*, op. cit. in note 3, p. 135.
49. C. Huygens, *Oeuvres Complètes*, 22 vols., The Hague (1888-1950), I, pp. 190-192.
50. G.B. Riccioli, *Almagestum Novum*, Bologna (1651), Part 2, p. 492.

fabrication of lenses was to have pure, unstained glass to start with. This glass was manufactured in the region around Venice and consequently enabled Italian lens-makers to produce top quality telescopes.[51] Around 1640, Francesco Fontana of Naples vulgarized a new type of telescope, the so-called 'astronomical' telescope. Although the image produced was inverted, the telescope had the advantage over Galileo's instrument of a larger field of view, a higher magnification and a positive focus, which enabled the insertion of measuring devices.[52] The position of Italy as the leading manufacturer of telescopes was not threathened until the end of the century. Yet another reason why Italian scientists were attracted to observational astronomy may have been the fact that this kind of research made it possible for them to avoid any discussion on cosmology or the system of the universe.[53] The mere technicality of the matter may have offered them a safe retreat from a more dangerous debate.

Through the work of Italian astronomers many of the anomalies in the movements of the Jovian satellites were pointed out. Without going into greater detail, it can be summarized that Hodierna studied the change in latitude of the planets and their eclipses in the shadow of Jupiter, while Borelli tried to incorporate the knowledge of the Moon's motion around the Earth in his theory of the Jovian system.[54] But the realisation of Galileo's project only came with Gian Domenico Cassini, who in 1668 published the first fairly accurate ephemerides of Jupiter's satellites that were to be used for several decades being corrected in 1676 and again in 1693. These tables were held in such high esteem by his contemporaries, and especially by Jean Picard at the *Académie des Sciences*, that Cassini was invited to come to France, to work in the Observatory. When Cassini arrived in Paris, he very soon assumed leadership of the astronomical activities of the *Académie*, among which figured the project of a geographical survey of France, and several world-wide expeditions with various astronomical aims. Cassini worked out a procedure for the exact determination of longitude, based on

51. For one of many affirmations of this claim, see e.g. S. Grillot, "L'emploi des objectifs italiens à l'observatoire de Paris à la fin du 17ème siècle," in: *Nuncius*, II (1987), fasc. 2, pp. 145-155.

52. A. van Helden, "The 'Astronomical Telescope', 1611-1650," in: *Annali dell'Istituto e Museo di Storia della Scienza di Firenze*, 1 (1976), pp. 13-36. Hodierna, however, still used the older 'Dutch-Galilean' telescope.

53. A. van Helden, "The Telescope in the Seventeenth Century," in: *Isis*, 65 (1974), pp. 38-58.

54. J. North, "The Satellites of Jupiter. From Galileo to Bradley," in: A. van der Merwe (ed.), *Old and New Questions in Physics, Cosmology, Philosophy and Theoretical Biology*, New York (1983), pp. 689-717.

the observation of the eclipses of the satellites around Jupiter.[55] Finally the invention propounded by Galileo, although still impractical to use at sea, was given full support by the end of the century as the most accurate method to determine longitude on land.[56]

The development of Galileo's heritage by his compatriots gives an indication of the seriousness of the still extant problems in the use of the Jovian satellites for the determination of longitude. Yet, this should not be taken as an explanation for the total rejection and the general lack of interest Galileo's proposal had encountered for almost three decades. Moreover, his case does not stand alone. While money was frequently spent on the most bizarre inventions, many of the more scientifically based proposals were hardly given any attention.[57] Perhaps due to the still low degree of institutionalisation of scientific enterprise during the first half of the seventeenth century, hindered even more by the aura of secrecy that surrounded their work, scientists could not lean on their personal prestige to encourage state officials to put their faith (and their money) into science. It is then no coincidence that Galileo's invention was finally completed by the first modern research institute of the new era, the Paris *Académie des sciences*.

55. On Cassini and his work, see B. Fontenelle, "Eloge de J.D. Cassini," in: *Eloges des academiciens*, 2 vols., The Hague (1740), I, pp. 273-312; J.F. Montucla, *Histoire des mathématiques*, Paris (1798-1799), II, 559-567; J.B.J. Delambre, *Histoire de l'astronomie moderne*, Paris (1821), II, 686-804; R. Taton, "Gian Domenico Cassini," in: *Dictionary of Scientific Biography*, New York (1971), III, pp. 100-104.

56. Incidentally, when John Harrison proposed his successful time-keeper "Number four" to the British Board of Longitude, the degree of accuracy was ascertained by comparison with the longitude established by the observation of Jupiter's satellites.

57. It must be kept in mind, on the other hand, that many scientists, such as Descartes and Beeckman, did not propose their solutions to the problem of longitude to any official body.

*Appendix*

Resolution of the States General on Galileo's first proposal.
5 December 1635
*Algemeen Rijksarchief Den Haag, Staten-Generaal, nr. 3194 f. 561v-562r*

(*in the margin: Galilerus Galilei*)

Gelesen de requeste van Galilerus (*sic*) Galilei opperste Mathematicus ende Philisophus (*sic*) vanden Grooten hertog van florencen, te kennen gevende hoe dat hij door neerstich ondersoeck vertrout gevonden te hebben een seecker middel omme op alle tyden ende in alle gewesten vande werelt soo te water als te landt te vinden de ware longitude vande plaets daermen sich bevint, ende hoe veel deselve oostelicker ofte westelicker gelegen is van eenig Stad ofte havens Meridiaen diemen sich selven soude mogen voorstellen, presenteerende de voorsz syne inventie in regard vande loffelicke reputatie ende Regieringe alhier als oock premie voorden eersten autheur daerop gestelt hare ho: Mo: te vertoonen ende opte dragen. Versoeckende alsoo sijn Suppliants ouderdom ende verheit van plaetse niet toeliet hier te commen dat mochten geordonneert worden eenige int voorsz. stuck ervaren om aen deselve schriftelick te mogen openinge doen ende d'examinatie van zijn werck te vervorderen. Waerop gedelibereert synde, Is goetgevonden ende verstaen mits desen te versoecken ende autoriseren doctorem hortensium praesidem Mathematicum tot amsterdam, Willem Jansz Blau mede tot Amsterdam Isaacq beeckmans Rector tot dordrecht om de opening van des Suppliants voorsz inventie schriftelick te ontfangen ende examineren, ende van alles hare ho: Mo: rapport te doen.

*Translation*:

Read the request by Galileo Galilei, chief mathematician and philosopher of the Grand Duke of Tuscany, announcing how through his zealous research he believes to have found a certain method to determine at every moment and in every place of the world, at sea as on land, the true longitude of the location, and how much more to the east or to the west this location is situated from the Meridian of any city or port, that may be chosen freely; presenting this invention with regard to the laudable reputation [to be gained for himself] and the government in this country, and also [with regard to] the premium offered to the first author who would show and dedicate [his invention] to Her Great Power. Begging, since the age of the Suppliant and the distance of location did not permit him to come [here], that some experts on this matter would be authorised to open a correspondence with the Suppliant and to go on with the examination of his work.
On which it was deliberated: agreed and accepted that would be invited and authorised doctor Hortensius as president, mathematician of Amsterdam, Willem Jansz Blaeu, also [living] in Amsterdam, Isaac Beeckman rector at Dordrecht to open [the negotiations on] the Suppliant's invention by letter and to examine it, and to report everything to Her Great Power.

# GIOVANNI FRANCESCO BUONAMICO AND THE FOSSILS: A FLOOD OF PROBLEMS

*Nicoletta Morello*

1. During the sixteenth and seventeenth centuries many naturalists turned their attention to the true nature of fossils, especially of the marine species, that were widely considered to be a special type of "figured stones" which had the shape of organisms. On the other hand, there had already been trends in the fifteenth century which favoured, encouraged by the Ancients, the supposition that the *lapides figurati* were the remains of organisms, buried in the soft ground that by petrifaction had turned into solid rock .

The position adopted and the attempts made to demonstrate one hypothesis or the other aroused rather than resolved the controversy. Independent of personal convictions and ties with traditional petrography, the naturalists who were faced with such a great variety of geological and paleontological phenomena, could make observations and choose evidence — in good faith or in bad — to support either the organic or the inorganic hypothesis. These two opposite images of inorganic nature were maintained by the impossibility of interpreting natural phenomena, such as the type of fossilization or rocks which presented marked local differences, outside of a scheme which presupposed constancy and uniformity of "causes".

It is thus possible to explain, for example, how the question of the *glossopetrae* survived well beyond the numerous 'demonstrations' of the organic nature of these fossils. On the other hand the hypothesis of the inorganic origin of fossils often had full scientific validity when it was the result, as it was in many cases, of an investigational accuracy often lacking in some of those who supported the explanation that was to win.

2. Among the fossils which were documented in the fifteenth and sixteenth centuries the *glossopetrae* (literally stone tongues; teeth of types of shark) aroused the greatest interest.

They were found in Malta and also in many places on the continent and they were quite a well-known and widespread remedy all over Europe because of the alexipharmic properties they were believed to have. This alexipharmic virtue was only possessed in the highest degree by the Maltese *glossopetrae* because it was conferred on the Maltese earth directly by St Paul as we can read in the First Book of the Acts of the Apostles.[1]

Some of the reasons which made the *glossopetrae* the fulcrum of the fossil problem were their fame as an alexipharmacon in comparison with other well-known figured stones in the form of marine creatures such as shells and whirls, their discovery also outside Malta, as well as the recognition of their similarity to shark's teeth owing to the development of the study of the morphology, anatomy and physiology of marine animals. Attempts were focused on the one hand on attributing them to the animal kingdom as shark's teeth, while on the other to leaving them in the category of stones, without depriving pharmacy of an important medicine and Malta of a peculiarity of its soil.

3. As we know, Steno in *Canis Carchariae*[2] and in *Prodromus*[3] closed the *lis necdum decisa* on the *glossopetrae* and fossils in general, merging the question of the origin of the 'figured stones' — in the form of vegetation or animals —and the codifying of the *naturae leges* which ruled the occurrence of geological and paleontological phenomena.

Before his stay in Italy, Nicolaus Steno (1638-1687) had already been in contact with scholars who studied fossils and at the same time were involved with research bordering on geology.

Naturalistic problems which were also dealt with by physicians, e.g. subjects as *balnea* and *fossilia*, were one of the probable reasons for the friendship that developed after Steno met Thomas Bartholin (1616-1680) at Copenhagen University.

When Steno arrived in Copenhagen in 1656, Bartholin had returned some time before from a study expedition in Europe during which he had visited the best-known natural history museums and private collections and had seen the *corpora petrefacta* of different types which for various reasons attracted the attention of the physician *curiosus naturae*.[4]

1. Cf. *Atti degli Apostoli*, 28, 1-10.

2. *Nicolai Stenonis elementorum myologiae specimen seu musculi dissertatio geometrica cui accedunt Canis Carchariae dissectum caput et dissectus piscis ex Canum genere*, Florentiae, sub signo Stellae, 1667.

3. *Nicolai Stenonis de solido intra solidum naturalister contento dissertationis prodromus*, Florentiae, sub signo Stellae, 1669.

4. *Thomae Bartholini Anatomicarum rariorum centuria I et II*, Hagae Comitum, ex typographia Adriani Vlacq, s.d., *Centuria II*, Historia IX — XX — LXIV — C.

*Thomae Bartholini Historiarum anatomicarum rariorum centuria III et IV. Eiusdem cura accessere observationes anatomicae Petri Pawii*, Hagae Comitis, ex typographia Adriani Vlacq, 1657, *Centuria IV*, Historia I.

Bartholin had stayed in Italy (Padua, Messina, Naples, Rome and Venice) and also in Malta. It was in Malta between 1643 and 1644 that he was the guest of Giovanni Francesco Abela, Vice Chancellor of the Order of Jerusalem in whose museum there was a wood fossil that he had found. Certainly the chance of a visit to Malta had stimulated Bartholin's interest in the famous Maltese *glossopetrae*. Furthermore his host was about to give the press a paper in which, among other things, he expressed his conviction of the inorganic origin of the *glossopetrae*. So faithful was he to the question which linked St Paul to the story of Malta and its "tongues" that he presented the picture of the bay in which St Paul was shipwrecked before landing on the island.[5]

Steno himself records in his *Canis Carchariae*[6] that Bartholin, whom he calls *praeceptor meus*, had seen in Malta (and had brought back specimens from there to Copenhagen) the famous *glossopetrae maiores* (teeth of *Carcharodon megalodon*) whose origin he was also near to dealing with.

Bartholin probably prepared a paper (which was never published) in favour of the organic nature of these specimens. In Padua, where he returned after the journey to Malta and Southern Italy, he instead gave the press a treatise on the unicorn[7] which was also considered a powerful antidote. For this reason and because of its interpretation as an "ivory fossil" it caught the attention of many scholars. Among them was Daniel Sennert (1572-1657), the writer of a "summary of natural history",[8] a quite widespread genre of 'popular' scientific literature. Another example of this genre was the *Historia Naturae* of Giovanni Eusebio Nieremberg, a Madrid Jesuit and professor of "physiology" at the Royal Academy of that city.[9] In these well-known works, as well as in the *Pharmacopoea spagyrica* by Pierre Potier,[10] a French physician who lived for a long time in Italy and who was quoted for example by Bartholin and

5. *Della descrittione di Malta isola nel mare Siciliano con le sue Antichita' ed altre notizie libri quattro del Commendator Fra Gio: Francesco Abela*, Malta, Paolo Bonacotta, 1647, p. 133 and pp. 222-224.

6. N. Steno, *Canis Carchariae*, op. cit. in note 2.

7. *Thomae Bartholini de unicornu observationes novae Accesserunt de aureo cornu Cl. V. Olai Wormii eruditorum iudicia*, typis Cribellianis, Patavii, 1645.

8. *Danielis Sennerti vratislaviensis epitome naturalis scientiae*, editio secunda, Witebergae, J.W. Fincelius, 1624.

9. *Joannis Eusebii Nierembergii* [...] *Historia Naturae maxime peregrinae Libri XVI distincta. In quibus rarissima Naturae arcana etiam astronomica et ignota Indiarum animalia, quadrupedes, aves, pisces, reptilia, insecta, zoophyta, plantae, metalla, lapides et alia mineralia, fluviorumque et elementorum conditiones etiam cum proprietatibus medicinalibus, describuntur novae at curiosissimae quaestiones disputantur ac plura Sacrae Scripturae loca erudite enodantur. Accedunt de miris ac miraculosis naturis in Europa Libri duo: item de iisdem in Terra Hebraeis promissa Liber unus*, Antverpiae, Balthasar Moretus, 1635.

10. *Pharmacopoea spagyrica Petri Poterii andegaunensis* [...], Bononiae, N. Tebaldino, 1622.

Holger Jacobaeus (1650-1701),[11] at least one brief chapter is dedicated to the Maltese fossils. These authors, together with Abela and Domenico Magri,[12] who represented the best of erudite Maltese culture, were the principal scientific sources of Giovanni Francesco Buonamico, a Maltese physician and naturalist still unknown among modern historians of science, especially of geological science.

4. Born in Valletta in 1639,[13] Buonamico left this city in 1657 for a study trip through Europe which lasted nine years. At Aix-en-Provence where he stayed for nearly a year he probably graduated in medicine. He spent more than eighteen months in Paris and from there he went to Worms for about six months. At the end of May 1661 he arrived at Utrecht and later, after several weeks in Amsterdam, he moved to Leyden where he stayed for two years, attended the Faculty of Medicine and took a second degree. On the 22nd September 1663 he left Leyden for Louvain, where he studied at the university for two more years. Afterwards he returned to Holland and he spent fifteen months in Amsterdam. In 1666, at the age of twenty-seven he returned via Italy and Greece to Malta and obtained a nomination from the Order of Jerusalem as General Doctor of the Galleys but after several months at sea in the Mediterranean he asked to be relieved of his post and retired to private life. He died, aged 41, on March 18th 1680.

Buonamico was an amiable conversationalist, with a knowledge of many languages, he was open to the discussion of naturalistic subjects in which he excelled as well as of humanistic subjects or other scientific questions such as mechanics. Coming from the island famous for its *glossopetrae*, he had presumably many opportunities to discuss the question of fossils during his years in Leyden. We do not know for certain if Steno was among his interlocutors. However, according to what Mangion[14] says, he did keep "rapporti di studio" with him after the Leyden period.

Among his published and unpublished works[15] there is the *Lettera missiva* in which Buonamico dealt with the theme of the organic origin of fossils. In the *Memorie di viaggio,* still handwritten, of which Mangion published the most

11. *Oligeri Jacobaei de ranis et lacertis observationes*, Hafniae, 1686, impensis Johannis M. Lieben.

12. Domenico Magri, *Viaggio al Monte Libano*, Roma, 1655.

13. The main bio-bibliographical source on Giovan Francesco Buonamico is Giovanni Mangion, *Giovan Francesco Buonamico scienziato e letterato maltese del Seicento*, "Studi secenteschi", Firenze, Olschki, v. XII (1971), pp. 285-321.

14. *Ibidem*, p. 287.

15. *Ibidem*, p. 288 and p. 300.

interesting parts[16], Buonamico gave a report of his European itinerary (1657-1666). This account, which was written in 1672, is incomplete because it stops at the year 1663, ending with the pages devoted to his stay in Holland, one of the most important stages in his intellectual life.

5. Buonamico entered the controversy which arose on the origins of *glossopetrae* and in particular the Maltese ones, because he was asked by Paolo Boccone (1633-1704), the Sicilian naturalist and botanist of the Grand Duke of Tuscany, then in Malta for research, to answer a letter from a friend in Messina, the painter and naturalist Agostino Scilla (1629-1700). Scilla had written bo Boccone because, before returning to Sicily, Boccone was to collect some examples of Maltese fossils for him, fossils both loose and from the parent rock. The reason for this request was the desire to examine closely these much-discussed fossils and the "casualness of their grouping" in order to sustain the hypothesis, that Scilla had recently arrived at, of their organic origin. Boccone, however, knowing that Buonamico had made a mental itinerary in the opposite direction to that of Scilla, asked the erudite Maltese to answer the letter which Buonamico did, attempting to reconvert the Sicilian painter by setting out the arguments that had led him to abandon the hypothesis of the organic nature of the fossils.

Buonamico's letter is dated 28th August 1668.[17] Buonamico was then 29 years old; shortly before Steno had finished the preparation of *Prodromus* which, however, he had not given to the printer. Far from being a scholastic treatment of the fossil theme — as many other writings in defence of the petrographic theories were to be — the *Lettera missiva* set out, first of all, the geological and paleontological objections that could be raised against the arguments adopted and the observations made in favour of the organic origin of fossils. Although the second part has the continuity and fluency of an epistle, it is dedicated to the development of Buonamico's philosophical assumptions.

6. The naturalistic treatment revolved around the theme of the universal Flood, whose validity as the only event responsible for the presence of fossil remains on the surface of the Earth, was firmly denied by Buonamico. At the

16. *Ibidem*, pp. 292-295; pp. 330-335 and pp. 312-321.

17. *Lettera missiva del signor Gio. Francesco Buonamico Maltese, dottore di medicina, filosofo e poeta, diretta ad Agostino Scilla messinese, pittore ed accademico della Fucina detto lo scolorito, Data sotto li 28 d'agosto 1668 ove si tratta dell'origine delle glossopietre, occhi di sepri, bastoncini detti di S.Paolo ed altre pietre figurate che si cavano dall'isola di Malta e dal Gozzo*, in *Opuscoli di Autori siciliani*, Palermo, 1770, tomo XI, pp. 105-200.

time of his return to Malta, Buonamico, who had seen rare examples of petrification and visited many famous natural history collections during his nine years abroad, was inclined to accept the hypothesis of the organic origin of fossils. Above all he was inclined to accept the "sedimentary" version, according to which the remains of marine animals, at one time alive, were enveloped in marine deposits which had emerged due to "various changes of place of the sea". In time, while these were consolidated into rock, the organic remains were transformed into the fossils we find today.

The "diluvialist" hypothesis moved away from this formulation attributing to a single catastrophic event, the Flood, the presence of recent fossil-bearing strata in the Earth's crust.

Many of the Ancients had believed that the fossils were the remains of living organisms and many modern authors, whose names Buonamico learned from Scilla's letter and from Boccone, defended this hypothesis; among them Steno, his "fellow-student at Leyden".

Buonamico thought that the conviction of the authors quoted by Scilla was based above all on the morphological similarities between the *figura externa* of fossils and analogous living creatures, something that seemed to him insufficient and anything but evidential, above all because many of these authors, in his opinion, associated the argument of the morphological analogy with the Flood, by means of which they explained, in a simplistic way, the problematical distribution of fossils upon and in the Earth's crust.

Buonamico's initial predilection for the sedimentary hypothesis was also linked to the conclusion that Athanasius Kircher (1602-1680) reached after numerous observations on land in the Maltese archipelago. Kircher, whom Buonamico often met in Malta, maintained that Malta was "by chance made up" of completely horizontal strata and that the whole archipelago had recently been formed, created from "various marine deposits"[18] and that it did not date from the Creation.

In order to "confirm by complete visual satisfaction those first mental impressions"[19] Buonamico made many excursions in the Maltese archipelago and above all in Malta, looking for phenomena that could confirm the affirmations about the organic origin of fossils and the recent formation of the island with which he agreed. In Malta and at Gozo, Buonamico also sought the "casualness of assembly" of the fossils and the overlapping in horizontal strata of sandstone, clay and gravel.

The "casual" amassing of fossils was one of the supporting arguments of the

18. *Ibidem*, p. 153.
19. *Ibidem*, p. 120.

diluvial hypothesis. It seemed in fact to indicate that a violent turbulence in the flood waters had upset the emerging land and the sea bed, dragging round the globe the remains of animals dragged from their natural environment and brought together. When the turbulence ceased, equally casually (a word that Buonamico uses to mean "confusedly") they were deposited at the bottom of those waters together with material in suspension, forming fossiliferous rocks in a short (Biblical) time.

The overlapping of horizontal marine sediments was on the contrary one of the arguments that supported the "sedimentary" hypothesis. The sea, abandoning temporarily or permanently the areas that it had covered, caused an *incrementum terrae habitabilis* (as Linnaeus was to call it) due to the deposits and the consequent consolidation of "rejettamenti" among which the remains of marine organisms were to be found. The advance of the coast, however, was aided by fluvial deposits of material brought down from inland. This formed, as in the case of Holland, "[...] clayey and sandy land [...] compounded of different soils" of which only some were fossiliferous.[20]

Malta and Gozo were composed, on the other hand, entirely of tertiary marine sediments, faulted and divided into five formations which contained, in regular distribution along the stratifications, those fossils whose true nature many European naturalists were investigating — *glossopetrae*, sea urchins, whirls, etc.[21] Instead of "a confused mixture of tongue stones, eyes, urchins, shells and bones [...]", Buonamico saw an ordered distribution of species: "[...] on the contrary I saw and marvelled that almost every species of our figured stones has its own special vein, that is the rock of very different quality where it is found [...]."[22] His initial marvelling induced Buonamico to further his paleontological investigations at the end of which he had reconstructed the map of the contemporary geographical distribution of the principal Maltese fossil forms. From his investigations it became clear that the local fossil types

20. *Ibidem*, p. 153.

21. The geology of the Maltese archipelago is relatively simple even if it is different from and more complex than the geology of the Netherlands, as Buonamico noted. The stratigraphic succession, ranging from Lower Miocene to Upper Miocene, is composed above all of calcareous rocks of different petrographic composition (recristallized biolitites, fine grained limestones, chalky limestones, etc.). The intermediate part of the succession (*Globigerina Limestone*) is characterized by the presence of phosphatic nodule beds full of fish teeth. The whole sequence is slightly folded and crossed by numerous faults, normal or transcurrent in character, originating frequent downworping. Cf. M.R. House, K.C. Dunham, J.C. Wigglesworth, *Geology and structure of the Maltese Islands* in H. Bowen-Jones, J.C. Dewney, W.B. Fisher, *Malta*, 1962, Newcastle upon Tyne, pp. 25-33.

22. G.F. Buonamico, *Lettera missiva*, op.cit. in note 17, p. 121.

did not reveal any confusion in their stratigraphic (ante litteram) or geographic distribution.

It was with this consideration that Buonamico opened the discussion of the "diluvio di difficultadi" (flood of problems) which engulfed, without providing a plausible answer, those who wanted to explain through the Flood the organic origin of fossils, and of the Maltese species in particular. The objections raised were largely the same as those which were to be raised, above all from the second half of the seventeenth century, by the sedimentarists against the diluvialists. According to Buonamico the Flood explained neither presence of fossils inside rock nor their absence between two fossiliferous areas; and not even their assembling in particular places instead of being spread at random over the Earth's crust. What is more, it did not explain the great quantity of fossils found in sediments compared with the very few found on the coast, caused by the retreat of the sea or after storms, bearing in mind that even during the Flood the molluscs had tried to anchor themselves onto the cliffs exactly as they did in storms at sea.

In addition to these there were other more general arguments that Buonamico put forward against the organic origin of fossils and in support of the opposite theory towards which he was more and more inclined. There were few sharks in the Mediterranean and none in the northern seas, but famous deposits of *glossopetrae* were found in these areas. Because each specimen exhibited a great number of teeth, and Malta's sediments were rich in tongue stones, an unprecedented number of sharks would have had to die — most of them near the island. In that case other parts of their skeletons should have been found, more solid and resistant than teeth; these, however, were never found in their original state (Buonamico gives an exact description of non-fossilized teeth but he was not able — or was unwilling — to recognize them as such) nor could they be considered "perfectly" similar to *glossopetrae*. Furthermore, sharks' teeth did not possess the alexipharmic virtue with which the tongue stones and especially those of Malta were invested. They should therefore be classified among stones, not different from those which were equally famous and discussed, whose *figura* had never been seen in *rerum natura*, namely the *cornua ammonis* (ammonites), the unicorns, the *astroites* and so on.

Buonamico did not believe at all that all the fossiliferous mounts were postdiluvian, formed as was usually believed by the "sand, mud and other muck haphazardly brought together by the waters."[23] On the contrary, "the

23. *Ibidem*, p. 152.

land of Holland and Zealand" could be considered recent and formed by repeated marine and river deposits "having remained not only very low and flat but also with not even a handful of rock [...]."[24]

The morphology of Malta and its diagenesized, subhorizontal, locally verticalized ("parallel to the earth's axis") rock types, close to faults, had nothing at all in common with the Dutch land or the parts of Northern Europe where the retreat of the Flood waters, according to seventeenth and eighteenth century naturalists, was still in progress. "There being always the same causes [...], from what we see nowadays, we should with every probability conclude that the same must have occurred in the past."[25] The "cause" which affected Holland and which had turned it into a "post-diluvian" land, with a flat morphology characterized by different kinds of rocks, parallel to the horizon and overlapping each other, did not affect the Maltese coasts which should "grow", but instead, were continually being eroded.

Furthermore, after paying rather closer attention, Buonamico noted that not all the strata in Malta were horizontal and that these placements were not regular but had a wide variability spectrum so as to create the impression that "the rocks of each piece of land also have their own arrangement of fissures."[26] In Malta there were not "those many soils tidily organized in distinct materials, one on top of the other" which were, as in Holland, together with a formation with marine fossils, a characteristic of alluvial deposits, considered by seventeenth-century naturalists to have followed — but not necessarily to have been caused by — the Great Flood. Malta "is nothing but a whole rock, solid and uniform from its surface to the deepest one can discover."[27] Deformed and faulted, covered not by sand but by fertile land, this "cristianissima" Malta was for Buonamico ancient, antediluvian and created as it was — contrary to what Kircher thought — by the hands of the Divine Architect. Its fossils, diverse for every piece of land, according to the types of sediment and their dislocation, were special as all Malta was special and endowed like the *glossopetrae* found there with the same alexipharmic virtue.

7. To explain the presence of fossils within the rocks, without having recourse to the Flood and still believing in their organic nature, some researchers of Nature, including Gassendi (1592-1655), hypothesized the existence of underground salt water lakes in which the fish and shells had lived.

24. *Ibidem*, p. 152.
25. *Ibidem*, p. 155.
26. *Ibidem*, p. 156.
27. *Ibidem*, p. 153.

Others, such as Agricola (Bauer, 1494-1555) and Neobrigense,[28] believed that the fossils were animals breeding within the rock or brought about by the planets or by the "fermentation" of some "material"; others like Terzago thought that they were the seminal fruit of marine substances that the waters, following their normal underground routes, had carried inside the mountains.

Sennert and De Boot, with whom Buonamico said he agreed, maintained instead that they were stones whose *figurae* did not reveal any morphological analogy with the *figurae* of living organisms and this was truc, as we remember, for unicorns, ammonitcs, and so on. Apart from some examples of animal generation within rocks, which Buonamico considered possible because he believed in spontaneous generation, fossils for him were "pure stones of their species formed by nature because it pleased God."[29]

The "final" reasons for the particular form of the stones had to do with a sort of analogical function: from the different *figurae* of the fossils man could gather the *virtus* concealed in every stone. And this was a sign of the Divine Benevolence which, "even though it had, after original sin, completely deprived us of that certain knowledge of the virtues of natural things, was however, pleased to leave in them certain obscure characteristics, or ciphers, namely the external figures, which give at least an indication that this or that stone is worthy of observation, because Nature could not have given them a certain figure by chance, since it never works at random but always regularly and in the same way."[30] Only the stones whose figures were "always various, irregular and completely different from each other" had been shaped at random and there was no possibility of attributing them to determinate and distinct species.[31] But for Buonamico it was not by chance if those stony objects, shaped like leaves, flowers and plant seeds, had "figures that are always regularly similar and uniform in the same rock species."[32] Thus the *glossopetrae*, shaped like serpent's tongues, reminded us that they were antidotes to poison. According to Buonamico even the non-Maltese *glossopetrae* had this quality, while it was not possessed by fossils in shell form

28. This is William of Newburg, a churchman from Bridlington, Yorkshire, who probably died in 1201. He was the author of a chronicle of events in England between 1066 and 1198. Among the many posthumous editions of this work, the most famous are: *Rerum anglicarum libri quinque, recens ceu e tenebris erutis et in studiosorum gratiam dati*, Antverpiae, 1567 and *Gulielmi Neubrigensis* [...] *de rebus anglicis* [...] *libri quinque* [...], Parisiis, 1610.

29. G.F. Buonamico, *Lettera missiva*, op.cit. in note 17, p. 165.

30. *Ibidem*, p. 168.

31. *Ibidem*, p. 175.

32. *Ibidem*, p. 177.

although the Maltese *conchitae* were composed of the same earth as the *glossopetrae*.

Many hypotheses were also advanced for the identification of the "cause prossime" of the figures of minerals, fossils or crystals (*lapides figurati*).

Some maintained that inanimate things were the products of Nature, not randomly but according to the form they took on or from the disposition of material or from the laws of God who, having created "all things *in numero, pondere et mensura*" had included in the measure or quantity even the forms "che sono modi dell'istessa". The proof of this would have been the spherical form of the sky, of the stars and of the Earth, the pyramid form of flames, the various figures of "stars" of snowflakes but above all the different crystalline forms of salts. Thus as the form of garnets, rock crystal, topaz etc., is constant, it is equally constant for each kind of salt. Every salt when it is crushed, dissolved in water and recrystallized reassumes its own particular form: "pyramidal the glass, cubic the mined salt, the form of a pyramid quadrangular blunt the sea salt [...]."[33]

The chemists who considered salt to be the rudiment of all "mixtures" attributed the forms of various bodies to the "predominance and mixture of the salts". But the interpretation did not explain why, for example, crystals were pyramidal nor how salts, which were mixed and not single bodies, always formed themselves into a determinate figure and after, according to the chemists, contributed to give shape to crystals and gems.

To respond to this and other questions, many had recourse, said Buonamico, to hypotheses such as the action of a formative faculty, of a heavenly idea which had produced an astral virtue, of a universal soul, of a vegetative soul for stones and minerals and also a formative and augmentative soul. These hypotheses did not, said Buonamico, allow us to "penetrate the essence of the truth", given that they explained "an unknown thing by another equally unknown and obscure one."[34] Democritus cast a little more light on the matter. Mixtures were formed, his theory explained, by "insensible and very minute particles" endowed with a great variety of different forms which, when combined, took a certain geometrical shape. The "cause" of the natural forms of salts and stones could be explained in such a way and, still more, the forms of their component particles could be deduced from their global shape.

"One can say in general, that the insensible particles that make up a grain of sea salt have a form which naturally enables them to unite into for example, a blunt, quadrangular pyramid, that of crystal into a hexagon, of garnet into a

33. *Ibidem*, p. 179.
34. *Ibidem*, p. 180.

globe, of topaz into a cube and so for thousands and thousands of other figures, thus also for our *glossopetrae*, serpents' eyes, shells, sea urchins and other more compound forms."[35]

Also, still according to Democritus, all stones with regular forms were "almost efflorescences of the different rocks" from which they came. The larger or smaller dimensions of these forms, and the forms themselves depended on the nature of the pores of the rocks which the lapidary saps had had to adept to; for this reason one type of rock generated a given type of stone with a given form. This, according to Buonamico, was "an ingenious way of philosophizing mechanically which clarifies obscure natural phenomena". A way which combined well in Buonamico with Aristotle's hypotheses of spontaneous generation and the pages of the Bible. The land and the sea were like Nature's womb, they contained the "principi materiali" of all animals. When these "principles of flesh, bones, nerves" met inside a rock, there was no obstacle to the generation of animals or parts of them. If this happened, as Buonamico believed it did, it could happen all the more in the Earth which had the "principles" of bones, isolated parts like teeth, shells and so on. Also God himself had said: *producat terra, producat aqua*, and by virtue of that *producat*, by virtue of the spontaneous generation of whole animals in the sea, by virtue of the analogy that united the animal, vegetable and mineral kingdoms there was no reason why stones shaped like plants or marine animals or bones of vertebrates could not be produced in the rock. The *Lettera Missiva* stopped at this point, since Buonamico probably thought that the matter had driven him too far from his original intentions.

8. In Naples in 1670 *La vana speculazione disingannata dal senso* was published.[36] This was Scilla's reply to Buonamico's letter and it contained a point by point criticism of the arguments of the Maltese scholar. The work did not mention the name of the person to whom this "lettera responsiva" was addressed, who was only referred to as "Dott. N.N.". Scilla thus wanted to

35. *Ibidem*, p. 183.

36. *La vana speculazione disingannata dal senso. Lettera responsiva circa i corpi marini che pietrificati si trovano in varij luoghi terrestri di Agostino Scilla pittore, accademico della Fucina, detto Lo Scolorito* [...], Napoli. Andrea Colicchia, 1670. There were at least three (partial) Latin translations of this work during the eighteenth century: Roma, De Rubeis, 1747?-1750?; Roma, De Rubeis, 1752; Roma, Monaldini, 1752. These editions, entitled *De corporibus marinis lapidescentibus quae defossa reperiuntur,* contain both Scilla's *Lettera* and the *Dissertatio de glossopetris* of Fabio Colonna (1567-1640). In the edition of which the date is uncertain, a third work is included (by Vitaliano Donati).

Mangion says that the publication of Buonamico's *Lettera missiva* was called for by the publication of the 1752 edition of Scilla. Cf. Mangion, quoted in note 13, p. 295-296.

save Buonamico from the fierce attacks to which he would have been subjected because of the opinions he had expressed. Clearly he had been struck by the fact that a writer who was so precise in his treatment of geological and paleontological subjects, had reached conclusions that seemed clumsy and even absurd.

The polemic between Scilla and Buonamico arose in fact during the years in which the 'proof' of the organic origin of fossils found its most complete expression in Italy, in the work of Steno. Besides, such organic origin had already been advanced in Italy at the beginning of the seventeenth century. If Scilla had been a little less impetuous, he would have realised that even if Buonamico had arrived at conclusions that were difficult to sustain because they were based on rather uncertain arguments (like the spontaneous generation of molluscs or the therapeutic virtues of figured stones), he had also produced solid arguments against the weak points both of the diluvial and the sedimentary theories, at least in the terms in which they were then presented.

It is also true that the *Lettera missiva* had the explicit aim of orienting criticism in favour of the inorganic origin of fossils and against the contrary hypothesis. This was something which, in the attempt to resolve a naturalist question about "objects" effectively linked in two otherwise distinct worlds —the organic and the inorganic — led to the association of biological arguments with geological, paleontological and mineralogical ones. Thus those who approached the question of fossils by a rigorous method, found that they had to face — regardless of the side they were on — the controversy about the generation of animals and the one which was emerging about the ways in which the Earth's crust was formed. Buonamico favoured the geological, paleontological and mineralogical arguments and his observations were those of a field naturalist. Thus the *Lettera missiva* was above all a short treatise on "Earth sciences" as it is called today. This was based, negatively with respect to fossils and the Flood, on themes around which Steno's writings revolved even though these came to different, if not opposite conclusions.

This thematic similarity — above all in the conception of the mechanics of the formation of sedimentary strata and the overlapping of those originally parallel to the horizon; in the treatment of the formation of the figures of salt and crystals, distinguished from other fossils because of their geometrical form but which like the others had a form which was constant and characteristic for each "species" — does not necessarily derive only from the particular nature of the problem nor does it indicate Buonamico's assiduity with Steno during his stay in Leyden. It rather suggests an analogous intellectual journey, the result perhaps of their travels through Europe, a journey that Steno after his stay in Leyden was to follow in the opposite direction to Buonamico's.

Above all it was the university studies carried out in Northern Europe that had a deep and positive influence on the young Maltese who came from an environment which, while enlivened by the presence of many foreign scholars, revealed itself to be, at least on the subjects of fossils, not very open to the great changes in scientific culture, even if news of such changes came from Southern Italy, which had close cultural ties with Malta. The climate of great liberty and the intellectual vitality of Holland, the richness of research and of naturalistic themes whose principal exponents were also university teachers, contacts with students from very different cultural backgrounds, all gave Buonamico, who was studying medicine, the opportunity to develop his special interests. However, Buonamico's scientific education does not seem to have been particularly oriented, as in the case of Steno, towards bio-medical studies. Rather, towards physico-mathematical themes in which he was sustained by a mechanistic view of nature that adopted the philosophical principles of Descartes but still linked Aristotle and the Bible. Buonamico was attracted by the research and experimentation of the iatrochemists and by chemists in general on crystal mineralogy — also in relation to the question of the origins of the *figura* of the crystals of snow and salts — as well as on the experiments on the mechanical deposition of substances in suspension in fluids. He also knew the observations made by many naturalists to explain the phenomenon of *decrementum aquarum* so manifest on the Swedish coasts, and from the many readings of geographical works as well as from the excursions he seems to have made, he learned to interpret the geological composition of Holland.

In Democritus' "filosofar mecanico" Buonamico found the unitary explanation of the great variety of phenomena of the inorganic world, from the geometrical forms of minerals and crystals to the varied morphology of the Earth's surface.

The Maltese cultural environment which was composed of erudite people intent on defending the image of Malta as a stronghold of Christianity against the Turks, as a land created directly by God and then marked by the miracle of a Saint, seems to have taken over Buonamico when he returned.

His Northern European experience was a defined and closed episode in his life. But he never forgot it. Whether involved in scientific reflection or in writing his *Memorie di viaggio*, Buonamico continually saw Holland as a counterpoint to sunny Malta.

The impression left on him by his journey can be seen perhaps in this touching nocturnal memory:

> "On nights when all is calm and quiet in the air with deep silence, by the light of the moon you see the water of the canals as quiet and unmoving as crystal but, at the

sound of the church bells nearby, on the surface of that so tranquil water suddenly, as if by the touch of a hidden hand, hundreds and thousands of circles are formed and these in their growing and decreasing continue in various guises to occupy all water, rippling it lightly rather as though an infinite number of extremely minute stones rained upon it. When the tocsin ceases so does the circulation and movement of the waters and tranquility returns. From all this, beautiful conclusions might be drawn to understand the nature of sound."[37]

37. "Nelle notti dunque quando è somma buonaccia e quiete nell'aria ed un profondo silenzio, al chiaro della luna vedrassi l'acqua che stagna ne canali non men quieta e immobile che se fosse un cristallo: ma non si tosto ai replicati colpi delle vicine campane risuona l'aria, che su la superficie di quell'acqua si tranquilla, veggonsi con celerità incredibile formarsi in un tratto da occulta mano centinaja e migliaja di circoli, e questi crescendo e disseccandosi continuamente in varie guise, occupar d'un subbito tutta la superficie dell'acqua, increspارla, ed agitarla leggermente, non altrimenti che se da infinito numero di minutissimi sassolini buttativi con impeto fosse stata percossa. Cessando poi il suono delle campane s'acquietano anco poco a poco quelle circolazioni e movimento nell'acque, e ritorna finalmente nella tranquillità di prima. Qui possono cavarsi bellissime conseguenze per comprendere la natura del suono." Cf. Mangion, op.cit. in note 13, p. 314.

# ITALIAN INFLUENCES ON ANTONI VAN LEEUWENHOEK

*L.C. Palm*

Leeuwenhoek remains an enigma, a colleague recently told me, when we briefly discussed the work of the microscopist from Delft. When we look at his achievements in microscopical science, e.g. the discovery of micro-organisms, bacteria, red blood cells, spermatozoa, certain aspects of wood structure, and many other parts of mineral, plant and animal life, one may say that there is some ground for the feeling that there is something enigmatic in his work. The explanation most offered is to state that he was a clear observer, had a sober mind, was an autodidact, acknowledged no authorities, and, above all, to point to the outstanding quality of his microscopes. Indeed, we do find in his letters scattered remarks which point in these various directions.[1] Since, to my understanding, these explanations are rather fragmentary and unsatisfactory, I would like to present with this paper a part of an attempt to analyze the influence which other scientists have exerted on Leeuwenhoek, both with their discoveries and their ideas, and to consider whether these influences can form part of an explanation for the stupendous number of original discoveries he made.

Leeuwenhoek has been ranked under the five great microscopists of the end of the seventeenth century: Nehemiah Grew (1641-1712), Robert Hooke (1635-1703), Marcello Malpighi (1628-1694), Jan Swammerdam (1637-1680), and the Delft microscopist himself. As a first step in this research I have concentrated on the Italian influences on Leeuwenhoek, for two main reasons. Firstly, because the names of Italian scientists are rather frequently mentioned in Leeuwenhoek's letters, especially in the early ones at times when he made his most important discoveries. Secondly, because of the similarities in types of research of Leeuwenhoek and Malpighi. I only need to point to the wood anatomical research, the research on blood and blood

1. See for example: C. Dobell, *Antoni van Leeuwenhoek and his "Little Animals"* (Amsterdam, 1932); A. Schierbeek, *Measuring the Invisible World* (London/New York, 1959); J. Heniger, "Leeuwenhoek", in C.C. Gillispie, ed., *Dictionary of Scientific Biography* VIII (New York, 1973) 126-130; L.C. Palm and H.A.M. Snelders, eds, *Antoni van Leeuwenhoek 1632-1723* (Amsterdam, 1982).

vessels and the conflicting ideas on generation and propagation. Moreover, both scientists had been stimulated in their work by the Royal Society for Improving of Natural Knowledge in London.

*Material*

The material which can be used to reconstruct the lines of influence is diverse and in some ways rather fragmentary.
We have:

1. His outgoing letters, more than 450, of which the majority was directed to the Royal Society in London.
2. The references Leeuwenhoek made to other scientists and their work in these letters, which, however, are scarce.
3. The minutes and the journal books of the council meetings of the Royal Society in which Leeuwenhoek's letters were discussed.

We do not have:

4. A collection of the letters he received. They are all lost; we only know them through other ways: in the correspondence of Leeuwenhoek's correspondents themselves or as mentioned in the archive material of the Royal Society.
5. A list of books he possessed or, even better, a catalogue of his library.

1. The letters Leeuwenhoek wrote are for the greater part well-known. Of the more than 190 letters he wrote to the Royal Society which had a scientific content 116 were published in fragmentary or complete English translation in the *Philosophical Transactions of the Royal Society of London*. Almost 200 letters, also partly published in the *Philosophical Transactions*, were printed contemporaneously in his own Dutch vernacular and in a Latin translation. These are the only publications Leeuwenhoek undertook. They can be characterized as being rather unsystematic. Leeuwenhoek often discusses more themes in one letter and the same theme may reappear in a very irregular way. So even the reconstruction of his own ideas is of necessity tentative. Leeuwenhoek wrote his letters in a period of fifty years from 1673 until his death in 1723. A few of them were abstracted in Dutch, German, French and Italian scientific journals.[2]

2. *Journal des Sçavans* (Amsterdam, 1675, 1678, 1679, 1683); *Acta Eruditorum* (Leipzig, 1682, 1683, 1685, 1689); *Bibliothèque Universelle* (Amsterdam, 1686, 1688); *Nouvelles de la République des Lettres* (Amsterdam, 1686); *Giornale de letterati del MDCXCII e primo di Modena* (Modena, 1692); *Collection Académique. Partie Etrangère* (Dijon, Auxerre, Paris, 1755, 1761, 1766); *Abhandlungen zur Naturgeschichte, Physik und Oekonomie* (Leipzig, 1719, 1780).

2. If we look at the references Leeuwenhoek makes to the work of other scientists then it is immediately obvious that he seldom does so. The bulk of these is to be found in the early years of his activity, when he was stimulated by the recurrent questions of the secretaries of the Royal Society Henry Oldenburg (ca 1618-1677) and Grew, to carry out new research and to explain the method by which he obtained his results. Undoubtedly the surprising results of Leeuwenhoek's work, especially the discovery of the microorganisms, led them to do so. The names Leeuwenhoek refers to in those early years are on the one hand those of fellow countrymen who stimulated him and brought him into contact with scientific circles, like Constantijn Huygens (1596-1687) and his son Christiaan (1629-1695), both of whom Leeuwenhoek counted among his supporters; Cornelis 's Gravesande (1631-1691), physician in Delft and a neighbour of Leeuwenhoek, who stimulated him to reveal his research and made him familiar with the anatomical work of Thomas Willis (1621-1675) and others, and who possibly introduced him to academic circles in the town of Delft;[3] Cornelis Bontekoe (1647-1685), a Cartesian physician who often visited him and undoubtedly made him familiar with the Cartesian theories of physics and medicine he later adopted[4]; Swammerdam, fellow microscopist, whom he had met at least once, but with whom he was not on friendly terms[5]; and Reinier de Graaf (1641-1673), town physician of Delft, a good friend of Leeuwenhoek, who introduced him to the Royal Society[6]. On the other hand we find the names of fellows of this society, like Grew, Hooke, Oldenburg and Robert Boyle (1627-1691), who apparently had asked him questions or commented on his results. It is, however, remarkable that Leeuwenhoek in these cases seldom referred to their scientific works, but to personal affairs. Nevertheless, only through these friends and correspondents did Leeuwenhoek become aware of some of the scientific works of his time.[7]

3. Leeuwenhoek's letters are published in *Alle de Brieven van Antoni van Leeuwenhoek/The Collected Letters of Antoni van Leeuwenhoek* (12 vols; Amsterdam/Lisse, 1939-1989; in progress), abbreviated as *Collected Letters*. The letters are numbered after the edition in the *Collected Letters*, between square brackets Leeuwenhoek's own numbering in the contemporary editions is given. See for 's Gravesande Letter 13 [8] of 4 December 1674 to Oldenburg, *Collected Letters* I, 193-195.

4. H.A.M. Snelders, "Antoni van Leeuwenhoek's mechanistic view of the world", in Palm and Snelders, *Leeuwenhoek* (o.c., note 1) 70-71.

5. G.A. Lindeboom, "Jan Swammerdam als microscopist", *Tijdschrift voor de geschiedenis der geneeskunde, natuurwetenschappen, wiskunde en techniek* 4 (1981) 107-110.

6. G.A. Lindeboom, *Reinier de Graaf. Leven en werken* (Delft, 1973) 57-59.

7. Leeuwenhoek refers, among others, to William Harvey (1578-1657) and René Descartes (1596-1650). See also Table 1.

3. From the journal books of the Royal Society we learn that particularly in the early decades of Leeuwenhoek's contact with the learned society in London he was stimulated by its members, notably Hooke, Grew and Oldenburg. In later years, when his results assumed more and more the character of an extension or repetition of earlier work, the interest of the Royal Society declined, to the extent that in the mid-1690's the council no longer discussed his letters; so Leeuwenhoek ceased writing to it. Only after Hans Sloane (1660-1753) had become more influential in the Royal Society, did the interest in Leeuwenhoek's rather Baconian collections of observations rise again and it lasted until his death in 1731.[8]

We know of two cases in which the Royal Society presented Leeuwenhoek with books. In 1686 they bestowed on him Francis Willughby's *Historia piscium*, and in 1697 Malpighi's *Opera Posthuma*, which had been published through the care of the Royal Society itself.[9] Apart from these he regularly received the *Philosophical Transactions* of the learned society in London of which he had been elected a fellow in 1680.[10]

4. Unfortunately the letters to Leeuwenhoek, which were once in his possession, are lost so that we cannot reconstruct the network of his correspondence. Only by chance do we have the text of letters to him, mostly from father and son Huygens, and we know of course that there must have been many letters from the Royal Society. One of the few conclusions we may draw from the existing material is that a real scientific correspondence in which facts and ideas were regularly exchanged only existed between the Royal Society and Leeuwenhoek, and to a lesser degree with both the Huygenses. All the other exchanges of letters we are aware of: with the Rotterdam editor Pieter Rabus (1660-1702),[11] the state pensionary Anthonie

8. See for Leeuwenhoek's contacts with the Royal Society L.C. Palm, "Leeuwenhoek and other Dutch correspondents of the Royal Society" (in press).

9. Leeuwenhoek acknowledged the receipt of Willughby's *Historia piscium* (1686) in Letter 93 [51] of 10 June 1696 to the Royal Society, *Collected Letters* VI, 127; and of Malpighi's *Opera Posthuma* (1697) in Letter 188 [110] of 10 September 1697 to the Royal Society, *Collected Letters* XII, 181.

10. Leeuwenhoek was elected a fellow of the Royal Society on 29 January 1679/80 (old style). See for the diploma J. Heniger, "Antoni van Leeuwenhoek en zijn diploma van de Royal Society", *Tijdschrift voor de geschiedenis der geneeskunde, natuurwetenschappen, wiskunde en techniek* 1 (1978) 157-169.

11. Rabus was editor of the *Boekzaal van Europe*, an influential Dutch literary journal which was published in Rotterdam. See J.J.V.M. de Vet, *Pieter Rabus (1660-1702). Een wegbereider van de Noordnederlandse Verlichting* (Amsterdam, 1980). The controversy between Leeuwenhoek and Rabus about the divining-rod is described in L.C. Palm, "Antoni van Leeuwenhoeks reactie op Pieter Rabus' problemen met de wichelroede", *Tijdschrift voor de geschiedenis der geneeskunde, natuurwetenschappen, wiskunde en techniek* 8 (1985) 1-14.

Heinsius (1641-1720),[12] the Florentine librarian Antonio Magliabechi (1633-1714),[13] and all those other "dignitaries and learned gentlemen", to use Leeuwenhoek's own words, are of a different nature. With Rabus he discussed a problem, certainly, but of a kind that could not give Leeuwenhoek the opportunity to make conclusive statements after microscopical research, namely the effectiveness of a divining-rod. The other addressees were used by Leeuwenhoek, as colours to be printed above the text of his letters. Some of them tried to pull Leeuwenhoek into learned discussions, in vain, however, as happened to Gottfried Wilhelm Leibniz when he tried to convince Leeuwenhoek of the necessity to establish a research-school, so that his skills might not pass away with the artisan himself.[14]

5. The last possibility of gaining more factual insights into the scientific influences he might have experienced is a reconstruction of his library, or to put it more accurately: the composition of a list of books he possessed and had read. The only useful sources for the composition of this list appear to be the references in his own letters and the confirmation of these from other sources, like the *Journal Books* of the Royal Society.

An examination of all published and unpublished letters of Leeuwenhoek and of the archives of the Royal Society on books Leeuwenhoek had seen brings us to a total of 33, some of which are uncertain.[15] The nature of these books is very miscellaneous. There are travel reports like those of Tavernier and Witsen, but also books on medicine by Blankaart, Overkamp, Ten Rhyne and Von Tschirnhaus, Hooke's *Micrographia* of 1666, and his *Lectures and Collections*, works by Swammerdam, De Graaf, Kircher, Jonston, Grew, Redi, and Malpighi.

A great problem in the evaluation of Leeuwenhoeks' perception of all these works is that he was unable to read other languages than Dutch. He was certainly helped by friends to grasp more of the reasoning in the books he had seen, but it is almost impossible to find criteria for an evaluation of these facts, since Leeuwenhoek seldom refers to them. We can be certain, however, that Leeuwenhoek studied the figures and the engravings. We know, for example, that already in 1675, when he was at the very beginning

12. Heinsius was descended from a Delft family of regents. He studied law and later entered the diplomatic service. From 1689 until his death he was Grand Pensionary of Holland and thus one of the most influential politicians of the Dutch republic. He always showed interest in Leeuwenhoek's work.

13. Magliabechi was librarian to Cosimo III de' Medici, Grand Duke of Tuscany.

14. See letter [XVIII] of 28 September 1715 to Leibniz, *Send-Brieven* (Delft, 1718) 168-169.

15. See Table 1 for a list of these books.

of his career, Leeuwenhoek studied the drawing of a louse in Redi's *Experimenta circa generationem insectorum* of 1671 (a Dutch edition), which he found rather inaccurate, and which he corrected.[16]

Apart from books Leeuwenhoek had access to the *Philosophical Transactions* of the Royal Society. The first issue mentioned by him is no. 106 of 21 September 1674, the last one no. 346 of December 1715. In total he refers to or acknowledges about 60 of these issues by their number spread regularly over the whole period of correspondence between him and the society. Since he often asked for the remittance of missing issues it is not improbable that in the end he had received them all.[17] Even if this is a too optimistic interpretation we nevertheless have to conclude, that the main source of scientific knowledge for Leeuwenhoek was the *Transactions*.

Moreover, there are indications that Leeuwenhoek himself considered this to be the case too. On 22 January 1676 he wrote to Henry Oldenburg, then secretary of the Royal Society:

> "I must tell you that I regret I do not understand any language but Dutch and when you write to me in French or Latin I can help myself all right, since I have enough friends who will translate it for me; but I cannot help myself with the English language since the death of a certain gentleman who was proficient in this language. I admit that usually Englishmen can be found everywhere, but not all are able to translate the Transactions from English into Dutch."[18]

Leeuwenhoek tried to face this problem, since he made an attempt to translate the *Transactions* with the aid of a dictionary as he wrote later in the same year to Boyle, but he found it difficult.[19] In 1679 he wrote to Christiaan Huygens that he was not able to read Hooke's *Lectures and Collections* which the latter had sent to him.[20] But with the same perseverance that characterizes his microscopical researches he tried to master the English language as appears from a letter of 1704 to the Royal Society in which he reacts to published research about cochineal and wrote down an abstract of

16. It concerns a drawing of the human head or body louse (*Pediculus humanus* L.), first described in the Florentine edition of Redi's work *Esperienze intorno alla generazione degl'insetti* (1668). Leeuwenhoek's drawing is to be found in Letter 16 [10] of 11 February 1675 to Oldenburg, *Collected Letters* I, 273. See also J.M.E.H. van Bronswijk, "Two fellow students of fleas, lice and mites: Antoni van Leeuwenhoek and Jan Swammerdam", in Palm and Snelders, *Leeuwenhoek* (o.c., note 1) 109-127.

17. For a list of these issues see Table 2.

18. Letter 20 of 22 January 1676 to Oldenburg, *Collected Letters* I, 343.

19. Letter 25 [17] of 28 July 1676 to Boyle, *Collected Letters* II, 45.

20. Letter 46 of 15 May 1679 to Christiaan Huygens, *Collected Letters* III, 49.

this article which is quite correct.[21] For the other languages Leeuwenhoek was confronted with, he continued to be dependent on the help of translators.

In this article the Italian influences on the work of Antoni van Leeuwenhoek will be examined as an example of the transmission of regular science on the work of the microscopist from Delft.

### *Italian scientists in Leeuwenhoek's letters*

In Leeuwenhoek's letters we meet the names of Malpighi, Redi, Ramazzini,[22] Buonanni,[23] Scaramucci,[24] Vallisneri,[25] Magliabechi, and four other Italians who are only important as bearers of letters and messages. Of all these Malpighi and Redi are no doubt the most important.

### *Marcello Malpighi*

Leeuwenhoek showed his knowledge of the existence of Malpighi for the first time in a letter of 19 October 1674 to Oldenburg. In the context of a description of his ideas about the structure of the gustatory corpuscles on the tongue he told Oldenburg that

> "several learned people who came to see him told him that Malpighi agreed with him on this point. (...) I thought it well to tell you this."[26]

Here we meet with a small sign of Leeuwenhoek's initial uncertainty when dealing with academic circles which later led him to an anti-academic attitude that has recently been described as a possible characteristic of self-taught scientists.[27] The next time we encounter Malpighi is in January 1680

21. Letter of 21 March 1704, *Philosophical Transactions* 24 (1704) 1614-1628. For Leeuwenhoek's relation with language see also B.C. Damsteegt, "Language and Leeuwenhoek", in Palm and Snelders, *Leeuwenhoek* (o.c., note 1) 13-28.

22. Bernardino Ramazzini (1633-1714) was professor of medicine in Modena and Padua.

23. Filippo Buonanni (1638-1725) was a Jesuit priest and the principal Italian vindicator of the theory of spontaneous generation. Leeuwenhoek criticized his ideas (*Collected Letters* X, 31, 91, 115, 311).

24. Giambattista Scaramucci (ca 1650-ca 1710) was a medical practitioner in Macerata and Urbino. Leeuwenhoek mentioned a book by him which he had received from Magliabechi (*Collected Letters* XII, 407).

25. Antonio Vallisn(i)eri (1661-1730) was professor of practical and theoretical medicine in Padua. He was mentioned by Leibniz in a letter to Leeuwenhoek as being not in agreement with Leeuwenhoek's animalculistic theory (*Send-Brieven*, o.c., note 14, 166).

26. Letter 12 [7] of 19 October 1674 to Oldenburg, *Collected Letters* I, 185.

27. See K. van Berkel, "Intellectuals and Leeuwenhoek", in Palm and Snelders, *Leeuwenhoek* (o.c., note 1) 187-209.

when Leeuwenhoek wrote to Hooke among other things about the vessels in different kinds of wood and in straw. Here Leeuwenhoek wrote:

> "And although I understand that the learned gentlemen, Malpighi and Nehemiah Grew, have very eruditely written about this, I yet take the liberty to indicate the vessels in the wood, as far as these are known to me ..."[28]

From the original Dutch text we can draw the conclusion that Leeuwenhoek indeed did not know the text of Malpighi's and Grew's works before he started his own researches in wood anatomy.[29] His first reports on wood anatomy date from 1673 and his first acquaintance with Grew's *The comparative anatomy of trunks* (1675) dates from 1676, when Constantijn Huygens showed the book to Leeuwenhoek. He did have it in his hands, but he was only able to study the figures. Malpighi's wood anatomical research perhaps never exerted much influence on Leeuwenhoek since, as far as we know, he did not possess any of Malpighi's works until 1697 when he was presented by the Royal Society with Malpighi's *Opera Posthuma*.[30] In the meantime Malpighi did not publish any wood anatomical work in the *Philosophical Transactions* but only in books. It remains possible that through discussions with visitors like the botanists Paul Hermann (1640-1695)[31] and Pieter Hotton (1648-1709)[32] Leeuwenhoek had been told about Malpighi's work, but there is no evidence for this.

Leeuwenhoek did read or study Malpighi's *Opera Posthuma*, since he derived arguments from it in a discussion about the ideas of the French scientist De Plantade who maintained that in the human spermatozoa a complete human being could be discerned.[33] Although both Leeuwenhoek and Malpighi were preformists in their theories of generation, they differed

28. Letter 54 [29] of 12 January 1680 to Hooke, *Collected Letters* III, 185.

29. See P. Baas, "Leeuwenhoek's contributions to wood anatomy and his ideas on sap transport in plants", in Palm and Snelders, *Leeuwenhoek* (o.c., note 1) 79-108.

30. After Malpighi's death the Royal Society took care of the publication of his remaining work resulting in the *Opera Posthuma* (Londini, 1697). The society sent this work to a number of its foreign correspondents and members, including Leeuwenhoek. See the *Minutes of Council* (London, Royal Society Archives) II, 95-96.

31. Hermann was professor of botany in Leiden. He visited Leeuwenhoek in 1685 or 1686 and provided him with plant seeds which Leeuwenhoek later examined (*Collected Letters* II, 89-91, 249).

32. Hotton was professor of botany in Leiden, successor to Hermann (*Collected Letters* IX, 81).

33. François de Plantade (1670-1741) published drawings of spermatozoa with a complete preformed human being in it, which Leeuwenhoek firmly criticized. See G.A. Lindeboom, "Leeuwenhoek and the problem of sexual reproduction", in Palm and Snelders, *Leeuwenhoek* (o.c., note 1) 146-149.

on the attribution of importance to the different sexes. Leeuwenhoek was an animalculist thanks to his discovery of spermatozoa and his life-long research in this field.[34] Malpighi was an ovist, no doubt because of his splendid embryological research into the development of the chicken within the egg.[35] From this research Leeuwenhoek borrowed an argument against the work of De Plantade:

> "Now if an Animalcule in the sperm of animals and birds were provided completely with all the limbs, so that they could be discerned through a Magnifying glass as people try to make us believe, these animalcules from time to time, as they grow bigger, ought to increase in such completeness. But this is not the case, we see from the observations which the late very learned Mr. Malpighius made, as well as he could, about the beginning of the chicken in the Egg."[36]

It is remarkable that both microscopists did not pay attention to their respective theories in the field of embryological research. Leeuwenhoek did study the ovaries, the Fallopian tube, the womb and the presence of male spermatozoa there, but he never found anything that resembled the bird's egg in mammalian specimens. Nor did he believe that the corpus luteum, which was considered to be the egg by some researchers, coud possibly pass through the small opening of the Fallopian tube to proceed to the womb. Leeuwenhoek was indeed too eager to explain away the existence of these female organs by analogical reasoning; a comparison with mammalian male nipples was enough for him. Nowhere in the rather long arguments about these themes Leeuwenhoek wrote down, did he show any insight into Malpighi's ideas.[37]

Malpighi, on the other hand, was quite aware of Leeuwenhoek's discoveries. The first reference to Leeuwenhoek which came under Malpighi's eyes is to be found in a letter of Grew to Malpighi of 10 July 1678 in which Grew told the Italian scientist that Leeuwenhoek had found more than a million animalcules in a single drop of pepper water.[38] Robert Hooke later told

34. For Leeuwenhoek's ideas on generation see E.G. Ruestow, "Images and ideas: Leeuwenhoek's perception of the spermatozoa", *Journal of the history of biology* 16 (1983) 185-224.

35. See for Malpighi's embryological research H.B. Adelmann, *Marcello Malpighi and the evolution of embryology* (5 vols; Ithaca, 1966).

36. Letter 200 [166] of 9 June 1699 to the Royal Society, *Collected Letters* XII, 299.

37. See for Leeuwenhoek's embryological researches and their context F.J. Cole, *Early theories of sexual generation* (Oxford, 1930), and Lindeboom, "sexual generation" (o.c., note 33).

38. Adelmann, *Malpighi* (o.c., note 35) I, 410. The letter is published in H.B. Adelmann, ed., *The correspondence of Marcello Malpighi* (5 vols; Ithaca and London, 1975) II, 780-781.

Malpighi that Leeuwenhoek, with this observation had found a new analogy between plants and animals, because the seeds of both types of organisms were filled with almost innumerable worms.[39] A year later Malpighi was told by a German friend that in gonorrhoea

> "animals like little toads, dead and deprived of motion, were to be found in the semen."[40]

Malpighi immediately communicated these discoveries to some of his friends among whom was Lorenzo Bellini (1643-1704). Bellini did not value Leeuwenhoek's discovery of the spermatozoa, as he wrote to Malpighi.[41] It is remarkable that, while Malpighi never was tired of discussing with Bellini, he left this opinion unchallenged. Nowhere in his works or in his letters does Malpighi attach any significance whatever to the spermatozoa, though it is certain that he knew of their existence at first hand because he possessed Leeuwenhoek's *Anatomia et contemplatio* of 1685[42] in which animalcules from male semen are described and their role discussed. Even in the *Opera Posthuma*, where Malpighi acknowledges Leeuwenhoek's descriptions of different types of red blood cells in different animals and Leeuwenhoek's confirmations of his own views on the reproduction of galls which were written when he had ample time to digest Leeuwenhoek's informations on spermatozoa, he makes no mention of "worms" in the semen, apparently because he did not believe they played any role at all in generation.[43] Here again we see how the preoccupation with one's own research, in this case the embryological development, obscures the perception of good ideas and new discoveries by others despite strong similarities in scientific method.

The other treatises of Malpighi, in the *Philosophical Transactions*, were published in volumes Leeuwenhoek did not possess or were treatises without illustrations in the Latin language, to which Leeuwenhoek did not react.

### *Redi*

Francesco Redi is important for his publications on vipers and their poison, his attacks on the theories of spontaneous generation through microscopical research on insect anatomy, and for his studies on parasitic worms in

39. Adelmann, *Malpighi* (o.c., note 35) I, 693. The letter is in Adelmann, *Correspondence* (o.c., note 38) II, 831-832.
40. Adelmann, *Malpighi* (o.c., note 35) I, 420, 503.
41. Adelmann, *Malpighi* (o.c., note 35) I, 422. The letter from Bellini to Malpighi is in Adelmann, *Correspondence* (o.c., note 38) III, 1219-1222.
42. Adelmann, *Malpighi* (o.c., note 35) I, 504.
43. Adelmann, *Malpighi* (o.c., note 35) I, 868.

different kinds of animals.[44] These three general themes were also covered by Leeuwenhoek, but his results were presented in a far less systematic way than Redi's. Leeuwenhoek possessed one book by Redi, the *Experimenta circa generationum insectorum*, published in Amsterdam in 1671, which was the Latin translation of the Italian original of 1668. Leeuwenhoek referred to Redi for the first time in a letter of 22 January 1675 to Oldenburg where he wrote that he had seen in the works of two different authors (who appear to be Redi and probably Swammerdam[45])

> "the drawings of scorpions as seen by them through the microscope, and I saw that they drew the sting of scorpions as sharp and pointed as if it were only designed to prick."[46]

Leeuwenhoek subsequently bought a few living scorpions on which he

> "observed the same and found the sting to be blunt and not, as others draw and describe it very sharp; and I noticed a hole in its extremity."[47]

After this correct observation he obtained some fluid from the sting which he analyzed further; he found it to consist of many small globules. In conclusion he imagined that not the sting of scorpions is lethal, but the fluid which is brought into the body of the victim. The correction by Leeuwenhoek of Redi's drawing of the louse has been mentioned above.

The most important reaction of Leeuwenhoek to Redi's work can be found in a letter to Thomas Gale (1635?-1702), then secretary of the Royal Society, written in 1680. Here Leeuwenhoek referred to Redi when he wrote:

> "Having become acquainted with the various opinions concerning the reproduction of little animals, and more especially with what a certain Gentleman has written, who maintains that no living being can originate in a firmly closed bottle, previously filled with meat-juice or meat, I had a mind to make some experiments upon this."[48]

He filled two glass tubes with rain-water; one of them he sealed and the other he left open. After a few days he found microorganisms in both tubes, no doubt because he did not heat the sealed tube with its contents enough, so that anaerobic microorganisms could generate there. In this letter Leeuwenhoek did not give an explanation for the difference between Redi's and his

44. F. Redi, *Esperienze intorno alla generazione degl'insetti* (Milano, 1970; reprint with an introduction and annotations by G. Montalenti), and P. Polito, *Scritti di botanica, zoologia e medicina di Francesco Redi* (Milano, 1975).
45. Letter 15 [9] of 22 January 1675 to Oldenburg, *Collected Letters* I, 227.
46. *Ibidem.*
47. *Ibidem.*
48. Letter 62 [32] of 14 June 1680 to Gale, *Collected Letters* III, 261.

own experiments, but later he stated that he was convinced that no living creatures could arise from putrefraction, a conviction he repeated again and again.[49] He could do so because he succeeded in proving the existence of spermatozoa in fleas and other very small organisms. By analogical reasoning he extended the existence and the role of spermatozoa to all living organisms, because of his conviction that provident nature always works in the same way. A repetition of the aforementioned experiment, but now over a period of several weeks, took away any doubts he still might have had, for now he could find no microorganisms at all in the tube.[50] Moreover, he discovered larvae in plant galls, which he rightly considered to be the cause of these excrescences, while Redi left these as one of the few possibilities where spontaneous generation might be the cause of their existence.[51] More references to Redi are not to be found in Leeuwenhoek's works.

*Other Italians*

About the other Italians we can be brief. Ramazzini and Scaramucci were mentioned only because Leeuwenhoek obtained books written by them; these covered themes, however, that were not of such great interest to him. Buonanni comes up because Leeuwenhoek came into possession of a summary of a book by the Roman Jesuit priest from which it appears that

> "the Author of this Book holds the opinion of the Ancients, and in particular of Aristotle, about the generation of shell-fishes, asserting that they are born spontaneously in mud or sandy earth ..."[52]

Leeuwenhoek rejected this opinion by writing a long report to the Royal Society on the researches he had made on the generation of mussels. The starting point of his research was as follows:

> "For my part, I am firmly convinced that just as it is impossible for a whale to spring from mud, it is also impossible for a little shell-fish to arise without generation,"[53]

and although in this letter he mistook other creatures for mussel eggs he later was able to demonstrate young embryonic mussels. To his disappointment he never succeeded in discovering the spermatozoa of this marine animal.[54]

49. See E.G. Ruestow, "Leeuwenhoek and the Campaign against Spontaneous Generation", *Journal of the history of biology* 17 (1984) 224-248.
50. Letter 92 [50] of 14 May 1686 to the Royal Society, *Collected Letters* VI, 63-65.
51. Redi, *Esperienze* (o.c., note 44) 109-111.
52. Letter 137 [83] of 30 April 1694 to the Royal Society, *Collected Letters* X, 91.
53. *Ibidem*, 93.
54. See for Leeuwenhoek's researches on molluscs L.C. Palm, "Antoni van Leeuwenhoek's

Vallisneri is mentioned in a Leeuwenhoek letter in a quotation from a letter of Leibniz. He had written to Leeuwenhoek that Vallisneri did not agree with the role Leeuwenhoek attached to the spermatozoa. The Delft microscopist was obviously not impressed by this since he replied with a Dutch saying that a hooded crow makes no winter;

> "if Mr Vallisneri opposes my theses, there are more than a thousand with me",

he optimistically remarked.[55]

The last Italian which should be mentioned here is the Florentine librarian Antonio Magliabechi. He was librarian to the house of the Medici, a great collector of books and always interested in news in this field. Therefore he maintained correspondence with many scientists and members of the Republic of Letters of the time. It is not very clear how Leeuwenhoek and Magliabechi had come into contact with each other, but the result was an exchange of more than 38 letters, some of which are now lost.[56] Many of them deal with practical matters and social talk, but there are two interesting aspects in the other ones. Magliabechi never wrote to Leeuwenhoek about scientific observations, theories or speculations, but corresponded about books. Leeuwenhoek on the other hand used this information to present his own ideas and results. The aforementioned book by Ramazzini, for example, was announced in a letter from Magliabechi to his Delft correspondent; it dealt with an epidemic of grass diseases in the vicinity of the town of Modena, probably a rust infection.[57] Leeuwenhoek replied with a story that a year earlier the meadows in the neighbourhood of Delft were covered with a rust-coloured powder. The population, suffering from fevers, supposed that the rust had fallen from heaven and was the cause of this fever. Leeuwenhoek denied this, he gave a description of the fungus and subsequently put forward his mechanistic theory of the cause of diseases.[58] Here we meet with another example of Leeuwenhoek's procedure when publishing his results of observation and speculation as we have seen in the case of Buonanni.

A second, but unintentional effect of this specific correspondence was the propagation of information on Italian books in the Netherlands. Magliabechi used his contact to enclose in his letters long lists of recent Italian books, which Leeuwenhoek passed to his friend, the Rotterdam editor of the

---

malacological researches as an example of his biological studies", in Palm and Snelders, *Leeuwenhoek* (o.c., note 1) 153-168.

55. See note 25.
56. See note 13.
57. Letter 115 of 18 September 1691 to Magliabechi, *Collected Letters* VIII, 175.
58. *Ibidem*, 173-189.

literary periodical *Boekzaal van Europe*, the notary public Pieter Rabus, who regularly published these lists under the heading of Italian Booknews ("Italiaansch Boeknieuws").[59]

*Conclusion*

No other works of Italian scientists are recorded by Leeuwenhoek (see Table 1). Of the numerous articles in the issues of the *Philosophical Transactions* Leeuwenhoek could have seen, 23 were written by Italian scientists, either as original contributions or as abstracts of articles in other journals. A number of them deal with astronomical topics in which Leeuwenhoek was not interested at all; while others, notably those by Malpighi, Redi, and the Leghorn apothecary Giacinto Cestoni (1637-1718),[60] dealt with subjects Leeuwenhoek did study too, but they were published a few years later than Leeuwenhoek's own studies or there could not be found any influence upon the work of Leeuwenhoek, nor did he refer to them.[61]

In sum the Italian influence on Leeuwenhoek's work seems rather eclectic in the sense that Leeuwenhoek used it only for his own, often limited purposes. He showed practically no interest in non-microscopical research, he made no attempts to get the more important scientific literature translated or otherwise made accessible to him, and he made no systematic treatment of research themes. He only reacted to the ideas of Italian scientists when they attacked the cores of his own thinking such as the ideas on the role of the spermatozoa and spontaneous generation. It confirms therefore the picture of Leeuwenhoek as a loner, a student of natural history working in his well-chosen isolation which was a prerequisite for the independent valuation of the wonders of nature he saw through his small lenses.

59. See note 11, and B. van der Saag, "Pieter Rabus en Antoni van Leeuwenhoek in de Boekzaal van Europe", in H. Bots, ed., *Pieter Rabus en de Boekzaal van Europe 1692-1702* (Amsterdam, 1974) 343-382.

60. See for example S. Baglioni, ed., *Giacinto Cestoni. Epistolario ad Antonio Vallisnieri* (2 vols; Roma, 1940-1941).

61. See my forthcoming *Invloeden op het werk van Antoni van Leeuwenhoek* (Amsterdam, 1990).

Table 1. *Books read or possibly seen by Leeuwenhoek*
(* indicates a book certainly seen by him)

1. Alonso Barba, *The first book of the art of metals*, translated by the Earl of Sandwich (London, 1679) [Letters of 8 December 1702 and 3 November 1703].
2. *Steven Blankaart, *Collectanea medico-physica; oft, Hollands Jaar-Register der genees- en natuur-kundige Aanmerkingen van gantsch Europa &c. ...* (Amsterdam: Johan ten Hoorn, 1680-1688) [*Collected Letters* V, 207].
3. Steven Blankaart, *De Kartesiaanse Academie ofte Institutie der Medicyne* (Amsterdam: Joh. ten Hoorn, 1684) [*Collected Letters* VIII, 131].
4. *Steven Blankaart, *Verhandelinge van het podagra en vliegende jicht* (Amsterdam: Jan ten Hoorn, 1684) [*Collected Letters* IV, 281].
5. *Hermanus Busschoff, *Het podagra nader als oijt nagevorst en uijtgevonden, mitsgaders des selfs sekere genesing of ontlastent hulpmiddel* (Amsterdam: Jacobus de Jonge, 1675) [*Collected Letters* II, 231; III, 377; IV, 283].
6. *[Dalenpatius], "Extrait d'une Lettre de M. Dalenpatius à l'Auteur de ces Nouvelles, contenant une découverte curieuse, faite par le moyen du Microscope", *Nouvelles de la République des Lettres* (1699) 552-554 [*Collected Letters* XII, 295].
7. *[Freijer], *Negenjaarige Reyse door Oostindien en Persien ... Begonnen met den jaare 1672 en geeyndigt met den jaare 1681. In agt Brieven beschreven door d'Heer Johan Freijer, M. Dr. tot Cantelberg, en lid vande Koninklijke Societeyt.* Uyt het Engels vertaalt ('s-Gravenhage: Abraham de Hondt, Jacobus van Ellinkhuysen & Meyndert Uytwerf, 1700) [Letter of 23 November 1711].
8. *Johannes Goedaert, *Metamorphosis naturalis, ofte Historische beschryvinghe van den Oirspronk, aerd, eygenschappen ende vreemde veranderingen der wormen, rupsen, maeden ...* (Middelburg: Jaques Firens, ca 1662-ca 1670) [Letter of 22 September 1711].
9. *Reinier de Graaff, *De mulierum organis generationi inservientibus tractatus novus: demonstrans tam homines et animalia caetera omnia, quae vivipara dicuntur, haud minus quam ovipara ab ovo originem ducere* (Lugduni Batavorum: Ex officina Hackiana, 1672) [*Collected Letters* IV, 5].
10. *Nehemiah Grew, *The comparative anatomy of trunks* (London: Printed by J.M. for W. Kettilby, 1675) [*Collected Letters* II, 5].
11. *Johann Franz Griendel von Ach, *Micrographia nova* (Nürnberg: Johann Ziegers, 1687) [*Collected Letters* VII, 37; IX, 291].
12. *Nicolaas Hartsoeker, *Proeve der Deur-zigtkunde* ('t Amsteldam: Jan ten Hoorn, 1699) [*Collected Letters* XII, 253].
13. *Robert Hooke, *Lectures and Collections: Cometa, Microscopium* (London: J. Martyn, 1678) [*Collected Letters* III, 49].
14. Robert Hooke, *Micrographia* (London: J. Martyn & J. Allestry, 1665) [*Collected Letters* I, 45; VII, 37; IX, 291].
15. *Robert Hooke, *Philosophical Collections* (London: Richard Chiswel, 1679-1682) [*Collected Letters* III, 191].
16. *Jan Jonston, *Beschryving van de Natuur der Vissen en Bloedloze Water-dieren* (Amsterdam: I.I. Schipper, 1660) [*Collected Letters* X, 177].

17. *Jan Jonston, *Beschrijving van de Natuur der Gekerfde of Kronkel-Dieren, Slangen en Draken.* Uyt 'et Latijn Vertaelt door M. Grausius (Amsterdam: I.I. Schipper, 1660) [*Collected Letters* X, 209].
18. *Athanasius Kircher, *D'Onder-aardse Weereld in Haar Goddelijk Maaksel en wonderbare uitwerkselen aller Dingen ... in XII Boeken naewkeurig beschreven* ('t Amsteldam: Erv. J.J. van Waasberge, 1682) [*Collected Letters* IX, 257, 291; XI, 213; Letter of 24 October 1713].
19. *[Lorenzo Magalotti], *Saggi di naturali esperienze fatte nell'Accademia del Cimento sotto la protezione del serenissimo Principe Leopoldo di Toscana e descritte dal segretario di essa Accademia.* Seconda edizione (Firenze: Gio. Filippo Cecchi, 1691) [*Collected Letters* XII, 201].
20. *Marcello Malpighi, *Opera Posthuma* (Londini: A. & J. Churchill, 1697) [*Collected Letters* XII, 181].
21. *Adriaan Metius, *Manuale arithmeticae et geometriae practicae* (Franeker: V. Balck, 1646; 2nd ed.) [*Collected Letters* XII, 221].
22. *Johannes d'Outrein, *De Roosendaalsche vermakelykheden, met een geestelijk oog beschouwd* (Amsterdam, 1700) [Letter of 16 June 1700].
23. *Heidentrijk Overkamp, *Nieuwe beginselen tot de Genees- en Heel-konst enz. steunende op de gronden der fermentatie en deze op die van Renatus Des Cartes* (Amsterdam: Jan ten Hoorn, 1681) [*Collected Letters* IV, 39; V, 25].
24. Bernardino Ramazzini, *De constitutioni anni M.DC.XXXX ac de rurali epidemia quae Mutinensis agri et vicinarum regionum colonos graviter afflixit* (Modena: Typis haeredum Juliani Cassiani, 1690) [*Collected Letters* VIII, 183].
25. *Francesco Redi, *Experimenta circa generationum insectorum* (Amsterdam: A. Frisii, 1671) [*Collected Letters* I, 227, 273; VII, 37].
26. *Willem ten Rhijne, *De Chineese en Japanse wijse om door het branden van moxa en het steken met een goude naald, alle ziekten en voornamelijk podagra te genezen* (Amsterdam: Jan ten Hoorn, 1684) [*Collected Letters* IV, 281].
27. Giambattista Scaramucci, *Meditationes familiares ad ... Antonium Magliabechium ... de Sceleto Elephantino* (Urbino: Litteris Leonardi, 1697) [*Collected Letters* XII, 163].
28. Jan Swammerdam, *Historis insectorum generalis, ofte, Algemeene verhandeling van de bloedeloose dierkens* (Utrecht: Meinardus van Dreunen, 1669) [*Collected Letters* II, 64; VII, 37, 345; VIII, 17; IX, 291].
29. *[Tavernier], *De zes reizen van de heer J. Baptista Tavernier, Baron van Aubonne, in Turkijen, Persiën, en in d'Indiën ... Door J.H. Glazemaker vertaalt* (Amsterdam: By de weduwe van Johannes van Someren, 1682) [*Collected Letters* VI, 57; VIII, 313].
30. *Ehrenfried Walther von Tschirnhaus, *Geneesmiddel der Ziele ofte Proeve van een oprechte Redenkonst. Waar in gehandeld word van het rechte Beleid, om onbekende Waarheden te ontdekken* (Amsterdam: Albertus Magnus & Jan Rieuwerts, 1687) [*Collected Letters* XII, 357].
31. Thomas Willis, *Cerebri anatome, cui accessit Nervorum Descriptio et Usus* (Londini: Typis Ja. Flesher, impensis Jo. Martyn et Ja. Allestry, 1664) [*Collected Letters* I, 217].

32. *Francis Willughby, *De Historia Piscium libri quatuor* ... (Oxford: E theatro Sheldoniano, 1686) [*Collected Letters* VI, 87, 127].
33. *Nicolaas Witsen, *Noord en Oost Tartarye, ofte Bondigh Ontwerp Van Eenige dier Landen, en Volken, Zo als voormaels bekent zyn geweest* (Amsterdam, 1692) [*Collected Letters* XI, 279].

Table 2. *Philosophical Transactions* mentioned by Leeuwenhoek.

| Number. | Comment. | Source. |
|---|---|---|
| 106 | mentioned | *Collected Letters*, I, 201 |
| 108 | mentioned | *CL*, IV, 211 |
| 113,117 | not yet received | *CL*, I, 345 |
| 120, 121 | mentioned | *CL*, II, 5 |
| 127 | mentioned | *CL*, II, 165 |
| 136 | mentioned | *CL*, II, 239 |
| 137, 138 | requested | *CL*, II, 353 |
| 137 | received | *CL*, II, 357 |
| 139, 140 | expected | *CL*, II, 411 |
| up to 138 | received | *CL*, IV, 11 |
| missing ones | received | *CL*, IV, 103 |
| 145, 146, 147, 148 | received | *CL*, IV, 281 [*CL*, IX, 327] |
| 159 | mentioned | *CL*, V, 319 |
| year 1686 | received | *CL*, VI, 193 |
| 142 | mentioned | *CL*, XII, 299 |
| 247 | read | *CL*, XII, 313 |
| 193 | read | Letter of 21 March 1704 |
| 283-289 | received | Letter of 22 July 1704 |
| 290-300 | received | Letter of 29 December 1705 |
| 301-309 | received | Letter of 12 July 1707 |
| 6 new ones | received | Letter of 14 January 1710 |
| 4 new ones | received | Letter of 22 September 1711 |
| 19 new ones | received | Letter of 29 December 1711 |
| new ones | received | Letter of 22 June 1714 |
| new ones | received | Letter of 21 August 1714 |
| 346 | read | Letter of 8 October 1717 |

# THE EDITORIAL FORTUNE OF BOLOGNESE SCIENTISTS IN HOLLAND (1669-1726)

*Marta Cavazza*

Erasmus of Rotterdam arrived in Bologna at the end of 1506 and remained there for a year, studying and debating with the best teachers of the University in Humanist Italy's climate of cultural liberty.[1] An historical event which took place at the time of his arrival was the victorious entry into the city of Pope Julius II and the transformation of Bologna into a legateship of the Papal state. So began the process which was to reduce the universal nature of *Alma Mater*, as the University of Bologna was called, to the size of the city.

The great events of the 16th century favoured this transformation: the Protestant Reformation, the Council of Trent and the Catholic Counter-Reformation, the religious wars which divided Europe in half and which made the circulation of books and the movement of scholars more difficult. Experiences like that of Erasmus became less and less conceivable. The decline of the Bolognese University was nevertheless slow: during the whole of the 16th century it continued to attract a large number of students, especially from the German area, and famous teachers like Andrea Alciato for law, Gerolamo Cardano, Scipione Dal Ferro, Lodovico Ferrari and Pietro Antonio Cataldi for mathematics, Pietro Pomponazzi for philosophy, Ulisse Aldrovandi and Luca Ghini for natural history, Berengario da Carpi and Gaspare Tagliacozzi for anatomy.[2] Relations with Europe were not completely lost, as Aldrovandi's extensive correspondence network shows[3], although their

I would like to express my sincere thanks to Cesare Maffioli who very kindly informed me from Holland of various texts and documents which were useful for my research.

1. Cf. A. Sorbelli - L. Simeoni, *Storia dell'Università di Bologna*, Bologna, Zanichelli, 1944-1947, 2 vols., II, pp. 1-2, and R. Newald, *Erasmus Roteradamus*, Darmstadt, Wissenschaftliche Buchgesellschaft, 1970, pp. 87-88. The recent work by S. Menchi, *Erasmo in Italia*, Torino, Bollati-Boringhieri, 1987, does not deal with Erasmus' stay in Italy, but examines the influence of his thinking on groups of Italian heretics in the XVIIth century.

2. See Sorbelli, *op.cit.*, II, *passim*.

3. On U. Aldrovandi see G. Olmi, *Scienza e natura nel secondo Cinquecento*, Trento, Libera Università degli Studi, 1976, and S. Tugnoli Pattaro, *Metodo e sistema delle scienze nel pensiero di Ulisse Aldrovandi*, Bologna, CLUEB, 1981.

significance changed radically. This became evident in the 17th century when the University's ability to attract students was greatly diminished; teachers were mostly recruited from among Bolognese doctors and the watchful eye of the Pope's delegate and the Inquisition grew even tighter and even more unquestioned. The upholders of modern ideas and methods found no room in the rigidly defined curricula of the university lectures, and readers frequently had to deal with attacks from their traditionalist colleagues and from suspicions aroused amongst the Ecclesiastic authorities.[4]

In the second half of the 17th century the "moderns" working in Bologna were followers of the Galilean school in close contact with the Florentine Academy of Cimento (1657-1667).[5] This is the case of Giandomenico Cassini (Perinaldo 1625-Paris 1711), who was subsequently to become the Director of the Observatoire Royale in Paris, of Marcello Malpighi (Bologna 1628-Rome 1694), one of the pioneers of microscopical anatomy, of the astronomer and "philosopher" Geminiano Montanari (Modena 1638-Padua 1687), who in 1665 founded the Traccia Academy in Bologna, with a syllabus similar to that of the Florentine Academy. These second generation Galileans were aware of the important contribution of the tradition of thought which they attached to the foundation of the new science, and they considered the Cimento Academy the "mother" of European scientific societies. Nevertheless, after its closure in 1667 they soon realised that the centre of research had moved elsewhere, to the north of Europe.

They immediately grasped the importance of the great official scientific societies which had grown up in England and France and which they tried to reproduce in Bologna, at first on a private level, with academies organised on the basis of the Royal Society and the Académie des Sciences. Above all they became aware of the dangers of remaining isolated through lack of informa-

4. For a picture of the development of the University of Bologna in the 17th century, besides Sorbelli, *op.cit.*, II, see G. Vernazza, *La crisi barocca nei programmi didattici dello Studio bolognese*, in "Studi e memorie per la storia dell'Università di Bologna", n.s. 2, 1961, pp. 95-177; H.B. Adelmann, *Marcello Malpighi and the Evolution of Embryology*, Ithaca, New York, Cornell University Press, 1966, 5 vols., I, pp. 15-106. On the control of the Inquisition in Bologna, see A. Battistella, *Il S. Officio e la Riforma religiosa in Bologna*, Bologna, 1905. On the decline in the number of foreign students, see in this volume, the contribution by H. de Ridder-Symoens, *Italian and Dutch Universities in the XVI and XVII centuries.*

5. On the methods and the events surrounding the introduction of Galilean and modern science in general in Bologna, I refer to my studies: *Verso la fondazione dell'Istituto delle Scienze: filosofia 'libera', baconismo, religione a Bologna (1660-1714)*, in *Sull'identità del pensiero moderno*, Firenze, La Nuova Italia, 1979, pp. 97-146; *Accademie scientifiche a Bologna: dal "Coro anatomico" agli "Inquieti" (1650-1714)*, in "Quaderni storici", n. 48, 1981, III, pp. 884-921.

tion on the most recent discoveries and theories and through the limited effects of their works printed in Italy. Therefore they tried to establish contacts with academies and scientists from "beyond the mountains" and enthusiastically welcomed those which were proposed to them. In particular, they did their utmost to get their works published in scientific or scholarly periodicals, which would assure their circulation in Europe: "Philosophical Transactions", "Journal des Savants", "Acta Eruditorum", "Mémoires des Trevoux", "Bibliothèque Universelle et Choisie". Those difficulties which they had met with at home in affirming their concepts and methods, whether it was a question of the experimentalism borrowed from the Cimento and the Royal Society, or of the Galilean and Borellian iatromechanical principles, would undoubtedly have been confronted with better possibilities of success thanks to the authority which they obtained from correspondence with major figures in European science, from being members of foreign academies, from the reports of their books published in the most important journals, or even just from their re-issues abroad.

In this context, did the relations between the Bolognese scientists and Holland play a significant and extensive role? The answer to this question is, I think, positive. I will try to show this by means of a series of episodes which, without any claim to being exhaustive, witness the reciprocal interest of Bolognese and Dutch scientists and illustrate some peculiar features in the relations between Holland and Bologna.

The most striking display of Dutch estimation of Bolognese science is shown by its effect on the book market. A notable number of works by authors in Bologna had their second editions in Holland. *Histoire Physique de la Mer* (1725) and, moreover, *Danubius Pannonicus-Mysicus* (1726) by Luigi Ferdinando Marsili were published for the first time there.[6] As far as the re-issues are concerned a simple list of authors' names is enough to show that it was a case not of casual choices, but of close-knit references to a precise theoretical trend and of a precise method of research. The Galilean and Borellian derived iatromechanical medicine, had for example one of its foremost representatives in Malpighi in Bologna. Works of Bolognese origin re-issued in Holland were, in fact, written by Malpighi, his colleague Carlo Fracassati, his immediate

6. *Histoire physique de la mer, ouvrage enrichi de figures dessinées d'après le naturel par Louis Ferdinand Comte de Marsili, membre de l'Académie Royale des Sciences de Paris*, Amsterdam, aux dépens de la Compagnie, 1725; *Danubius pannonico-mysicus observationibus geographicis, astronomicis, hydro-graphicis, historicis, physicis perlustratus et in sex tomos digestos ab Aloysio Ferd. Com. Marsili socio Regiarum Societatum Parisiensis, Londinensis, Monspeliensis*, Hagae Comitum, apud P. Gosse, R.Chr. Alberts, P. de Hondt — Amstelodami, apud Herm. Uythwere & Franc. Changuion, 1726, 6 vols.

pupils Anton Felice Marsili, Domenico Guglielmini, Anton Maria Valsalva, and a follower, who was none other than Giambattista Morgagni.

The first Dutch reprints of Malpighi's works date back to 1669. In that year *Epistolae Anatomicae*, written in collaboration with Fracassati[7], and *De Viscerum Structura* (with *De Polypo Cordis* in the appendix)[8] all appeared in Amsterdam. These Amsterdam reprints are just one of the signs of the renown which Malpighi was gaining in Europe: his correspondence with Henry Oldenburg began in 1667[9]; in 1669 he became a member of the Royal Society and his *De Bombyce* was first published in London.[10] Malpighi's correspondence shows that the mediators at that time between him and Dutch culture were Nicolas Steno and Oldenburg himself. Through Borelli and Bellini, Malpighi had known, since 1662, the work *Disputatio Anatomica de Glandulis Oris* (Leyden 1662)[11] by the Danish scientist, who had studied in Holland, had then been converted to Catholicism, and so had moved to Italy. In a letter of 1669 Steno established a Dutch contact for Malpighi in the person of Johannes Druyvenstein, a Venetian bookseller, probably of Dutch origin.[12] Together with a letter dated 6 December 1670, he sent Malpighi a book from Holland entitled *Spicilegium Anatomicum* by Theodor Kerckring (Amsterdam

7. *Epistolae anatomicae ... Marcelli Malpighii et Caroli Fracassati ...*, Amstelodami, apud Casparum Commelinum, 1669. Including: *De cerebro, De lingua, De externo tactus organo, De omento, pinguedine et adiposis ductibus*, by Malpighi, *De cerebro, de lingua*, by Fracassati.

8. M. Malpighi, *De viscerum structura exercitatio anatomica, cui accedit dissertatio de polypo cordis*, Amstelodami, apud Petrum Le Grand, 1669. The book, the first edition of which was issued in Bologna in 1666, was fully reviewed in the Roman "Giornale dei Letterati" in 1668 (pp. 45-50) and was reprinted several times, apart from the Dutch edition: in London (1669), Jena (1677), Frankfurt (1678 and 1683), Bologna (1680), Toulouse (1682), and was translated into French in 1683 (1687[2]). This information was gleaned from the introduction by V. Busacchi and P. Gelmetti to the Italian translation of *De viscerum structura*, in *Celebrazioni malpighiane. Discorsi e scritti*, Bologna, Azzoguidi, 1966, pp. 77-83. The only element in some way linking this work with Holland is the fact that in the section on the liver Malpighi reveals that he is familiar with the ideas on bile secretion held by Frans dele Boë Sylvius, who held the chair in Medicine from 1652 at the University of Leyden.

9. See *The Correspondence of Marcello Malpighi*, H.B. Adelmann ed., Ithaca-London, Cornell University Press, 1975, 5 vols., I, pp. 181-183: Oldenburg's letter from London dated October 2nd or December 28th (cf. *ibid.* p. 182, note 2). Together with the reply from Bologna, dated April 1st 1668 (*ibid.* pp. 184-191), Malpighi sent a copy of *De viscerum structura*.

10. M. Malpighi, *Dissertatio epistolica de Bombyce*, Londini, Martin & J. Allestry, 1669.

11. The work had already been published in Leyden in 1661 and had been reprinted in 1662. A letter from G.A. Borelli to Malpighi dated February 15th 1662, from Pisa, shows that L. Bellini had tried to establish the existence of the salivary ducts discovered by Steno, through anatomical sections and preparations: cf. *The Correspondence of M. Malpighi*, op. cit., I, pp. 147-148.

12. *Ibid.* I, pp. 429-430: letter dated October 27th 1669 from Vienna.

1670).[13] In those years Malpighi was engaged in the preparatory research of *De Formatione Pulli in Ovo*[14], and Kerckring's theories on procreation, together with those of the other young Dutch anatomist Regnier de Graaf, were to be the subject of intense discussions with his correspondents who included Francesco D'Andrea, Giovanni Battista Capucci and Thomas Bartholin.[15] News of De Graaf's theories, and in particular a report in the book *De Mulierum Organis Generationi Inservientibus*, which had just appeared in Leyden, had reached Malpighi by means of a letter from Oldenburg in the spring of 1672.[16]

Malpighi's acquaintance with Jan Swammerdam was also through Steno and Oldenburg. These interesting circumstances contain at least two small mysteries, one old and one more recent. Both scientists had a common interest in the anatomy of the silkworm. In a letter written at the end of 1670, Malpighi asked Oldenburg to send him books by Boyle, Lower and Swammerdam.[17] He particularly wished to see works by the latter on insects, since he had learnt from Oldenburg that the author made honourable mention of his *De Bombyce*. He also affirmed to have seen the book's "extremely elegant" illustrations, i.e. those of *Historia Insectorum Generalis*, which was published in Utrecht in 1669.[18] This is the first mystery, to which I can only repeat Howard B. Adelmann who says, "I am unable to explain how Malpighi came to see these figures before the work came into his hands."[19] In fact Swammerdam was not completely convinced by *De Bombyce* and at the end of 1671, through Steno, he told his Bolognese colleague that he had again carried out the experiments described in it and had "met with some differences in some of them."[20]

Another letter by Steno in summer 1675 informed Malpighi that Swammerdam had decided to abandon nature study in order to commit himself exclusively to the search for God. He destroyed the treatise he had written on the silkworm, but kept the illustrations in order to give them to Malpighi.[21]

13. *Ibid.* II, pp. 492-493: letter dated December 6th 1670 from Florence.
14. M. Malpighi, *Dissertatio epistolica de formatione pulli in ovo*, Londini, J. Martyn, 1673.
15. *The Correspondence of M. Malpighi*, op. cit., I, *passim*. The discussion of Kerckring and De Graaf's procreation theories took place between 1671 and 1674.
16. *Ibid.* II, pp. 615-616: letter dated April 24th (May 4th) 1672 from London.
17. *Ibid.* II, pp. 483-484: letter dated November 20th 1670 from Bologna.
18. *Ibid.* II, pp. 467-468: letter dated July 15th (25th) 1670 from London.
19. *Ibid.* II, p. 486, note 17. Cf. also H.B. Adelmann, *M. Malpighi and the evolution of Embryology*, op.cit., in note 4, I, p. 715.
20. *Ibid.* II, pp. 596-598: letter from Steno dated November 24th 1671, from Florence. Swammerdam published his critical observations in the work *Miraculum naturae sive uteris muliebris fabrica*, Leida, 1672.
21. *Ibid.* II, p. 715: letter from Steno dated July 18th 1675, from Florence.

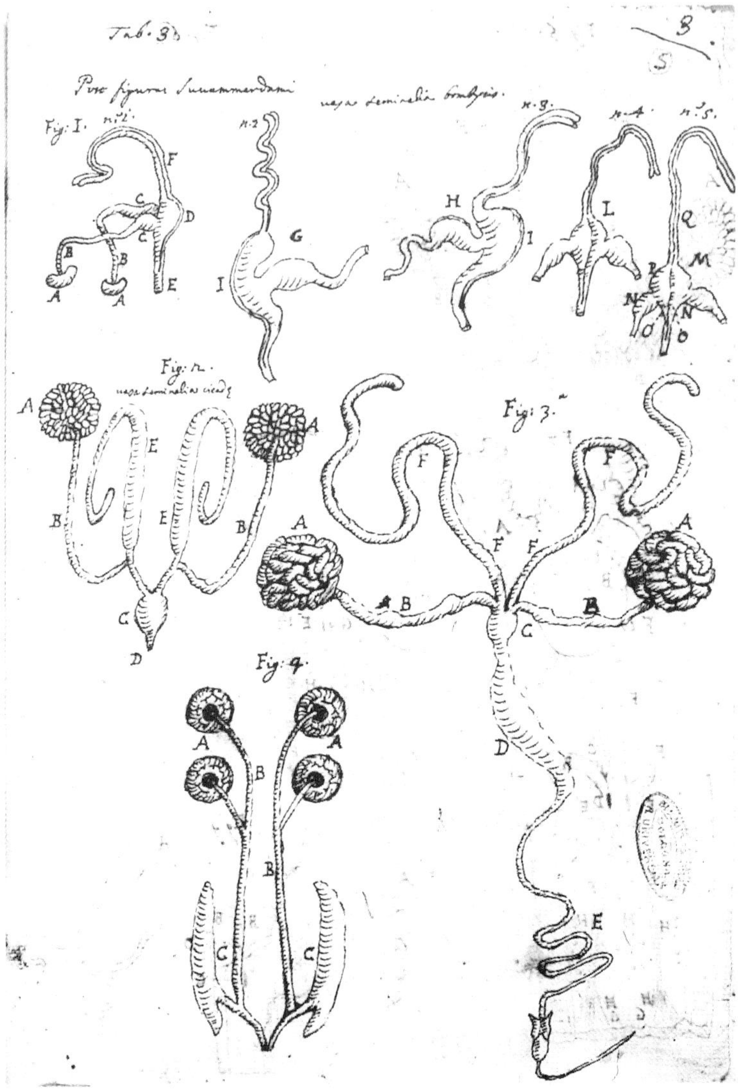

Fig. 8 Bologna, Biblioteca Universitaria, ms. 936, K, table III, pen drawings relative to the silkworm with the indication "prout figuras Swammerdami".

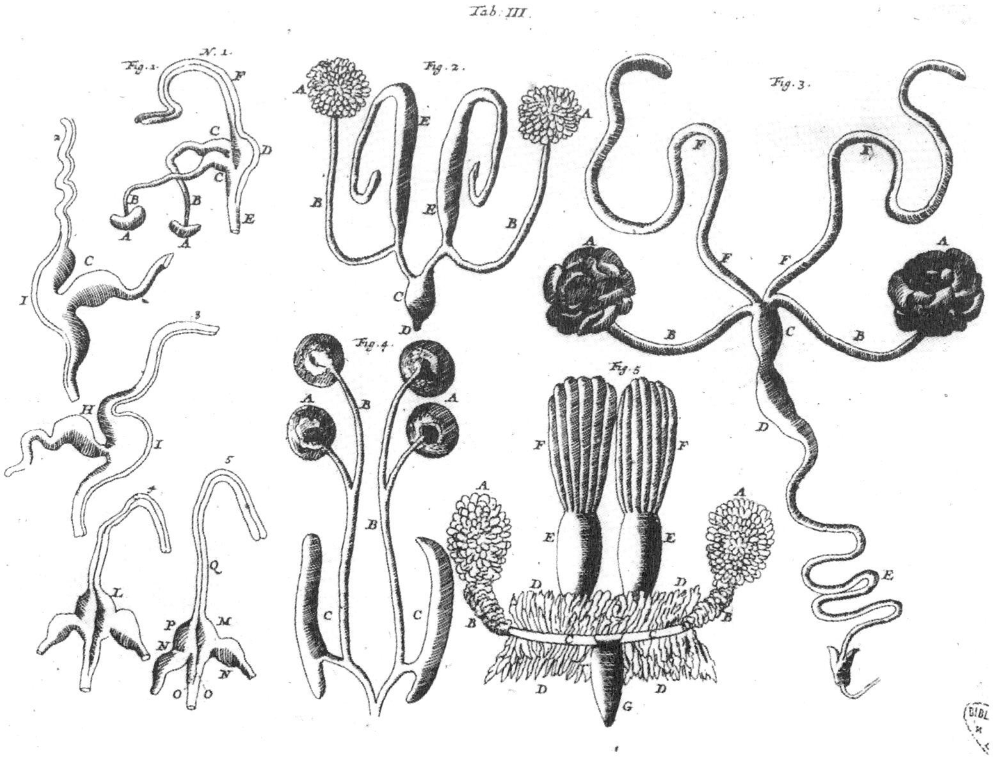


**Fig. 9 M. Malpighi, *Opera Posthuma,* Amsterdam, 1698, table III, on the silkworm, with drawings taken from tables III and IV of the ms. 936, K, of the Biblioteca Universitaria of Bologna (fig. 8).**

Here we have the second mystery: Adelmann states (1966 and 1975) that those illustrations — 28 colour plates to be precise — are still kept among Malpighi's manuscripts in the Bologna University Library.[22] Unfortunately, under the classification given by him, there are now only the illustrations, some in pen, some in sanguine, which were destined by Malpighi to illustrate his *Opera Posthuma.*[23] Only plate III deals with the silkworm and has the note "Prout figuras Swammerdami" (fig. 8), which is missing in the corresponding plate of the edited work (fig. 9). Here Malpighi gives ample room to his observations on the silkworm after *De Bombyce*, and to his differences of opinion with

22. Cf. *ibid.* II, p. 486, note 17, and p. 715, note 2. Cf. also Adelmann, *M. Malpighi and the evolution of Embryology*, op.cit., I, p. 399.

23. Bologna, Biblioteca Universitaria, ms. 936, K. Adelmann (places mentioned in note 22) also refers to ms. 2085, which contains the originals of all the Malpighi texts collected together in the *Opera posthuma*, published first in London (1697) and then in Amsterdam (1698).

Swammerdam.[24] He grudgingly points out that his *historia* had had more admirers than critics. Only Swammerdam in a book entitled *Naturae Miraculum*, "for the sake of contradiction", attempted to take up his observations on the silkworm once more, in an extremely rigorous and severe way. According to Malpighi the Dutchman was moved by the spirit of emulation, in that he had shown the external metamorphosis of the insects "in an elegant book", but had not dealt with the internal organs, "where the force of nature shines through more wonderfully."[25] Almost completely reproducing the quoted letters of Steno, Malpighi tells of the illustrations given to him by Swammerdam, and of his decision to publish them in order that the learned world could compare the studies of both, and so judge whether Swammerdam's discoveries were more important than his.[26] Malpighi also asks to take into account the fact that while he had penetratcd new and unknown ground, Swammerdam had had the advantage of already knowing of his observations. Nevertheless, since some of the criticisms made too harshly did in fact point out some actual mistakes, it was Malpighi's intention to take a long pause in order to analyse them, not only to honour the gift of a very important man, but also to let the truth emerge more easily.[27]

Apart from these differences, Malpighi found a stimulating affinity in Kerckring, De Graaf and Swammerdam's research, with objects and methods which were difficult to come across in Italy, except in the iatromechanical trend, and least of all in Bologna. Malpighi, however, did not understand the importance of the studies of another great Dutch researcher, a pioneer, like himself, of microscopic observation of nature: Antoni van Leeuwenhoek. In particular he seemed sceptical and indifferent with regard to the fundamental discovery of spermatozoa[28], which he had already heard about in 1678 through letters from Nathaniel Grew[29] and Robert Hooke.[30]

Even if there was some indirect contact between Malpighi and the Dutch researchers, as we have seen, there does not seem to be any relationship with

24. M. Malpighi, *Opera posthuma in quibus Excellentissimi Authoris vita continetur, ac pleraque quae ab ipso prius scripta aut inventa sunt confirmantur et ab adversariorum objectionibus vindicantur*, Amstelodami, apud Donatum Donati, 1698, pp. 97-185 of the Malpighi autobiography.

25. *Ibid.*, p. 81.

26. *Ibid.*, and see notes 20 and 21. The originals of Steno's letters are in ms. 2085 in the Bologna University Library.

27. *Ibid.*

28. Cf. Adelmann, *M. Malpighi and the evolution of Embryology*, op.cit., I, p. 868.

29. *The correspondence of M. Malpighi*, op.cit., II, p. 780: letter dated July (20th) 1678, from London.

30. *Ibid.* II, pp. 831-832: letter from London which may be dated 1678 or 1679.

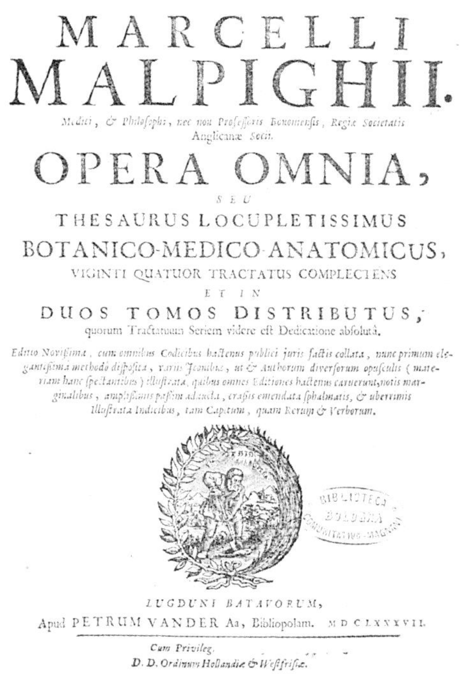
MARCELLI
MALPIGHII.
Medici, & Philosophi, nec non Professoris Bononiensis, Regiæ Societatis Anglicanæ Socii.
OPERA OMNIA,
SEU
THESAURUS LOCUPLETISSIMUS
BOTANICO-MEDICO-ANATOMICUS,
VIGINTI QUATUOR TRACTATUS COMPLECTENS
ET IN
DUOS TOMOS DISTRIBUTUS,
quorum Tractatuum Seriem videre est Dedicatione absolutâ.
Editio Novissima, cum omnibus Codicibus hactenus publici juris factis collata, nunc primum elegantissimâ methodo disposita, variis Iconibus, ut & Authorum diversorum opusculis (materiam hanc spectantibus) illustrata, quibus omnes Editiones hactenus caruerunt, notis marginalibus, amplissimis passim adaucta, crassis emendata sphalmatis, & uberrimis illustrata Indicibus, tam Capitum, quam Rerum & Verborum.
LUGDUNI BATAVORUM,
Apud PETRUM VANDER Aa, Bibliopolam. MDCLXXXVII.
Cum Privileg.
D. D. Ordinum Hollandiæ & Westfrisiæ.

Fig. 10 M. Malpighi, *Opera Omnia*, Leyden, 1687: a) title-page; b) frontispiece.

the two-volume edition of his *Opera Omnia*, issued in Leyden in 1687[31] (fig. 10). In fact Malpighi was at this time sufficiently well-known in Europe to be able to represent a secure commercial proposition: all his works written after 1669 were printed in first edition in England, where in 1686 a volume was issued which contained all his printed works up to that time.[32] The London edition of Malpighi's *Opera Omnia*, which did not receive the Royal Society's official sanction, was not very accurate and contained several glaring mistakes. The editor of the Leyden edition, Peter Vander Aa, states in the letter of dedication to have decided to republish Malpighi's medical-botanical and anatomical works above all with a view to amending them. As well as being more correct, provided with indexes and margin notes, the Leyden edition was more complete with the addition of plates and of some pamphlets of authors associated with Malpighi: Johann Jacob Harder, Thomas Bartholin, Carlo Fracassati, Anton Felice Marsili. The last two are Bolognese: the first, a colleague of Malpighi and his successor to the Messina professorship, has two dissertations (*De Cerebro* and *De Lingua*) contained in the work[33]; the second, Marsili, a pupil who had begun an ecclesiastical career, and who was later to be accused by Malpighi of protecting his adversaries, has his epistle *De Ovis Cochlearum* included, a contribution in support of Malpighi's ovistic thesis against upholders of spontaneous generation.[34]

The Leyden edition of the *Opera Omnia*, as well as competing well with the London edition on the Continental market, also made Holland more aware of Malpighi's methods and discoveries. These had a great influence on the works of the two important exponents of Dutch medicine and anatomy at the beginning of the 18th century, Frederick Ruysch and Herman Boerhaave. In *Institutiones Medicae* by the latter, Malpighi's most controversial and methodological writings in his *Opera Posthuma*, first published in London in 1697, are also frequently quoted. Here again, Dutch editorial organisations

31. M. Malpighi, *Opera omnia, seu thesaurus locupletissimus botanico-medico-anatomicus, viginti quatuor tractatus complectens*, Lugduni Batavorum, apud Petrum Vander Aa, 1687, 2 vols.

32. M. Malpighi, *Opera omnia*, London, R. Scott, 1686, 2 vols., reprinted in 1687 by Littlebury. The works on anatomy were also published in *Bibliotheca anatomica* by Leclerc and Manget published in Geneva in 1685.

33. Cf. note 7; on G. Fracassati, see G. Fantuzzi, *Notizie degli scrittori bolognesi*, Bologna, Stamperia S. Tommaso d'Aquino, 1781-1794, 9 tomes, III, pp. 357-359, and Adelmann, *M. Malpighi ...*, op.cit., *passim*.

34. On A.F. Marsili, see Fantuzzi, op.cit., V, pp. 276-278; Adelmann, *M. Malpighi*, op.cit., I, *passim*; M. Cavazza, *Riforma dell'università e nuove accademie nella politica culturale dell'arcidiacono Marsili*, in *Università, accademie e società scientifiche in Italia e in Germania dal '500 al '700*, Bologna, Il Mulino, 1981, pp. 245-282.

MARCELLI
MALPIGHII
PHILOSOPHI ET MEDICI
BONONIENSIS
E
SOCIETATE REGIA LONDINENSI
OPERA POSTHUMA.

In quibus Excellentiſſimi Authoris vita continetur, ac pleraque quæ ab ipſo priùs ſcripta aut inventa ſunt confirmantur, & ab adverſariorum objectionibus vindicantur.

*Supplementa neceſſaria, & Præfationem addidit, innumeriſque in locis emendavit* PETRUS REGIS *Monſpelienſis, in Academiâ patriâ Medicinæ Doctor.*

Editio ultima figuris æneis illuſtrata, priori longè præferenda.

AMSTELODAMI,
Apud GEORGIUM GALLET.
M. DC. XCVIII

Fig. 11 M. Malpighi, *Opera Posthuma,* Amsterdam, 1698, title-page and frontispiece.

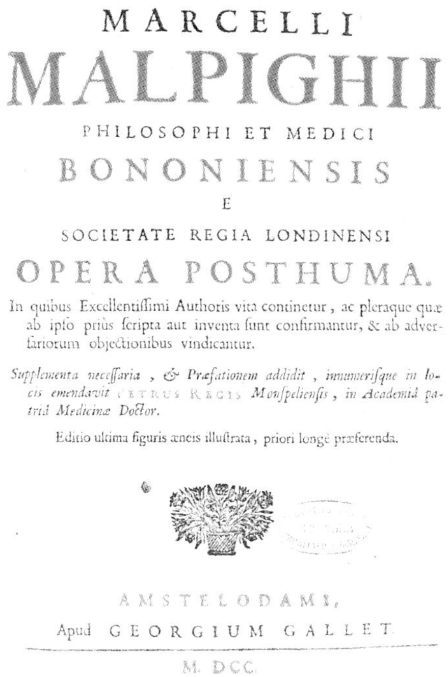

MARCELLI
MALPIGHII
PHILOSOPHI ET MEDICI
BONONIENSIS
E
SOCIETATE REGIA LONDINENSI
OPERA POSTHUMA.

In quibus Excellentiſſimi Authoris vita continetur, ac pleraque quæ ab ipſo priùs ſcripta aut inventa ſunt confirmantur, & ab adverſariorum objectionibus vindicantur.

*Supplementa neceſſaria, & Præfationem addidit, innumeriſque in locis emendavit* PETRUS REGIS *Monſpelienſis, in Academiâ patriâ Medicinæ Doctor.*

Editio ultima figuris æneis illuſtrata, priori longè præferenda.

AMSTELODAMI,
Apud GEORGIUM GALLET.
M. DCC.

Fig. 12 M. Malpighi, *Opera Posthuma,* Amsterdam, 1700, title-page.

were of great service to Malpighi, and an undoubtedly commercially-inspired operation took on a notable cultural role. In 1698 two editions of Malpighi's *Opera Posthuma* were issued in Amsterdam, differing only in their printers and in their dedications: one printed by Georges Gallet (fig. 11) with a dedication by Pierre Régis, the editor, to Nicolaas Witsen, the most important magistrate in the city of Amsterdam[35]; the other, perhaps intended for the Italian market, printed by Donato Donati and dedicated to *Mario Fiorentinio Patricio Lucensi et Medicinae Doctori.*[36] However, the frontispiece is the same and the letter to the "benevolent reader" signed by the physician Pierre Régis[37] is identical. Both versions were reprinted in 1700 (fig. 12). In both versions the Dutch edition was presented as *editio ultima ... priori longe praeferenda*. And rightly so, in that the London edition was so incorrect that it could have become an arm in the hands of Malpighi's adversaries. Particularly untrustworthy was the Latin translation of the *Risposta* to the epistle *De Recentiorum Medicorum Studiis*, i.e. Malpighi's defence of the rational methodology in medicine, against the criticisms of the Bolognese physician Giovanni Gerolamo Sbaraglia.[38] In fact, the latter was to take his cue, in his counter-reply,[39] from the ambiguous or equivocal phrases in the Latin text in order to produce further attacks on the credibility of Malpighi's methods.

In the preface to the Amsterdam edition the Huguenot Pierre Régis not only shows complete agreement with Malpighi's reasoning, but also interprets the controversies and hostility, which he had to deal with in Bologna, as an episode in the struggle for *libertas philosophandi* and likens his case to those of Galileo, Harvey, Gassendi and Descartes.[40] Régis significantly pauses on the first, who, when almost seventy years old, was forced by the harassments of the

35. N. Witsen was, together with J. Basnage de Beauval, Chr. Huygens and A. van Leeuwenhoek, one of the four Dutchmen or Dutch residents to whom the Royal Society had decided to send a copy of the London edition of the Malpighi *Opera posthuma*: cf. Adelmann, *M. Malpighi ...*, op.cit., I, p. 709.

36. Cf. note 24.

37. Pierre Régis (Montpellier 1656 - Amsterdam 1726) was a French huguenot who took refuge in Holland when the edict of Nantes was revoked. In his youth he had followed the lessons of the Cartesian doctor Pierre Sylvain Régis in Montpellier and those of the anatomist Duverney and the chemist Lemery in Paris (cf., *ad vocem, Nouvelle Biographie générale*, Paris, 1862).

38. On G.G. Sbaraglia (Bologna 1641-1710) and on the struggle in favour of "empiric" medicine against Malpighi and his followers, see Fantuzzi, *Notizie degli scrittori bolognesi*, op.cit., VII, pp. 332-341.

39. G.G. Sbaraglia, *Oculorum et mentis vigiliae ad distinguendum studium anatomicum et ad praxin medicam dirigendam*, Bononiae, P.M. Monti, 1704.

40. *Lectori benevolo Petrus Regis M.D.*, in M. Malpighi, *Opera posthuma*, op.cit. in note 24, 4 pp. not numbered.

Inquisition to forsake, against his will, his ideas on the universe and the Earth's movement. As had happened with other great men, despite all the obstacles, Malpighi's ideas were also finally accepted and now "bloom" everywhere, while those of his adversaries were condemned to eternal oblivion, "if it were not for the Spanish, the Portuguese and the Muscovites who still lie in darkness, favouring this nonsense". Through the figure of Mini, one of Malpighi's adversaries who had struck Régis the most, he grasps well the basic philosophical contrast between "empirical" and "rational" medicine: on the one hand the nostalgics such as Mini, who used the scholastic *facultates* and *virtutes* as categories in interpreting biological and physiological phenomena; on the other hand the upholders of a mechanical explanation of the same and of the usefulness of instruments such as the microscope in their investigation. On his part Régis has no doubts: the "faculties" are by now completely defeated and are only discussed in "some squalid schools."[41] Pierre Régis is certainly not the only or even the most important upholder of "rational" medicine in Holland. In 1703 Herman Boerhaave published in Leyden a famous academic introductory lecture on the use of mechanical reasoning in medicine and on the renewal of the medical studies curriculum in this direction.[42] He was also to show a steady interest in the theories of Italian iatromechanists, in Giovanni Alfonso Borelli and Lorenzo Bellini, as well as in Malpighi.[43] This favourable climate for mechanistic viewpoints is probably the major reason for the success in Holland of the physical-medical works of Domenico Guglielmini.[44] Among the immediate pupils of Malpighi, he is in fact the one who was most engaged in the defence of his teacher's memory and in an attempt to develop his ideas, presenting a project for the mathematicisation of medicine, which won the encouragement of Leibniz.[45] He first thought of this project when he was still teaching mathematics and hydrometry in Bologna but he dedicated himself to its realisation in Padova, particularly

41. *Ibid.*

42. H. Boerhaave, *De usu ratiocinii mechanici in medicina*, Leiden, 1703.

43. See G.A. Lindeboom, *Herman Boerhaave. The man and his work*, London, Methuen & Co., 1968. See also, by the same author, the item *Boerhaave* in *Dictionary of Scientific Biography*, ed. C.C. Gillispie, New York, 1970-1980, 16 vols., II, pp. 224-228.

44. For information on the life and works of D. Guglielmini (Bologna 1655-Padua 1710) it is still necessary to refer to the biography written by G.B. Morgagni and foreword to D. Guglielmini, *Opera Omnia*, Genevae, Cramer & Perachon, 1719, 2 vols., and to B. de Fontenelle, *Eloge de M. Guglielmini*, in *Oeuvres*, Amsterdam, 1754, V, pp. 178-196, as well as to Fantuzzi, *Notizie degli scrittori bolognesi*, op.cit., IV, pp. 320-328.

45. Cf. M. Cavazza, *La corrispondenza inedita tra Leibniz, Domenico Guglielmini, Gabriele Manfredi*, in "Studi e memorie per la storia dell'Università di Bologna", n.s., VI, 1987, pp. 51-79.

after he had gained the professorship in theoretical medicine in 1702. He began teaching with an introductory lecture in defence of rational medicine entitled *Pro Theoria Medica Adversum Empiricam Sectam*, published that same year in Venice. The short work was reprinted in Utrecht in 1704,[46] together with *De Sanguinis Natura et Constitutione*, whose first edition was issued in Venice in 1701.

In those years Guglielmini enjoyed notable prestige in Europe for his works on hydraulics which earned him admission both to the Académie des Sciences and the Royal Society.[47] If his fame is sufficient to explain the interest of the Dutch editors in the new books which he was publishing, it is also true that academics could find original contributions to various fields of research in them. This is certainly the case with *De Salibus,* issued in Venice in 1705[48] and in Leyden in 1707,[49] a work which takes up and elaborates research by the young Guglielmini. The demanding, methodological preface ends with the statement that none of the physical sciences, among which medicine can undoubtedly be included, can be dealt with without the help of geometry, mechanics and arithmetic.[50] The contribution which Guglielmini made with this book to the foundation of modern crystallography, is strongly emphasized by R. Hooykaas. He maintains that all merit is due to the Bolognese scientist for the construction of "the first general theory on crystals on the basis of the principle that crystalline polyhedra consist of polyhedric molecules of the same shape."[51] Guglielmini's explanation gave a completed theoretical form to a hypothesis put forward in 1675 by A. van Leeuwenhoek who is quoted in *De salibus* regarding the trapezoid shape which he observed in alum crystals.[52]

46. *Apud Guilelmum Vande Water.*

47. On Guglielmini's contribution to the scientific foundation of hydraulics, see the studies by C. Maffioli, *Guglielmini vs. Papin (1691-1697). Science in Bologna at the end of the XVIIth Century through a debate on Hydraulics*, "Janus", LXXI, 1-4 (1984), pp. 63-105; *D. Guglielmini, G. Rondelli e la nuova cattedra d'idrometria nello Studio di Bologna (1694)*, in "Studi e memorie per la storia dell'Università di Bologna", n.s., VI, 1987, pp. 81-124.

48. D. Guglielmini, *De salibus – Dissertatio epistolaris physico-medico-mechanica*, Venetiis, apud Aloysium Pavinum, 1705. The topics of this work had already been announced in *Riflessioni filosofiche dedotte dalle figure de sali*, Bologna, Heirs of A. Pisarri, 1688.

49. *Apud Federicum Haaring*. The "Acta Eruditorum" (1708, February, pp. 72-80) and the "Journal des Sçavans" (XLI, 1708, pp. 289-294) reviewed the *De salibus* in the Leyden edition.

50. Guglielmini, *De salibus*, op.cit., preface *Lectori benevolo.*

51. R. Hooykaas, "Domenico Guglielmini et le développement de la cristallographie", Estratto dagli 'Atti delle Fondazione G. Ronchi', VIII, 1953, p. 5.

52. Guglielmini, *De salibus*, op.cit., pp. 26-28. Guglielmini had already shown that he knew Leeuwenhoek in *De sanguinis natura* (1705). In this work the Dutch researcher's observations and experiments on blood, together with those by other neoterics, are often cited to support the author's theses.

Some indications lead us to think that the intermediary between Guglielmini and the world of Dutch scholars and editors was Pieter Hotton (1648-1709), professor of botany in Leyden, who was at the centre of a vast network of scientific correspondence. For example, in the introductory letter to the "benevolent reader" in *De Salibus*, Guglielmini affirms that the encouragement to complete his doctrine on salts, published sixteen years before, had come from this character and from another two, who had in fact been in contact with the former: Felice Viali, botanist at Padua University and a correspondent of Hotton, and the recipient of the work, Cristino Martinelli, Venetian nobleman and patron of the sciences.[53] At the end of the book Guglielmini asks the latter to send a copy to Hotton, together with the manuscript of a dissertation entitled *De Febrium Natura, et Causis*.[54] *De Salibus* also met with opposition in Holland from G.C. Schelhammer, who criticised the text in a treatise on saltpetre printed in Amsterdam in 1709.[55]

In order to complete the picture of Guglielmini's editorial fortunes in Holland it is necessary to remember that in Leyden in 1709 *De Idearum Vitiis, Correctione, et Usu ad Statuendum et Inquirendum Morborum Natura* was also published, a work dedicated to the determination of rational as opposed to empirical diagnoses of diseases (first edition, Padua 1706).

Another immediate pupil of Malpighi who had his fundamental work republished in Holland within a very short time is Anton Maria Valsalva. His treatise *De Aure humana*, which had come out in Bologna in 1704, was reprinted in Utrecht in 1707, by Van de Water, and in this edition it was read by Jean Leclerc, who praisingly reviewed it in "Bibliothèque Choisie" in 1710.[56] It was effectively a fundamental text for the anatomy of the ear, which enormously improved the knowledge of the organ's structure and immediately aroused interest in Europe, as well as being opportunely recommended with lengthy reports in "Philosophical Transactions", "Acta Eruditorum" and "Mémoires de Trevoux". Boerhaave was to include and to quote it in the bibliography of his *Institutiones Medicae*.[57]

53. *Ibid.*, preface *Lectori benevolo*.
54. *Ibid.*, p. 268.
55. G.C. Schelhammer, *Tractatus de nitro, vitriolo, alumine et atramentis*, Amstelodami, 1709. I have taken the information of Schelhammer's criticisms of Guglielmini from Fantuzzi, *Notizie degli scrittori bolognesi*, cit., IV, p. 327.
56. "Bibliothèque choisie pour servir de suite à la Bibliothèque universelle", by Jean Leclerc, 1710, t. XX, first part, pp. 140-145.
57. "Philosophical Transactions", n. 299, May 1705, pp. 1978-1988; "Acta Eruditorum", 1705, pp. 462-474; "Journal de Trevoux ou mémoires pour servir à l'histoire des sciences et des arts", VII, 1707, pp. 415-417. The work by Valsalva is widely quoted in *Institutiones medicae* by Boerhaave at least from the second edition onwards (Lugduni Batavorum, apud Johannem vander Linden, 1723).

The news of the second edition of Valsalva's book reached Bologna through Pieter Hotton, who wrote about it in the same year, 1707, to the botanist Lelio Trionfetti, whom he had been in correspondence with since 1699. It was, incidentally, a correspondence which dealt above all with the exchange of news about books and in which Hotton was interested not only in botanical texts but also particularly in developments in the controversies between Sbaraglia, the leader of the "empirical" physicians, and the followers of Malpighi. *De Aure humana* had already entered his list of required books in February 1705 and he discussed it in a letter the following year.[58]

The credit for having taken care of the distribution of *De Aure humana* among European scholars by means of letters in the name of the author, but which he in fact wrote himself, is claimed by Giambattista Morgagni in one of his autobiographies.[59] He was a pupil of Valsalva and his collaborator in the final draft of the book and in its preparation for printing. Rough copies of letters written in 1705 are in fact preserved in Bologna University Library among Valsalva's papers. These letters were to be sent with the volume, and were addressed to various scientists and learned men including the Dutch anatomists Govert Bidloo and Frederick Ruysch.[60] A letter of reply by Bidloo also exists, in which he says he has received the book and compliments the author[61] (fig. 13). Morgagni did actually have a strong awareness from a very young age of the vital importance of scientific information in order to learn in time of other discoveries and to arouse interest in his own. Born in Forlì, he arrived in Bologna in 1698, gaining his degree in medicine there. He then played a fundamental role in the reform of the Accademia Filosofica degli Inquieti, which in 1704 clearly defined its tendency as experimental and mathematical and reorganised itself on the lines of the Académie des Sciences in Paris.[62] Significantly, academics were, in turn, obliged to read to their colleagues reports on recently issued books and on articles appearing in scientific periodicals. Some of these reports are preserved in the archive of the

58. Bologna, Biblioteca Universitaria, ms. 1072, VI, 7. These are nine letters written between 1699 and 1707. The references to Valsalva are in the ones dated February 2nd 1705, February 12th 1706 and May 31st 1707.

59. Cf. G.B. Morgagni, *Opera posthuma*, 6 vols., Rome 1964-1982; vol. I, *Le autobiografie*, 1964, p. 14.

60. Bologna, Biblioteca Universitaria, ms. 4031, caps. CXXI.

61. Bidloo's letter is dated *Lugduni Batavorum postridie Kal. mart. 1705*. Bidloo (Amsterdam 1649-Leiden 1713) then held the chair of medicine and surgery at the University of Leyden.

62. On Morgagni's participation in the Accademia degli Inquieti, see M. Cavazza, *L'impegno del giovane Morgagni per la riforma dell'Accademia degli Inquieti e in difesa della tradizione malpighiana*, in *De sedibus, et causis. Morgagni nel Centenario*, eds. V. Capelletti and F. Di Trocchio, Rome, Istituto della Enciclopedia Italiana, 1986, pp. 91-103.

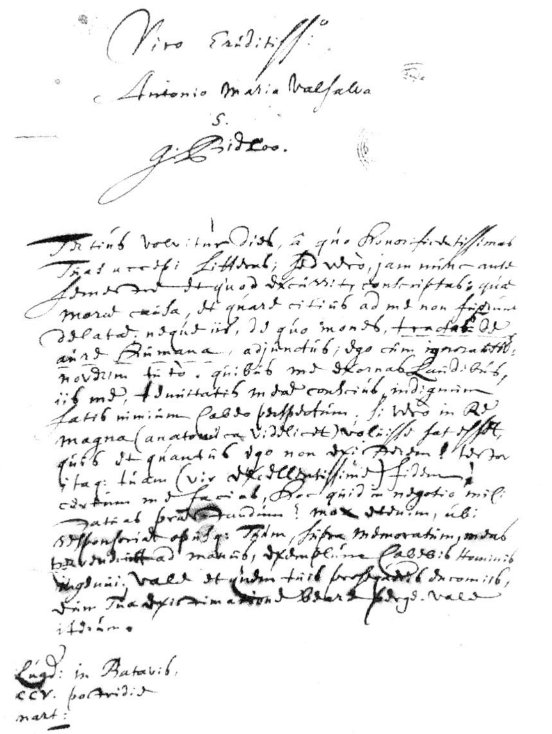

**Fig. 13 Bologna, Biblioteca Universitaria, ms. 4031, caps. CXXI, Letter from G. Bidloo to A.M. Valsalva, dated Lugd. Batav. postridie Kal. Mart. 1705.**

"Accademia delle Scienze" in Bologna, such as, for example, the lengthy report by Eustachio Manfredi[63] on the *Discours de la Cause de la Pesanteur* by Christiaan Huygens (Leyden, P. Vander Aa, 1690), while others, including those by Morgagni on the "Memoires" of the Académie des Sciences, indicated in an eighteenth century catalogue, are nowhere to be found.

When Morgagni published his own research in 1706 with the title *Adversaria Anatomica Prima*,[64] he adopted the same distribution plan already tested with the work by Valsalva: that is, he sent it to the major European journals and to the most distinguished anatomists. To use an effective expression by Paolo

63. The manuscript, entitled *Relazione della dissertazione d'Ugenio sulla causa della gravità* is kept in the archives of the Accademia delle Scienze of Bologna, with the pressmark *Academia vetus*, titolo IV, sez. I, memorie anni 1690-1723, M-Z.

64. *Joannis Baptistae Morgagni forolivensis, Philos. & Medici, ac Philosophicae quae Bononiae est Academiae Physicomed. & Anatomes Professoris, Adversaria anatomica prima, ab eo nuper in eadem Academia publicae lecta, multis deinde accessionibus novisque iconismis adaucta et Viris praestantissimis ejusdem Academiae Principi ac Sodalibus D.D.D.*, Bononiae, typis Ferdinando Pisarri, 1706.

Casini, he gained the *Adversaria*, "the European circuit which rendered influential in Italy books written in Italy" through the "recognition of the international community of experts."[65] Initially, however, the distribution of *Adversaria* met with several difficulties, especially on a practical level. Despite all his efforts Morgagni did not manage to get a copy to Hans Sloane or to assure himself a review in "Philosophical Transactions". The copies sent to Hotton in 1708, with a view to distributing them to various people, were apparently lost, including those sent to the Leyden anatomist Johannes Jacobus Rau end the Fleming Philippe Verheyen.[66] In spite of this, the book rapidly became famous throughout Europe thanks to the reviews which appeared in the "Journal des Sçavans", in the "Acta Eruditorum", in the "Giornale de' Letterati d'Italia" and finally also in Leclerc's "Bibliothèque choisie".[67] Morgagni's work began to penetrate Dutch university medical culture, and we know that even as early as 1711 Boerhaave recommended his students to study it. But he only inserted it in the bibliography and footnotes of the third edition of his *Institutiones*.[68] In 1714 the "Giornale de' Letterati" gave news that *Adversaria Anatomica Prima* had been reprinted in Leyden by Conrad Wishoff, with two very flattering letters to the author by the printer.[69] Morgagni was quite rightly convinced that such an editorial undertaking was due to the intervention of Boerhaave. He was to ask his confirmation of this in 1718 in a letter sent with *Adversaria Altera et Tertia*.[70] Boerhaave replied that his supposition was correct and promised to arrange that subsequent *Adversaria* would also soon be reprinted "in Academiae usum".[71]

The appraisal for the young Italian anatomist by Boerhaave (and by

65. P. Casini, *Morgagni nella stampa periodica, 1706-1722*, in *De sedibus, et causis*, op.cit., pp. 199-207, esp. p. 200.

66. Cf. A.M. Luyendijk-Elshout, *Morgagni, Ruysch and Boerhaave, three Musketeers for Anatomia subtilis*, in "Scienza e cultura", Padua, 1983, pp. 312-320, in particular p. 313.

67. "Journal des Sçavans", XXXVIII (1707), pp. 226-232; "Acta Eruditorum", January 1708, pp. 29-72; "Giornale de' Letterati d'Italia", I (1710), pp. 222-240; "Bibliothèque choisie", t. XX (Première partie), 1710, pp. 140-145.

68. Luyendijk-Elshout, *Morgagni, Ruysch and Boerhaave*, op.cit., p. 313.

69. "Giornale de' Letterati d'Italia", XX (1714), pp. 419-420.

70. Letter dated January 13th 1718, from Padua, published in L. Belloni, *Contributo all'epistolario Boerhaave-Morgagni. L'edizione delle Epistolae anatomicae duae, Leida 1728*, "Physis", XIII (1971), pp. 81-109, in part. pp. 82-83.

71. Letter dated December 23rd 1718 from Leyden, *ibid.*, pp. 83-84. The letter had already been published by G.A. Lindeboom, *Boerhaave's Correspondence*, Leyden, E.J. Brill, 2 vols., 1962-1964, II, pp. 79-80. The 1714 edition was probably also a reprint for university use, which could explain why there is not even one copy remaining in Holland, at least judging from the *Centraal Catalogus*. The NUC (National Union Catalog) lists a copy kept at the Medical School Library in Connecticut.

Ruysch, who was his host at the moment of the arrival of the second and third *Adversaria*) was in fact so great that he was asked, right from the first letter, to consider the idea of moving to the University of Leyden.[72] Here a vacant professorship in anatomy was anticipated, which did in fact happen (1719) due to the death of Rau. Morgagni did not accept this highly prestigious offer,[73] but his relations with Boerhaave remained strong, as is shown in their correspondence, now entirely published by G.A. Lindeboom and Luigi Belloni.[74] In this picture of the fortunes of Bolognese science in Holland, further developments in the relations between Morgagni and Boerhaave are not described since they concern works conceived and written in Padua, where from 1712 Morgagni taught firstly theoretical medicine and then anatomy.[75] However, what is underlined is the acceptance in Holland of the *Adversaria Anatomica Prima*, a completely Bolognese work, and the original fruit of the university teachings received from his tutors Valsalva, Albertini and Sandri, who in turn were all pupils of Malpighi. In the frontispiece Morgagni presents himself as a professor of anatomy of the Accademia Physicomedica of Bologna, i.e. of the Accademia degli Inquieti, where the paper had been read and approved before being printed.[76] Despite this Morgagni had no chance of a career in Bologna. Due to the presence of a strong anti-Malpighi group the doors of Bologna University remained closed to the young Morgagni, who, albeit under a pen name, entered the scene in 1705 to defend the method and discoveries of Malpighi and of the moderns in general, including Steno, De Graaf, Swammerdam and Leeuwenhoek, from the accusations of the powerful Sbaraglia.[77]

However, it was precisely in this period that, thanks also to Morgagni and to the group of young mathematicians and physicians who with him took part in the activities at the Accademia degli Inquieti, favourable conditions were being created for the institutional recognition in Bologna of the importance of new experimental-mathematical and naturalistic methods. In 1714, after long negotiations, a public Istituto delle Scienze e delle Arti was inaugurated, promoted by the city Senate with the approval of the Pope.[78] The institute

72. *Ibid.*, p. 84.

73. Morgagni, *Opera posthuma*, op.cit., I, p. 53.

74. Cf. notes 70 and 71.

75. For further developments of Morgagni's editorial fortunes in Holland, see the works of Lindeboom, Belloni, Luyendijk-Elshout cit.

76. Cf. note 64.

77. *Horatii De Florianis Epistola qua centum et quinquaginta errores ostenduntur in recenti libro inscripto Oculorum et mentis vigiliae* (...), Rome, Buagni, 1705.

78. On the foundation of the Istituto delle Scienze in Bologna, as well as to the works cited in notes 5 and 61, I take the liberty of referring to M. Cavazza, *La casa di Salomone realizzata?*, in *I Materiali dell'Istituto delle Scienze*, Bologna, CLUEB, 1979, pp. 42-54.

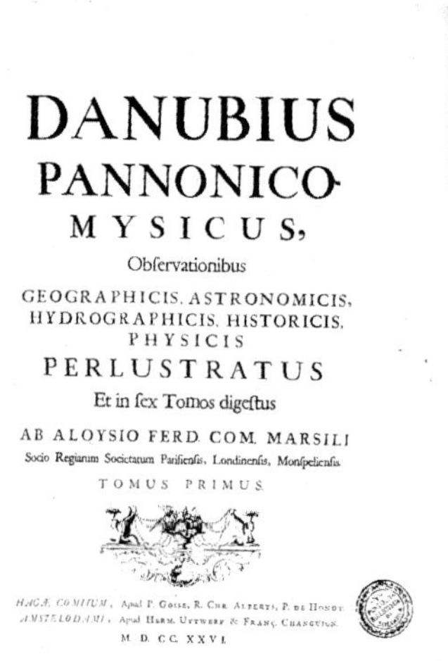

DANUBIUS
PANNONICO-
MYSICUS,
Obſervationibus
GEOGRAPHICIS, ASTRONOMICIS, HYDROGRAPHICIS, HISTORICIS, PHYSICIS
PERLUSTRATUS
Et in ſex Tomos digeſtus
AB ALOYSIO FERD. COM. MARSILI
Socio Regiarum Societatum Pariſienſis, Londinenſis, Monſpelienſis.
TOMUS PRIMUS.
HAGÆ COMITUM, Apud P. Gosse, R. Chr. Alberts, P. de Hondt.
AMSTELODAMI, Apud Herm. Uytwerf & Franç. Changuion.
M. D. CC. XXVI.

**Fig. 14 L.F. Marsili, *Danubius Pannonico-Mysicus*, The Hague, Amsterdam, 1726, title-page.**

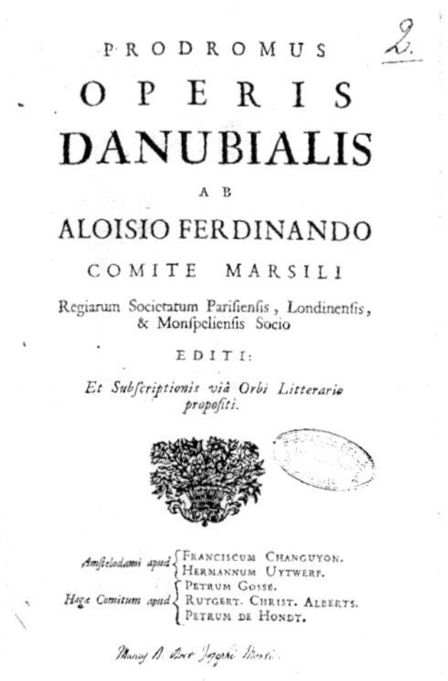

PRODROMUS
OPERIS
DANUBIALIS
A B
ALOISIO FERDINANDO
COMITE MARSILI
Regiarum Societatum Pariſienſis, Londinenſis, & Monſpelienſis Socio
EDITI:
*Et Subſcriptionis viâ Orbi Litterario propoſiti.*
*Amſtelodami apud* Franciscum Changuyon. Hermannum Uytwerf.
*Hagæ Comitum apud* Petrum Gosse. Rutgert. Christ. Alberts. Petrum de Hondt.

SUBSEQUENTES, OPUS DANUBIALE TYPIS MANDATURI, BIBLIOPOLÆ CONDITIONES OFFERUNT.

CONDITIONS DE LA SOUSCRIPTION.

**Fig. 15 L.F.Marsili, *Prodromus Operis Danubialis*, The Hague, Amsterdam, s.d.: a) title-page; b) enclosed sheet with the subscription conditions of the work on the Danube (fig. 14).**

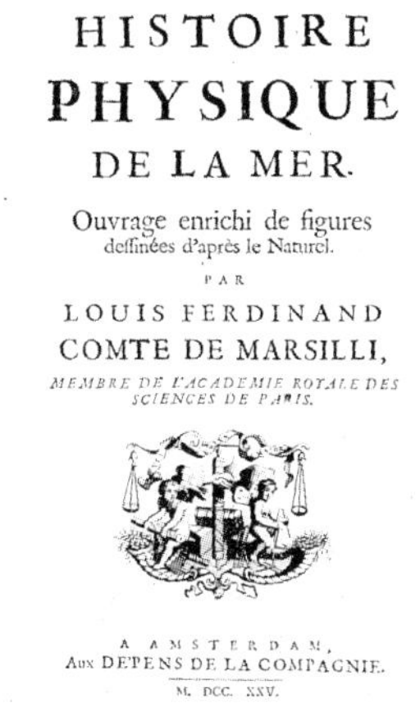
HISTOIRE
PHYSIQUE
DE LA MER.
Ouvrage enrichi de figures
dessinées d'après le Naturel.
PAR
LOUIS FERDINAND
COMTE DE MARSILLI,
MEMBRE DE L'ACADEMIE ROYALE DES
SCIENCES DE PARIS.
A AMSTERDAM,
Aux DEPENS DE LA COMPAGNIE.
M. DCC. XXV.

**Fig. 16 L.F. Marsili, *Histoire physique de la mer*, Amsterdam, 1725, title-page.**

included the Accademia degli Inquieti which had become Accademia delle Scienze. The deviser and creator of the undertaking was Count Luigi Ferdinando Marsili, who had previously been host to the Inquieti in his palace and had then given the Senate the basic materials and apparatus necessary for the running of the institution. As well as being a soldier who had had a brilliant career in the Imperial army, which ended in an unfortunate way, Marsili was himself a scientist and not even in the middle of campaings against the Turks did he stop gathering data for his research in natural history and geography.[79] In fact, it was during that time that he planned and wrote his great work on the Danube basin, which under the title *Danubius Pannonico-Mysicus* was published in 1726 in The Hague and in Amsterdam, in six tomes of atlas-size pages[80] (fig. 14 and 15). After his enforced withdrawal from military life Marsili continued his research in France, near Marseilles, into the structure of coasts, tides and sea currents, sea sponges and corals, which he maintained were plants. The fruit of this research is *Histoire Physique de la Mer*, printed in Amsterdam in 1725 (fig. 16). The dedication to the Académie des Sciences in Paris reveals that the publication in Holland of this work is due to the

79. For information on the life and works of L.F. Marsili, see B. De Fontenelle, *Eloge de M. le Comte de Marsigli*, in *Oeuvres*, Amsterdam, Aux dépens de la Compagnie, 1754, t. VI, pp. 272-286; G. Fantuzzi, *Memorie della vita del gen. co. L.F. Marsili*, Bologna, L. dalla Volpe, 1770.

80. See note 6. To promote the work, the *Pròdromus*, already published in Nuremberg in 1700, was reprinted without the dedication to the Royal Society and with the new title: L.F. Marsili, *Prodromus Operis Danubialis editi et subscriptionis viam Orbi Litterario propositi*, Amstelodami et Hagae Com., apud J. Changujon et alios, s.a. A French translation of *Opus danubialis* came out in The Hague, care of J. Swart in 1744.

insistence and the help of a "renowned professor of Leyden". It was, of course, Boerhaave, who wrote a long and enthusiastic preface to Marsili's text.[81]

Marsili had met Boerhaave during his long stay in Holland (from the beginning of 1722 to the beginning of 1723), and attended his chemistry lectures and worked with him in the laboratory.[82] During that time he also met other revered characters, such as Leeuwenhoek in Delft and Ruysch in Amsterdam, and became very interested in many aspects of the country, from its water-defence system to its religious problems.[83] The main aim of his journey was to successfully carry out negotiations for the printing of the manuscripts, maps and plates relative to the history of the Danube. He agreed with the printers and the editors, who took on the task, on a payment in books and scientific material, which he then donated to the Istituto in Bologna.[84] The city Senate was informed of these negotiations, as they were of others carried out by him with a view to the annual printing in Amsterdam of the history and the facts of the Accademia delle Scienze,[85] which were anticipated in its constitution, but which were late in appearing. Such a solution, according to Marsili, would guarantee a rapid and widespread distribution of the acts to the major European centres, all to the benefit of the prestige of the Istituto in Bologna. The idea gave rise to several uncertainties, above all with the concern

81. See note 6. Boerhaave's preface, in Latin with a French translation opposite, follows the dedication *A Messieurs de l'Academie Royale des Sciences de Paris* and a *Préface*, both by Marsili.

82. Cf. G.A. Lindeboom, *Boerhaave and his Italian Correspondents*, in "Atti del XXI Congresso Internazionale di Storia della Medicina, Siena 1968", pp. 49-61. On the relationship Boerhaave-Marsili see pp. 54-56.

83. Various documents relative to Marsili's visit to Holland are kept in the University Library of Bologna, ms. Marsili 87, F. These are sketches and notes and, above all, copies and drafts of letters to: Newton, Halley, Mead, Woodward, Derham, Boerhaave, Leeuwenhoek, 's Gravesande , De Groot and others. The visit to Ruysch which is mentioned in the letter to Newton is also documented in the anatomist's *Album amicorum*, which shows Marsili's name with the date Amsterdam October 21st 1722 (Amsterdam University, H. Ruysch, *Album amicorum (1710-1722)*, I, E, 36, f. 87). More information on Marsili's stay in Amsterdam can be found in three letters dated 1722 and 1723 sent by him to P. De la Court van der Voort in Leiden (University of Amsterdam, Diederrichs, 121 Ad 1-3). For further documents on Marsili's visit to Holland, see the ample contribution in this volume, by Anita Mc Connell, *A profitable visit: Luigi Ferdinando Marsigli's studies, commerce and friendships in Holland, 1722-23*.

84. The contract between Marsili (represented by the procurator Cesare Sardi) and the Dutch printers was drawn up in Holland on March 6th 1723: see *Strumento di una seconda donazione fatta dall'Illustriss. ed Eccellentiss. Signor Generale Conte Luigi Ferdinando Marsili all'Illustrissimo, ed Eccelso Senato di Bologna a favore, e comodo dell'Instituto delle Scienze, ed Arti liberali*, Bologna, Stamperia Bolognese di San Tommaso d'Aquino, 1728.

85. See Marsili's autographed letter to the Senate of Bologna dated March 19th 1723, in the Archivio di Stato di Bologna, *Archivio del Reggimento*, files year 1723, vol. 49, ff. 161-164.

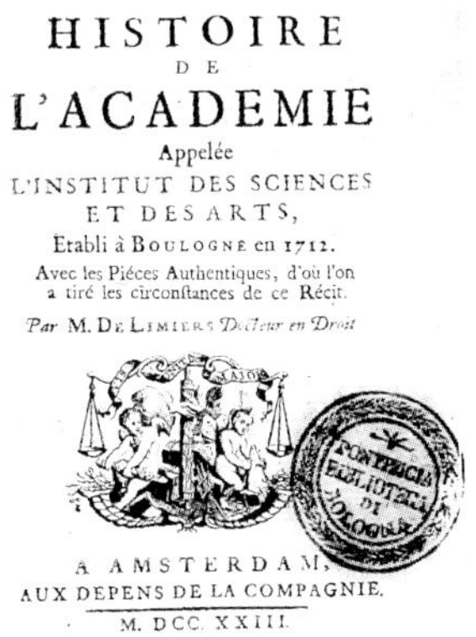
HISTOIRE
DE
L'ACADEMIE
Appelée
L'INSTITUT DES SCIENCES
ET DES ARTS,
Etabli à BOULOGNE en 1712.
Avec les Piéces Authentiques, d'où l'on a tiré les circonstances de ce Récit.
Par M. DE LIMIERS Docteur en Droit

A AMSTERDAM,
AUX DEPENS DE LA COMPAGNIE.
M. DCC. XXIII.

**Fig. 17** ***Histoire de l'Académie appelée l'Institut des Sciences et des Arts, Etabli à Boulogne en 1712*, Amsterdam, 1723, title-page.**

that the publication in Holland of the academic acts could appear in the eyes of the ecclesiastic hierarchies as a means of taking them out of their control. Marsili's initiative served, however, as a stimulus for the publication *in loco* of the *Commentarii* of the Accademia delle Scienze, the first volume of which was edited by the Bolognese printer Lelio dalla Volpe in 1731.[86]

Nevertheless, the first history of the Istituto delle Scienze of Bologna appeared in Holland. In fact, during his stay in Amsterdam, Marsili gave the historian Henry Philippe de Limiers all the information and documents necessary for the compilation of a short volume entitled *Histoire de l'Académie appellée l'Institut des Sciences et des Arts etablie à Bologna en 1712*, which was issued in Amsterdam as early as 1723[87] (fig. 17). With this modest short volume in octavo, as with the splendid *in folio* of the *Danubius* and the *Histoire Physique de la Mer*,[88] Dutch editors and culture once more made a significant contribution to a more widespread knowledge of Bolognese science, and to the

86. *De Bononiensi Scientiarum et Artium Instituto atque Academia Commentarii*, Bononiae, Ex typographia Laelii a Vulpe, 1731. The volume was widely reviewed in "Journal Literaire", t. XIX$^{2}$ (A la Haye, chez P. Gosse & J. Neaulme, 1732), pp. 289-321.

87. Limiers' *Histoire* came out "aux dépenses de la Compagnie". The Dutch public had already been informed in 1715 of the foundation of the Bologna Institute through the translation of an article from the "Giornale de' Letterati", t. XVII, 1714, pp. 148-169, in the "Journal Literaire", t. VII$^{1}$ (La Haye, chez T. Johnson, 1715), pp. 179-193.

88. In 1732 another work by L.F. Marsili, *Stato militare dell'imperio ottomano, incremento e decremento del medesimo*, was published in The Hague (eds. P. Gosse, J. Neaulme, P. de Hondt and A. Moetiens) and in Amsterdam (eds. H. Uythwere and F. Changuion), with Italian text and French facing it.

formation of that mixture of interest, admiration and curiosity with which 18th century Europe looked upon the Istituto of Bologna.

In the 18th century Bologna was no longer a leading European centre of culture as it had been during the Middle-Ages and the Renaissance period. It had become a "European province,"[89] certainly on the periphery but with strong ties with the new leading centres of culture. If the risk of remaining isolated was avoided, part of the merit is undoubtedly due to the Dutch issues and reprints which made the more important works of Bolognese scientists "visible" and available to European scholars.

89. The expression and its positive significance was borrowed from E. Raimondi, *Settecento bolognese: antichi e moderni*, in *Padre Martini. Musica e cultura nel Settecento europeo*, Florence, Olschki, 1987, pp. 1-13, p. 6.

# A PROFITABLE VISIT: LUIGI FERDINANDO MARSIGLI'S STUDIES, COMMERCE AND FRIENDSHIPS IN HOLLAND, 1722-23*

*Anita McConnell*

Luigi Ferdinando Marsigli (1658-1730), founder of Bologna's Institute of Arts and Sciences, decided in 1721 to pay his first visit to his many friends and correspondents in London and Holland. He travelled by ship from Leghorn (Livorno), and after a few weeks in London, he crossed over to Holland, arriving early in 1722 and departing just over a year later. This visit was not merely social; Marsigli intended to extend his knowledge of the sea, to purchase natural history specimens, and to arrange for the publication of his own writings.

Marsigli's autobiography[45] does not extend to this later part of his life; Quincy[46] and Fantuzzi[47], both of whom knew him, had to report on this voyage to Holland from what Marsigli later told them and their accounts are sketchy and formal. My reconstruction of events has been drawn from the lively letters that Marsigli wrote at the time, and soon afterwards, back to his close friends in London, from other letters written by the people that he met in Holland, and from official documents preserved at Amsterdam and Bologna. The Marsigli Archives in Bologna holds many notes and drawings apparently made in Holland, some in Marsigli's hand, some written for him, in various languages; many of these are undated and poorly sorted.[48] Marsigli also kept a scientific journal of his voyage, with additional notes on some of his observations made in Holland.[49] These sources all demonstrate that Marsigli, widely travelled as he was, had never encountered the landscape and way of life that was practised so skilfully in 18th century Holland, and he was fascinated and impressed by what he saw.

Marsigli, who was 64 years old when he reached Holland, had spent a lifetime passionately devoted to science. From 1682 to 1703 he was employed as a military engineer to the Emperor Leopold, then engaged against the Ottoman army. Thus his service had taken him the length and breadth of the Danube, with the requirement to map the natural features and the fortifications where the armies marched and fought. He had usefully occupied his

*The numbering in this article refers to file numbers in the Bibliography at the end of the text.

spare time in an exhaustive study of the lands bordering that great river —their archaeology and history, their economy, and every aspect of their natural history.

The Royal Society of London had encouraged this study, probably for political reasons; indeed Marsigli had been granted permission to dedicate it to the Society.[1] A prodomus was published in 1700, in Nuremberg[33], but in 1703 Marsigli had been dismissed from the army and the full Danube study had never seen the light of day. In his enforced retirement Marsigli went to the south of France where he turned his attention to the sea, its waters and the marine life which it contained. His original ambition, to which he clung throughout his life, was to write a treatise on the organic structure of the Earth, but although his enquiries were nominally directed to this end, he becamc obsessed with the study of coral and other sessile marine invertebrates. One day, having put a branch of coral in a jar of water and seen it 'flower', Marsigli became convinced that these were plants. Thereafter, he was always searching for evidence of other marine flowers and seeds; he died resolutely unconverted by all evidence to the contrary.

Never one to keep his findings to himself, Marsigli corresponded with learned men throughout Europe. On Newton's recommendation he had been elected a Fellow of the Royal Society in 1691; he had been a corresponding member of the Montpellier Royal Academy of Sciences since helping to found it in 1706, and was elected to the Académie Royale des Sciences in Paris in 1715. In 1711 he had published a prodomus to a book on the sea[35], and over the years other letters and articles on a diversity of natural history topics. The manuscript of his book about the sea — without the final section promised in the prospectus — had been sent to Paris where it received mention in the Memoirs of the Académie, but nothing more. By 1715 Marsigli had retrieved it.[47] He was then summoned from France to assist the Papal States' troops in their defence of the Adriatic coast. It was an opportunity for further marine investigations which could confirm his earlier ideas and provide new information to go into his text.

In 1711 Marsigli saw his dream of a new institute of Arts and Sciences realised. The institute was formally constituted, alongside the ancient University, and endowed by Marsigli with a library, and collections of fine arts, antiquities, natural sciences, and scientific apparatus.[34] Most of these items came from Marsigli's own holdings, garnered over many years of travel through war-torn Europe, where such items could be freely, or at least cheaply, acquired. Consequently the natural history collections lacked specimens from beyond Europe and Marsigli hoped to remedy this by purchasing in Holland

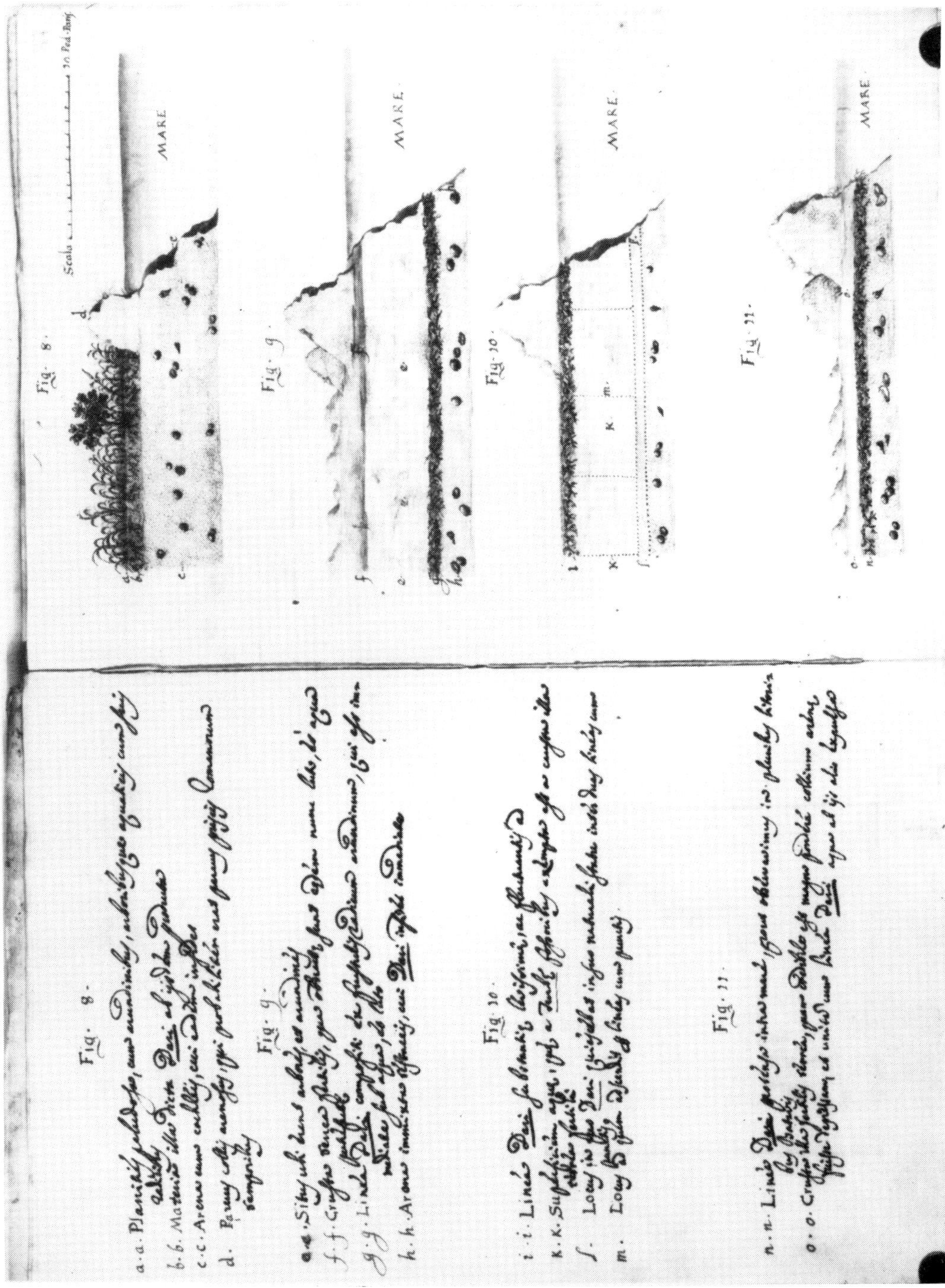


**Fig. 18** Drawings of sections through coastal deposits showing the height of vegetation, peat, and shelly layers relative to sea level (from: *Epistola ... ab Aloysio Ferdinando Marsili, ... scripta ad eruditiss. virum Hermannum Boerhaave, ...*, Biblioteca Universitaria, Bologna, Mss. Marsili, 97)

such exotic items as could be bought from collectors, or which ships' captains brought back for sale.

The long sea journey offered Marsigli the opportunity to compare conditions in the ocean to those in the Mediterranean, which he could add to his 'Natural History of the Sea', and so he kept a journal of the voyage.[49] Travelling as a passenger, he could only take daily samples of surface sea water to measure its salinity, observe wind and weather, the waves, and occasionally the geological structure of the passing coastline. He resumed the daily salinity measurements, from 23 January when he left London, until 28 January when, approaching the Meuse, the ship ran into a mass of ice, then grounded on sandbanks. This misfortune was, as Marsigli was to complain bitterly in several of his letters, due in part to bad weather, but far more to the stupidity of the captain. He landed at Hellevoetsluis, took carriage to Brielle, and came at last to Rotterdam, 38 days having elapsed since he left London.

By mid-March he was lodging in Amsterdam, having broken his journey at Leyden and Delft. From here the first of many letters were despatched, back to old friends and new that he had met in London (biographical notes follow the Bibliography). To Sir Hans Sloane,[2] after apologising for the delay in thanking him for hospitality, he wrote about the natural history collections that he had already seen in Holland, all of which consisted merely of coloured shells and insects, chosen for their appearance and financial value, without their owners knowing or caring for their scientific importance. He had met Frederik Ruysch, and was favourably impressed, although the latter's deafness inhibited conversation. Natural history, he confided to Sloane, was less well-received in Holland than in England.

A friendly letter went off to William Sherard,[3] with whom he had corresponded and exchanged material for many years. He spoke of tarrying a few days at Leyden, meeting another long-time correspondent: Herman Boerhaave, and attending one of Boerhaave's chemistry lessons which he found excellent in all respects. Boerhaave had shown him every kindness, and had invited Marsigli to attend his botany class in May. Boerhaave had written the letter of introduction to Ruysch, and they had met, although Marsigli had not yet seen Ruysch's collection.

Marsigli explained to Sherard that science came second to commerce in Holland. Learned men were few in number, unlike England which was a veritable Athens. Nonetheless, it is clear that Marsigli was impressed by the Amsterdam Botanic Garden; here, and in the Leyden Garden, he discovered species which were not cultivated in Italy, and he hoped to take plant material back to his Institute in Bologna. He also told Sherard that a natural history collection was to be sold on 15 April. He had seen the catalogue, and if prices were reasonable he intended to buy some marine plant specimens.

Fantuzzi says that in Amsterdam Marsigli met with the rich and famous and was invited to fourteen balls. His life was not entirely smooth, however, for news reached him from Bologna that the 15000 scudi intended to complete his Institute had been appropriated by the city authorities to pay off state debts and that the building works would be delayed until times were more favourable. On 26 March a strongly-worded diplomatic letter was drafted to Cardinal Paolucci,[43] who put matters right, so that Marsigli was promised that the building would be completed by 1725.

The letter drafted in May to Sir Isaac Newton,[8] President of the Royal Society, was a formal expression of thanks and obligation. After brief comments on the natural history collections, and the botanic gardens, Marsigli told how he had called on Antoni van Leeuwenhoek on his way through Delft, and how upset the old man was at the Royal Society's lack of response to four of his letters. Marsigli hoped that Newton would instruct the Secretary to reply, as Leeuwenhoek had five more letters ready to send.

This is a curious report, for the Royal Society Journal records several of Leeuwenhoek's letters being read to the Society at this time, and thanks ordered. Either the strained relationship between Leeuwenhoek and conventional academic world, or the confusion of old age, may be responsible.[52]

Marsigli had in fact called on Leeuwenhoek during March, soon after an event which they both found noteworthy. An iron cross, which had topped the steeple of Delft's New Church for two hundred years, was brought down for repair and a 'foreign gentleman' told Leeuwenhoek that such permanent iron fixtures often became magnetic. Leeuwenhoek acquired a sample of the iron but it was non-magnetic. Later, he was given part of the base of the cross, which although rusty, proved to be strongly magnetic, and also extremely hard. He communicated this information to the Royal Society on 21 April; his letter was translated and read to the Society on 3 May.[4]

On 13 May Marsigli wrote to Edmond Halley.[15] He had seen those rusty pieces of hard magnetic iron at Leeuwenhoek's house, but although his curiosity was aroused, had not dared to ask for one of the pieces. He also told Halley that he had come across a man who prepared good marine charts, based on the reports of ships' captains and pilots and that he himself sometimes went along to their meetings.

Other letters were drafted at this time: to William Derham,[6] John Woodward,[11] to the King's physician[12] (Marsigli had forgotten his name and had to send the letter via Sherard), to booksellers William and John Innis[7] who also acted as a 'mailbox' for packets and letters between Holland and London, and to Richard Mead.[14] He thanked each for the hospitality afforded him in

London, and commented briefly to each on topics of common interest. To Mead he spoke of having met a young man well-versed in physics, named 's Gravesande.

A warm friendship between Marsigli and 's Gravesande is explicit in Marsigli's draft of 23 May.[13] He thanked 's Gravesande and his wife for their kindness during his recent visit to the University of Leyden, and said that he was looking forward to attending the wedding of 's Gravesande's sister-in-law, in a fine wig expressly brought from London. He asked 's Gravesande to supply readings of the mercury barometer at Leyden, relative to sea level, to add to those he himself had previously made in central Europe and thus enable him to draw a profile of height of the continent. He ended by congratulating 's Gravesande on his physics treatise, promising to send a copy of this work to Bologna the following week.

Brief notes were drafted to two acquaintances in Rotterdam[9,10], with the request that they ask ships' captains to bring back from the Arctic parts of whales and examples of the food that whales ate.

Around this time, judging from internal evidence, Marsigli wrote a long letter to another correspondent of earlier years at the Académie des Sciences: the Abbé Bignon.[5] Marsigli's relationship with the Paris Academy was never cordial, despite his communications with them in the early years of the century, his visit to Paris in 1705, and his election as a Foreign Member in 1715 (a seat which coincidentally passed to Boerhaave on Marsigli's death). Now he was able to give Bignon a glowing account of the Royal Society in London, and his own reception thereat. He was also able to report that Boerhaave had asked him to send to Bologna for the two manuscript volumes of his natural history of the sea, written fourteen years previously and dedicated to the Paris Academy. This would now see the light of day, translated into Latin by a Flemish priest. The French language, Marsigli wrote, had become universal, and in Amsterdam they preferred to publish books in French, rather than in Latin. (Marsigli's book was in fact published in French.) After his usual remarks on the natural history collections, and the botanic gardens, he added that the Dutch gave nothing for nothing, not even directions to a poor stranger who had lost his way. On a more positive note he remarks on the wind, which is Holland's saviour, its main source of power driving industrial machinery and sawmills, and the pumps without which the land would be waterlogged. He explained to Bignon that he was planning a visit to north Holland to see if the drainage works there might be applicable to his own country in time of flood.

He remained in Amsterdam throughout June, where he spent considerable time and effort investigating the matter of peat cutting at Amstervene (now Amstelveen),[43] and he set himself to learn about the techniques of draining and

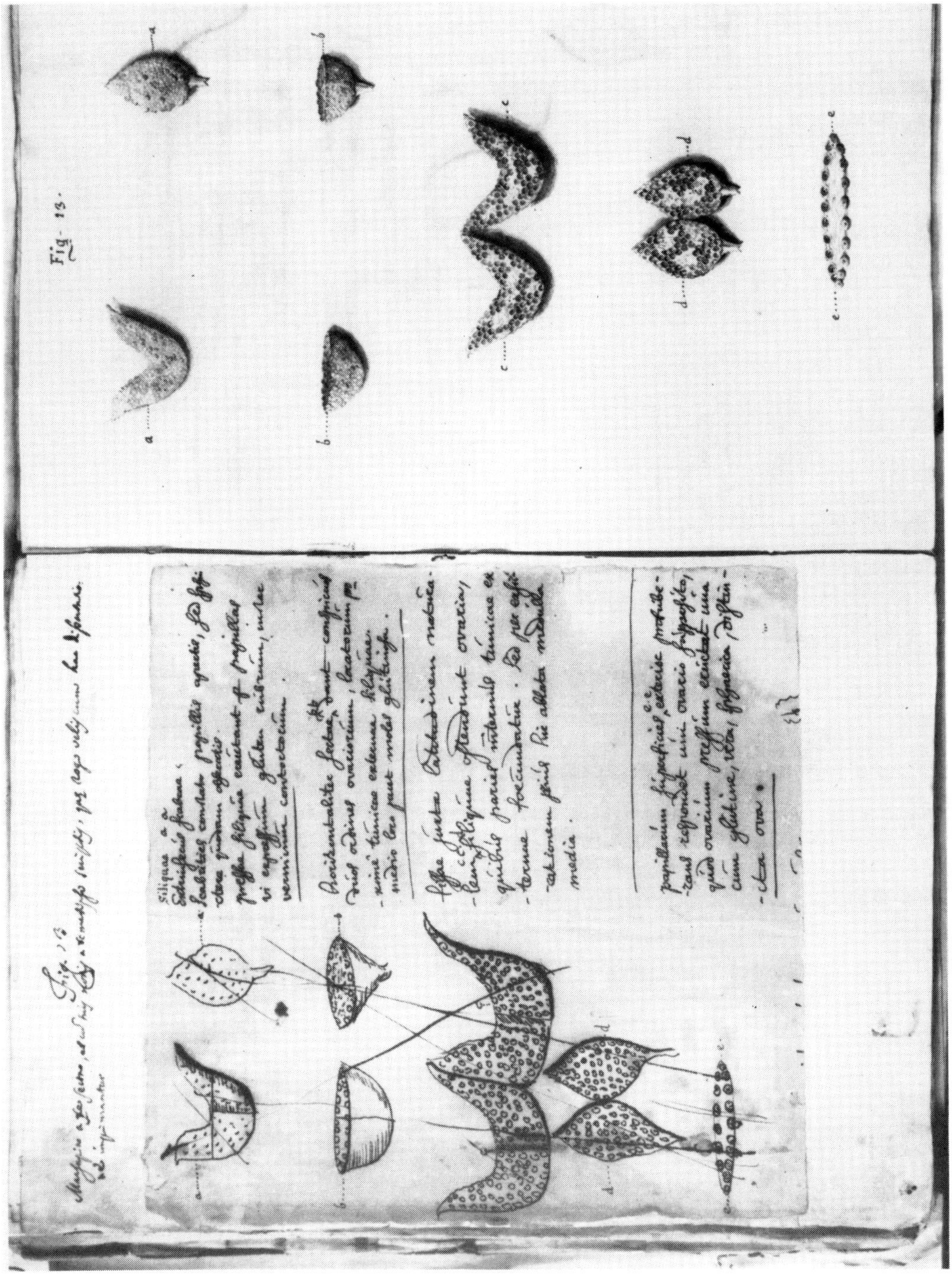

**Fig. 19 Drawings of seaweed husks sectioned by Marsigli and with his notes: Boerhaave's identification and description on the left; fair drawings on the right (from: *Epistola ... ab Aloysio Ferdinando Marsili, ... scripta ad eruditiss. virum Hermannum Boerhaave, ...*, Biblioteca Universitaria, Bologna, Mss. Marsili, 97)**

working the land in Holland. In Italy there had been disastrous floods in the valleys of the Reno and Po rivers a few years previously and Marsigli thought that he would learn how the Dutch were able to manage their land, much of it below sea level, while at the same time extracting a useful fuel from below the ground. The dykes fascinated him — those man-made barriers which alone protected the land from the raging storms. Yet the wind, which could create such destruction, was also the means of pumping the land dry. He was impressed by the ability of the diligent citizens to build canals, industrial works, and entire cities on such insubstantial foundations of peat and sand. For clearly the sea had formerly rolled over this land, depositing the layers of sand and shells now covered by soil. Watching the peat diggers, and looking at the sites where building was in progress, he sketched the layers of sediments, inspected the tools, and even obtained some to send home as examples for his countrymen.[31,43] He spent a considerable amount of his time with the younger Commelin,[17] studied Dutch books on the history of the towns and the countryside, acquired maps, and with the help of translations, gleaned as many facts as he could.

During the summer months Marsigli devoted himself again to marine science. Boerhaave wrote to Sherard[16] that Marsigli was once more out on the shore hunting for marine plants. He travelled north along the coast, collecting seaweeds that had been cast on shore, taking samples of seawater, and the water in wells; generally inspecting the dunes, their vegetation, looking at birds and animals, and enquiring about the way of life in the region. On 16 July he was at Heide, above the Meuse; on 18th opposite The Hague; on 20th at Katwijk; on 22nd at Pappen; 24th at Callantsoog. On the 25th he reached the northernmost point of his tour: Eijerland — the 'Island of the Birds', from whence he returned via Wieringen, on 28th, crossing the Pampus near Amsterdam the following day.[49] The seaweeds that he had gathered were sent to Boerhaave for identification but two days later Boerhaave fell ill, and was not to leave his bed until January 1723.[19]

Marsigli, resident again in Amsterdam, fretted over his failure to discover what ailed Boerhaave. Rumours abounded, even suggesting that Boerhaave had lost his senses. Marsigli sent his servant across to Leyden, but the man returned without having got an answer from the household and he was no wiser.[18] The illness was severe and debilitating, and it kept Boerhaave in his bed for some months, weak and in pain.

As he had agreed with Boerhaave, Marsigli took his samples of damp peat into the public laboratory and with an assistant occupied himself during the latter part of August with various analyses,[20] although he could not identify the seaweeds without Boerhaave's help. He did however manage to complete his analyses of the water collected on his coastal journey, and he teased out various plant fragments from the unconsolidated peat, which he identified and set aside for transport to Bologna.

We hear nothing of Marsigli's activities during September and October, but it seems probable that he was gathering information on the dykes, drainage, windmills, peat digging and land management, and pondering how this information might be used to good effect in dealing with the floods that periodically ravaged the lowland areas of the Papal States, of which he considered himself a citizen. Interestingly, it appears from earlier letters that he wrote to Halley and 's Gravesande that he intended to cross over to England once more, but possibly Boerhaave's illness forced a change of plan — certainly it delayed some of the analyses and identifications that he required. The next clue to his activities comes in a long letter to Sherard[18] written in November, with news from Italy where Marsigli had commissioned some paintings for, or on behalf of, Sherard. In this letter he announced that he was negotiating for the publication of his Danube survey, which for many years had been gathering dust in Bologna. It was to have a new Prodromus, and a prospectus which offered subscribers a reduced pre-publication price.[36] The negotiations were set on a formal basis when on December 8 Marsigli went before a Notary Public to authorise his agent Cesare Sardi to negotiate on his behalf with a company of Amsterdam booksellers for the publication of his work on the Danube, and other works which he could give them for printing.[44a]

The winter of 1722 was mild, with standing water not frozen,[53] in contrast to the previous year when extensive river ice had delayed Marsigli's arrival in Holland. Boerhaave resumed his lectures in January 1723, and the weather became more severe in the early months of 1723. There is no news of Marsigli during the first two months, other than a report from Boerhaave in a letter to Sherard[22] that Marsigli had been ill but hoped to leave soon for Leghorn. Marsigli himself had mentioned some ailment which afflicted him in a rather illegible letter to Sherard[17] written at the beginning of November, but wrote later that month of plans for his departure from The Hague to Lisbon.[19] He did not wish to sail in winter, but spoke of being back in Italy by Lent —Easter Sunday fell on March 23 that year — and it seems probable that the illness of Boerhaave and Marsigli himself forced a change of plan.

Final arrangements for his publications were set in hand. On 16 March he authorised Sardi to contract with a named group of publishers to print the Danube study.[44b] Sardi signed the contract on the same day.[44c] It was to be a royal work, six double-folio volumes in Latin, with over three hundred engraved plates. Only 625 copies would be printed, 20 of these going to the author. For this work Marsigli was to receive not money but books of equivalent value. The outcome of this bargain was indeed a happy one; the publishers bought 302 books from the famous library of Cardinal DuBois, recently deceased, auctioned at The Hague in 1725,[51] which Marsigli was able to present to the Institute at Bologna[38] where they were valued at 2000 Dutch florins. The British Library has a copy of the 4-volume sale catalogue with the prices of each book written in the margin.

In addition, and for the same group of publishers (who had used it to print the Prodromus), there was a collection of typefounts with Latin, Greek and Arabic letters, which Marsigli needed for the press which he had established in Bologna, which from 1719 had occupied its own substantial premises. This press and the founts were subsequently consigned to the Dominican fathers.[38]

Besides his Danube survey, Marsigli also arranged for two other books to be printed: one was a modest history of the foundation of his Institute, which he commissioned from De Limiers, a French historian who was living in Holland at the time. Working from available published material, and probably amplified by conversations with Marsigli, De Limiers' book was published in 1723.[37] The other text which finally saw the light of day after its earlier rejection by the Académie des Sciences, was Marsigli's study of the sea. This was to be published by another group of booksellers, at their own expense,[28] and its production was committed to the care of Jean LeClerc.[18] Boerhaave was persuaded to write its Preface for which Jean LeClerc provided a parallel French translation. The engravings were made by Mathias Poole, a Frenchman working in Holland at that time. Marsigli was not able to include his report of his recent voyage since his identifications of seaweeds and other marine plant material had been delayed by Boerhaave's illness. There is no evidence for any contract or exchange of money or goods, the publishers presumably met cost of engraving from their expected profits.

The book came out in 1725, published in French under the title of *Histoire Physique de la Mer*.[27,40] From an unknown print run, 11 copies were to be coloured, following the one painted by Buchoz in the French Royal Print Cabinet. Later, Marsigli tried to get his "Voyage to London" published separately; Boerhaave offered it to the publisher Pieter van der Aa, but the latter refused it, saying that the publishers of the *Histoire* should undertake it, and this they were apparently unwilling to do.

By now Marsigli had completed his natural history purchases. The collection consisted of marine animals and their parts (103 items); birds, with nests and eggs (76 items); 20 boxes containing many specimens of butterflies, beetles and lizard skins; 26 assorted animal parts, including bezoar stones, musk and beaver glands, wasps' nests; glass jars of snakes and other reptiles,

Fig. 20 Drawings of two seaweeds, with the pods sectioned and annotated (from: *Epistola ... ab Aloysio Ferdinando Marsili, ... scripta ad eruditiss. virum Hermannum Boerhaave, ...*, Biblioteca Universitaria, Bologna, Mss. Marsili, 97)

insects and marsupials in spirit; the embalmed snakes; land and marine vegetation (47 items); minerals and fossils (100 items). It contained specimens from the Arctic and Asia, the East and West Indies, as well as Holland and England.[38] A summary of the collection stands at the end of this text.

Marsigli's date of departure and his actual route home are uncertain. Against his recorded intention to travel by sea, Quincy tells us that he wished to see the mineral waters of Aix-la-Chapelle, and the mines of Switzerland, and that he arrived in Italy via Lake Garda, which so attracted him that he returned later to make a survey of it.

Back at Bologna, with his contract signed, Marsigli now prepared the Danube study for printing, a task which, including overseeing the engraving, took the best part of a year. Fantuzzi says that Marsigli had an excellent engraver for anatomical work, and an equally good secretary, Dr Girolamo Melani, whose Latin was marvellous and who corrected Marsigli's works. Boerhaave told Sherard in November 1723 that Marsigli was 'drowning in his Danube',[21] but in May 1724 Marsigli was able to write to Newton[23] that the work which the Royal Society had encouraged so long ago, and to whom it was dedicated, was finally on its way to Amsterdam to be printed.

The natural history collection was carefully stowed into numerous packages; the books were crated up by Hyguenz & Co., and the whole consignment was shipped to Leghorn under the watchful eye of Cesare Sardi, and at the expense of the Institute.[38] When eventually his goods arrived at Leghorn Marsigli was there to watch them unloaded. He escorted them to Bologna, and when they were unpacked in his house, was relieved to find that they had survived the journey quite well. The collections were admired; they were certainly exotic, and they were also vast. It was 1727 before the legal deed of transfer was signed, and the collections finally laid out in the correct manner.[50]

While he was in Holland Marsigli kept in touch with affairs in Italy. The death of Cardinal Trionfetti, a notable botanist and long-time friend was reported to Sherard, who knew many Italian scientists. Envoys from Italy brought him news when they arrived in Holland or passed through on their way to England. He wrote to friends in Britain and elsewhere begging for specimens for himself, or for his colleagues. He listed books that might be obtained from England for his Institute, and considered the matter of their translation. Slips of paper with such lists and jottings are to be found among his papers on peat digging, draining and other records of his travels through Holland.

Similarly, long after he returned to Italy, he continued to think about what he had seen in Holland, and to deal with the outcome of his visit. In January and March 1728 he addressed letters to Pope Benedetto XIII[31] suggesting that dykes and hydraulic works on the Dutch pattern might profitably be constructed along the Reno. This advice was misplaced, because the Reno floods in spring when melting snow overloads its channel, whereas the land surface in Holland was perpetually below sea level. However, in the course of these letters Marsigli describes many of the features of Holland that had impressed him. He refers to the industrious people, whose labour and ingenuity had drained the land and made it possible to construct buildings and canals in an otherwise waterlogged landscape. He illustrates a metal scoop that he brought back to Bologna, and how the canals in Holland were dredged and cleansed annually.

In May 1726 Boerhaave sent two copies of the *Histoire* to Sherard[30] — one was to be forwarded to the Royal Society. Marsigli supplied a letter to accompany it, dated October 1725 and written from Chiozza (now Chioggia),[28] which was read before the Society on June 23,1726. Marsigli found his own copies of his books awaiting him on his return from Lake Garda and visits to M. Baldo, Chioggia and Venice.[46] He was displeased at the errors that LeClerc had overlooked in the *Histoire*: the early plates were unidentified and many of the later figures were incorrectly keyed to the text.

The manuscript of the Ottoman military establishment had been put aside while Marsigli pursued the more congenial work of surveying Lake Garda, as he wrote from there to Gagliardi in January 1725.[32] It was to be published only in 1732;[42] its manuscript text, an addition to his study of the Bosphorus, and the voyage to London and Holland were among the documents by Marsigli's bed when he died in 1730, as we learn from Giuseppe Monti's oration on Marsigli's works published in the Institute's Commentaries (tome II, 1745-1746).

* * *

*Summary of the natural history specimens sent from Holland.*[38]

*PART I* *Animals and parts of animals*

103 items or groups of items, from all parts of the world, representing whales, walrus and other amphibians, goats, elephants, turtles, fish, oysters, mussels, fish eggs, shellfish, freshwater and marine nacre and pearls, sea snakes.

*Birds from the Indies*

76 items: mostly entire specimens, nests, eggs. From the West Indies, Philipines, Cape of Good Hope, Ceylon.

*Insects from the Indies*

20 boxes containing many butterflies, beetles and other insects. Dried skins of American lizards wrapped in paper, porcupine spines.

*Bezoar stones of great value, and other things*

26 items: models of whales and their parts, musk glands from Tonquin, Bengal and Siberia, beaver glands from Siberia, Poland and Canada, armadilloes, fish, wasps' nests, snakes.

*Glass jars containing animals in balsam spirit*

10 items: snakes from America, butterflies from Guinea, male and female toads, the female giving viviparous birth. A large box containing a wood mouse from Surinam — 'it carries the fetus in a pouch and the young are carried and fed in the same pouch — a most fine and very rare object'. Rhino horn, shells.

*Serpents and other creatures in spirit in glass jars*

'bought in Amsterdam, their latin names being copied from the sale catalogue ...'
100 items: snakes, birds, shells, fish, lizards, rats, cockroach, ostrich egg, insect nest.

*PART II* *Plants and parts of them*

14 items, consisting of nine woods, palm leaves and fruit.

*Marine vegetation*

33 items: sponges, alcions, fucus, some dry, others in balsam spirit.

*PART III* *Fossils or minerals*

7 items or groups of items: lava, iron ore, glass from fused minerals.

*Petrifactions*

21 items or groups of items: marbles showing crystals, belemnite or ammonite or other fossils.

*Fish in stone*

28 items: fragments or impressions of fossil fish.

*Other fossils*

27 items: minerals.

* * *

## BIBLIOGRAPHY

Locations of unpublished primary source material are abbreviated as follows:

BL = British Library
CU = Cambridge University
MA= University of Bologna, Marsigli Archive
RS = Royal Society

*Primary sources* (i) Correspondence

1. BL SLOANE 4068 f29. London, 2 October 1701, draft, Sloane to Marsigli.
2. BL SLOANE 4046 f215. Amsterdam, 17 March 1722, Marsigli to Sloane.
3. RS SHERARD 757. Amsterdam, 17 March 1722, Marsigli to Sherard.
4. RS Journal, 1720-5, p. 243. Delft, 21 April 1722, Leeuwenhoek to RS. (Trans. J. Jurin.) Published in *Phil. Trans.* 32, (1722-3); 72-75.
5. MA FRATI 87F 37-8. [Amsterdam], nd, Marsigli to Bignon.
6. MA FRATI 87F 29v. [Amsterdam], 13 May 1722, draft, Marsigli to Derham.
7. MA FRATI 87F 29v. [Amsterdam], nd, draft, Marsigli to Innis Bros.
8. MA FRATI 87F 27. [Amsterdam], 13 May 1722, draft, Marsigli to Newton.
9. MA FRATI 87F 31. [Amsterdam], 23 May 1722, draft, Marsigli to Leewenhoofd.
10. MA FRATI 87F 32. [Amsterdam], 23 May 1722, draft, Marsigli to Baron de Goot.
11. MA FRATI 87F 28, 28v. [Amsterdam], 13 May 1722, draft, Marsigli to Woodward.
12. MA FRATI 87F 30. [Amsterdam], nd, draft, Marsigli to NN, 'Medicin du Roy'.
13. MA FRATI 87F 30v. [Amsterdam], 23 May 1722, draft, Marsigli to 's Gravesande.
14. MA FRATI 87F 27v. [Amsterdam], 13 May 1722, draft, Marsigli to Mead.
15. MA FRATI 87F 27v, 28. [Amsterdam], 13 May 1722, draft, Marsigli to Halley.
16. RS SHERARD 39. Leiden, 16 August 1722, Boerhaave to Sherard.
17. RS SHERARD 681 and 653. np, 'beginning of' November [1722], Marsigli to Sherard.
18. RS SHERARD 709. Amsterdam, 20 November 1722, Marsigli to Sherard.
19. RS SHERARD 654. Leiden, 1 January 1723, Marsigli to Sherard.
20. RS SHERARD 48. Leiden, 18 March 1723, Boerhaave to Sherard.
21. RS SHERARD 46. Leiden, 12 November 1723, Boerhaave to Sherard.
22. CU Keynes 94(E). Bologna, 29 February 1724, Marsigli to Newton.
23. CU Keynes 94(F). np, May 1724, Marsigli to Newton.
    The Marsigli-Newton letters are published in HALL, A R & TILLING, L (Eds) *Newton's Correspondence* VII, pp. 264-266, 283-286.
24. RS SHERARD 59. Leiden, 12 April 1724, Boerhaave to Sherard.
25. RS SHERARD 56. Leiden, 4 August 1724, Boerhaave to Sherard.
26. RS SHERARD 80. Leiden, 3 April 1725, Boerhaave to Sherard.
27. RS SHERARD 62. Leiden, 9 July 1725, Boerhaave to Sherard.
28. RS M2.2. Chiozza sur le Venetien, 14 October 1725, Marsigli to Sloane.
29. RS SHERARD 64. Leiden, 6 March 1726, Boerhaave to Sherard.
30. RS SHERARD 63. Leiden, 6 May 1726, Boerhaave to Sherard.
    The Boerhaave-Sherard letters are published in LINDEBOOM, G A (Ed) (1962-79) *Boerhaave's correspondence* (2 parts), Leiden, 1962-4.
31. Two letters from Marsigli to Pope Benedetto XIII: Bologna, 28 January 1728 and 24 March 1728. Published in CARDINALI, F (1829), *Nuova Raccolta d'autori*

*italiani che trattano del moto dell'acque 6*; 245-283 and 284-315 + pl. 1.
32 [Lettera 80]. Maderno, 27 January 1725, Marsigli to Gagliardi. Published in *Memorie istorico-critiche intorno all'antico stato de'cenomani ... raccolte e pubblicate dall'Abate Antonio Sambuco.* (Brescia) 1750.

*Primary sources* (ii) Other documents.

33. MARSIGLI, L F (1700) *Danubialis operis prodromus* (Nuremberg).
34. ANON. [1711] *Instrumentum Donationis Illustrissimi et Excellentissimi Viri Domini Comitis Aloysii Ferdinandi de Marsiliis ...* (Bologna).
35. MARSIGLI, L F (1711) *Brieve ristretto del saggio fisico intorno alla storia del mare &c* (Venice).
36. MARSIGLI, L F [1725] Prodromus Operis Danubialis. (Amstelodami Hagae Comitum). A copy in the British Library, *ex libris* Joseph Banks, has ms insert: 'Conditions de la souscription' (single folio, latin and french text, with list of European booksellers accepting orders).
37. DE LIMIERS, H P (1723) *Histoire de l'Académie appelée Institut des Sciences et des Arts établi à Boulogne en 1712.* (Amsterdam).
38. ANON. (1727) *Strumento di una seconda donazione fatta dall'Illustrissimo & Eccellentissimo Signor Generale Conte Luigi Ferdinando Marsili all'illustrissimo, e eccelso Senato di Bologna a favore e comodo dell'Instituto delle Scienze in essa Città eretto.* Printed at Bologna, but only copy known to me is at MA FRATI 146a.
39. ANON (1728) *Atti Legali per la fondazione dell'Instituto delle Scienze ed Arti Liberali ...* (Bologna).
40. MARSIGLI, L F (1725) *Histoire physique de la mer.* (Amsterdam).
41. MARSIGLI, L F (1726) *Danubius Pannonico-Mysicus.* (Amsterdam).
42. MARSIGLI, L F (1732) *Stato Militare dell'Imperio Ottomano.* (The Hague and Amsterdam).
43. MA FRATI 87F, 99A and 97 contain notes and sketches in Marsigli's and other hands, relating to his travels and observations in Holland.
44. Amsterdam City Archives, Not. Arch. 1829:
(a) 8 December 1722, Marsigli authorises Sardi to negotiate with the booksellers' company for the printing of the work on the Danube, and other works which he can give to them.
Amsterdam City Archives, Not. Arch. 5830:
(b) 16 March 1723, Marsigli authorises Sardi to negotiate with the booksellers' company for the printing of *Opera Danubiale.*
(c) 16 March 1723, Contract between Sardi and the booksellers, concerning the printing and publication of *Opera Danubiale.*
Contract (b) is summarised in van EEGHEN I H, *De Amsterdamse Boekhandel* (Amsterdam 1960-67) *III*; 65-66.

(iii) Biographies

45. LOVARINI, E (ed) (1930) *Autobiografia di Luigi Ferdinando Marsigli.* (Bologna).
46. QUINCY, L D C H D (1741) *Memoires sur la vie de M. le Comte de Marsigli.* (Zürich).
47. FANTUZZI, G (1770) *Memorie della vita del generale Co. L F Marsigli.* (Bologna).

*Secondary sources*

48. FRATI, L (1925-28) 'Catalogo dei manoscritti di L F Marsigli conservati nella Biblioteca Universitaria di Bologna' *La Bibliofila* (Florence) *27*, 185-215, 263-285, 370-385, 447-451; *28*, 50-56, 193-200, 291-301, 405-413; *29*, 46-54, 298-312, 404-419; *30*, (index) 120-139.
49. McCONNELL, A (1986) 'L F Marsigli's voyage to London and Holland, 1721-22' *Notes and Records of the Royal Society 41*; 39-76.
50. NEVIANI, A (1932) 'L F Marsigli e la seconda donazione del 1727' *Atti della Pontificia Accademia Romana dei Nuovi Lincei, 85*; 49-72.
51. [DUBOIS Guillaume] (1725) *Bibliotheca Duboisiana: ou catalogue de la bibliothèque du feu ... le cardinal Dubois, receuillie çi devant par ... l'Abbé Bignon.* 4 parts (La Haye).
52. van BERKEL, K 'Intellectuals against Leeuwenhoek' in PALM, L C & SNELDERS, H A M (1982) *Antoni van Leeuwenhoek, 1632-1723* (Amsterdam); 187-209.
53. FAHRENHEIT, D G (1724) 'Experimenta et Observationes de congelatio ...' *Phil. Trans. 33*; 78-84.

*Biographical Notes on Marsigli's contacts*

BOERHAAVE, Herman, 1668-1738. Professor of chemistry, botany & medicine, and clinical practice at Leyden, and curator of its Botanical Garden. (Elected FRS in 1730.)

DERHAM, William, FRS, 1657-1735. Held various ecclesiastic posts.

GRAVESANDE, William Jacob 's, FRS, 1688-1742. Had spent a year in England. Professor of mathematics at Leyden.

HALLEY, Edmond, FRS, 1656-1742. Astronomer and geographer. Secretary RS.

INNIS, John and William. Booksellers. Printers and publishers to Newton and the Royal Society.

LEEUWENHOEK, Antoni van, FRS, 1632-1723. Devised microscopes for his study of micro-organisms. His many visitors included royalty.

MEAD, Richard, FRS, 1673-1754. Physician. On Royal Society Council.

NEWTON, Isaac, FRS, 1642-1727. Mathematician. President of the Royal Society.

RUYSCH, Frederik, FRS, 1638-1731. Professor of anatomy and botany at Amsterdam and noted for his anatomy and natural history collections.
SARDI, Cesare, ? — ?. Originally from Lucca, Sardi and his descendants were bankers and agents in Amsterdam.
SHERARD, William, FRS, 1659-1728. Studied in Europe and travelled widely throughout Europe and Near East. Close friend of Boerhaave.
SLOANE, Hans, FRS, 1660-1753. Physician. Studied botany in France, founder of the Chelsea Physic Garden. His collections became the British Museum.
WOODWARD, John, FRS, 1665-1728. Geologist and physician.

* * *

*Acknowledgements*

I am grateful for permission to consult manuscripts at the Royal Society, the University Library of Bologna, and Amsterdam City Archives, and would like to thank their staff, and those of the Science Museum Library in London, for patient help and advice. I am also indebted to Mrs van Eeghen, Mrs Carpine-Lancre and Dr John Stoye for their assistance.

# TRACTIONAL MOTION 1692-1767
# THE JOURNEY OF A MATHEMATICAL THEME FROM HOLLAND TO ITALY[1]

*H.J.M. Bos*

§ 1. In 1765-67 the Italian mathematicians Vincenzo Riccati and Girolamo Saladini published a textbook on mathematical analysis entitled *Institutiones analyticae.*[2] In its third volume, the book contained a 60 page section on what was called the "construction of differential equations by means of a tractional curve."[3] Riccati was responsible for that section; the subject had captured his interest already in 1750. In 1752 he had published a monograph about it, entitled "On the use of tractional motion in constructing differential equations,"[4] and in 1753 the Florentine periodical *Symbolae Litterariae Opuscula Varia* carried a letter on tractional motion which Riccati had written to Maria Gaetana Agnesi in 1750.[5] I take these publications of Riccati as the end of the journey through mathematics of a curious theme: tractional motion.

Tractional motion[6] occurs when a heavy object is dragged over a horizontal resisting surface by a cord (whose motion is frictionless) of constant or variable length, the free end of which is moved along a straight line or a curve in the plane (see fig. 21). It is supposed that, because of the friction, the direction of the motion is always along the cord. The curve along which the free end of the cord is moved is called the base. The path described by the heavy object is called a tractional curve. If the end Q of a cord PQ is move along a

1. In the present article I give a non technical survey of the story of tractional motion, leaving out many of the mathematical details. In my article [Bos 1988] I have dealt more extensively with the early part of the story (till c. 1710). I hope to give a detailed description of the later part of the story elsewhere.

2. [Riccati-Saladini 1765].

3. Chapters 14-16 of book 2 of part II, vol. 3, pp. 470-531. Chapter 14 is called "Quaenam aequationes construi possunt ope tractoriae praedita tangente constante, methodo directe investigatur"; chapter 15 deals with the corresponding "inverse method" and chapter 16 gives the solution of some further inverse problems.

4. [Riccati 1752].

5. [Riccati 1753], the letter was dated Nov. 3, 1750.

6. For a more detailed explanation see [Bos 1988] pp. 10-12.

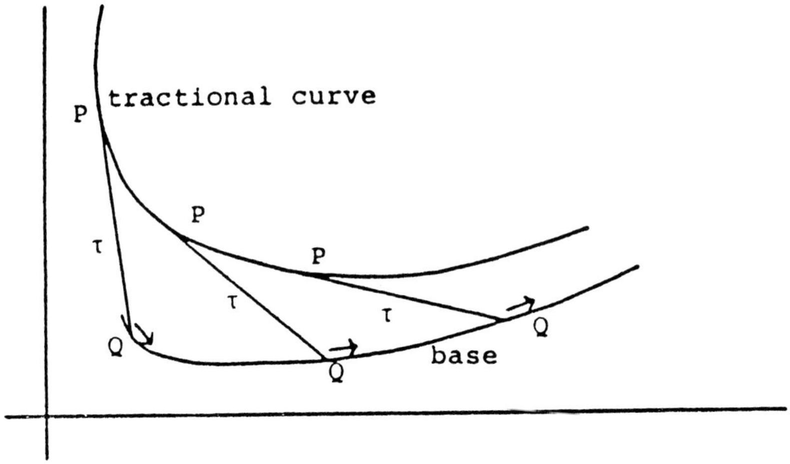


Fig. 21 **Tractional Motion.**

straight or curved base as QQQ, then the other end P of the cord (where the heavy body is assumed to be) describes the pertaining tractional curve PPP which has the property that in each of its points the cord is tangent to it.

Riccati studied the use of tractional motion in solving differential equations. Differential equations involved infinitely small quantities, the so-called differentials; such equations were at the centre of the mathematical interest in the eighteenth century. The creation of the differential and integral calculus had enabled scientists to translate mechanical problems about motion and deformation in mathematical terms. In most cases the result of such a translation was a differential equation whose solution provided the description of the mechanical process that was studied. The majority of the mathematical investigations in that period had the ultimate purpose of solving differential equations. Differential equations were problems; their solutions were functions or curves. Eighteenth century mathematicians endeavoured to solve these differential equations by deriving the equations of the solution curves or the formulas of the solution functions. Being solutions, these equations or formulas should no longer involve differentials, they should be "normal" equations concerning finite quantities only. However, mathematicians encountered types of differential equations whose solutions they could not express in such finite equations or formulas. Riccati's study concerned certain such types. He argued that, if finite equations for the solutions could not be found, the solution curves might perhaps be traced by tractional motion. To be more specific, he suggested a method for tracing the solution curves of certain classes of differential equations, and that method involved tracing auxiliary curves by tractional motion. The details of his procedure are mathematically too technical to be discussed here.

§ 2. Seemingly Riccati offered instrumental solutions of problems that could not be solved directly in a mathematical way — a timehonored procedure in applied mathematics, witness a rich history of computational instruments from abacus, via astrolabe, slide rule and mechanical integrator to the electronic computer. But did Riccati seriously consider tractional motion as a practicable means of solving differential equations? His terminology suggested actual execution of the curve tracing process, and he even referred to instruments that had earlier been made for tracing tractional curves.[7] On the other hand, in the majority of his examples the base curve was so complicated that, although its equation was known, it was hardly practicable to draw it — and tractional motion of course presupposed the base to be drawn. Despite the imagery of movement and machines, Riccati's approach has to be understood as belonging to abstract mathematics. He considered the abstract physical process of tractional motion as a solution of certain problems in pure mathematics, namely a certain class of differential equations. In other words, he considered tractional motion as a *means of construction* in abstract mathematics in the same way as the use of compass and straightedge is considered as means of construction in pure geometry; the mechanical imagery of "tracing", "using the compass", etc. serves as metaphoric description of abstract processes. The essential construction occurs not on paper but, if anywhere, in the mind of the mathematician.

This, then, is the theme whose journey through time and place I want to follow: the use of tractional motion as an abstract means of construction in solving problems equivalent to differential equations. I find it an intriguing theme. One would expect tractional motion to be seen as a problem in mechanics: what is the curve traced by such and such a process of dragging an object? That mechanical problem would, if translated into mathematical terms, lead to a differential equation, which in its turn might be solved; its mathematical solution providing the answer to the original mechanical problem. Here, however, by a notable 'Gestalt' switch, the roles are reversed: tractional motion, rather than being a problem itself, serves as a (means of) solution. Besides this unusual role division between mathematics and mechanics, solution and problem, my theme also concerns a remarkable

7. [Riccati 1752] pp. 28 and 42. The references were to Poleni's instruments which I discuss below.

interplay of geometrical procedures (construction of lines within figures) and analytical ones (derivation of equations). We see Riccati accepting an abstract mechanical process in a construction that should serve as an alternative for such a convincingly satisfying analytical procedure as writing down the equation of the solution.

In following the theme's journey from Holland to Italy I will try to show how and why tractional motion for some time acquired such a special status in mathematics.

§ 3. Riccati's work constituted, as far as I know, the theme's journey's end — no mathematician after him seems to have studied tractional motion as an abstract theoretical procedure for solving differential equations. The theme started its journey in Holland in 1692. In November of that year Christiaan Huygens studied tractional motion and wrote down his thoughts, calculations, designs and experimental results on nine manuscript folios which he gave the general title

> The quadrature of the hyperbola by a new quadratrix of it, which is traced in one single motion.[8]

The title reflects a structure of argument similar to Riccati's: a mathematical problem, the quadrature of the hyperbola, is solved, constructed, by means of a geometrical object, its quadratrix, which is a curve that in its turn is obtained by a mechanical process, namely tractional motion.

It appears from the manuscripts that Huygens wanted to design and make an instrument that accurately traced the simplest kind of tractional curve. This was the curve that arose when the base was a straight line and the cord had constant length (see fig. 22). Huygens called it the *Tractoria*; the curve has become generally known under the name *Tractrix*. It has the property that at each of its points the distance along the tangent from the point to the base is constant.[9] Figure 23 shows the design of a tractrix instrument which ultimately satisfied Huygens best. A block B was moved along the straight rim (serving as the base) of a horizontal surface. The beam DC was connected to B by a pin at

8. [Huygens ms 1692], cf. [Bos 1988] pp. 25-32.

9. With the X-axis along the base and the initial position of the cord of length a along the Y-axis, the tractrix has the equation

$$x = a \log [(a+\sqrt{a^2-y^2})/y] - \sqrt{a^2-y^2}.$$

The curve enjoys fame in mathematics not so much for being the simplest tractional curve as for the fact, discovered by Beltrami, that its surface of revolution around the X-axis provides a model for the hyperbolic type of non-Euclidean geometry.

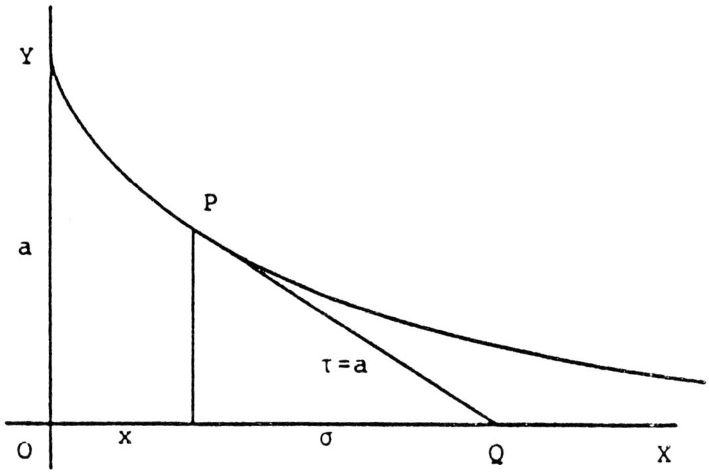


Fig. 22 The Tractrix.

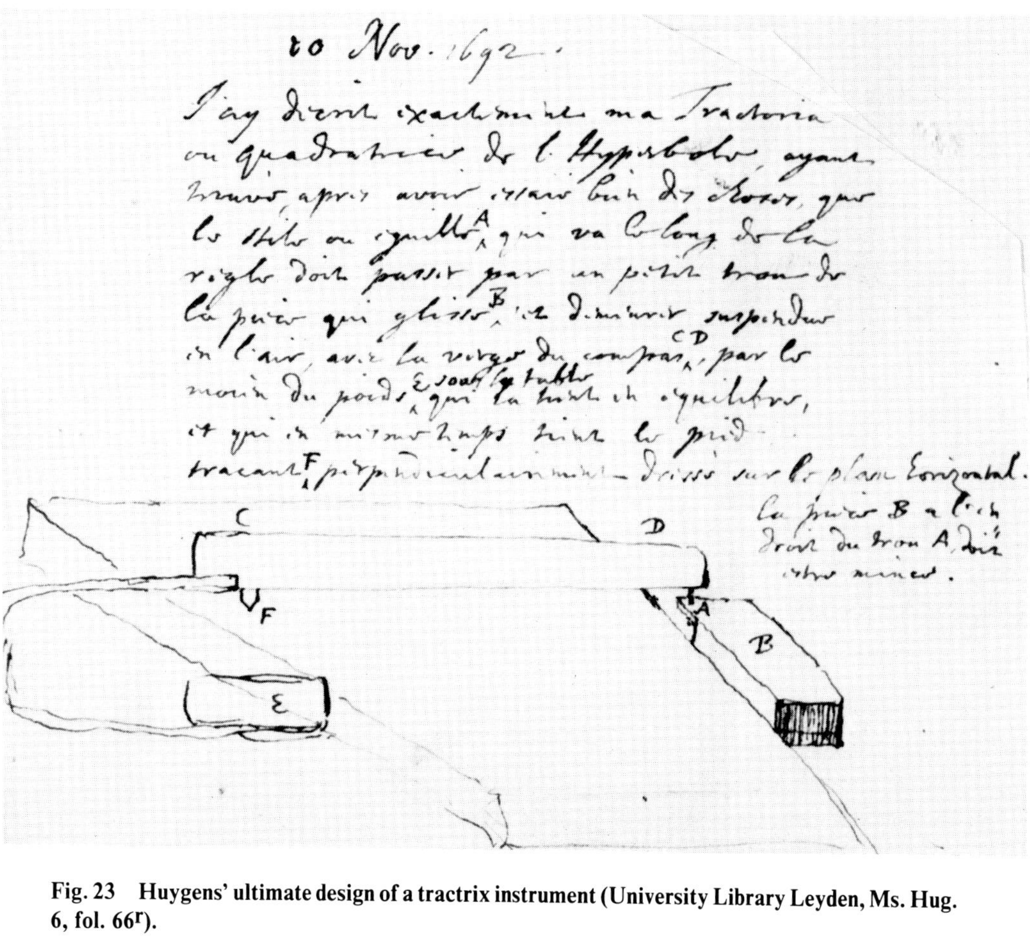


Fig. 23 Huygens' ultimate design of a tractrix instrument (University Library Leyden, Ms. Hug. 6, fol. 66r).

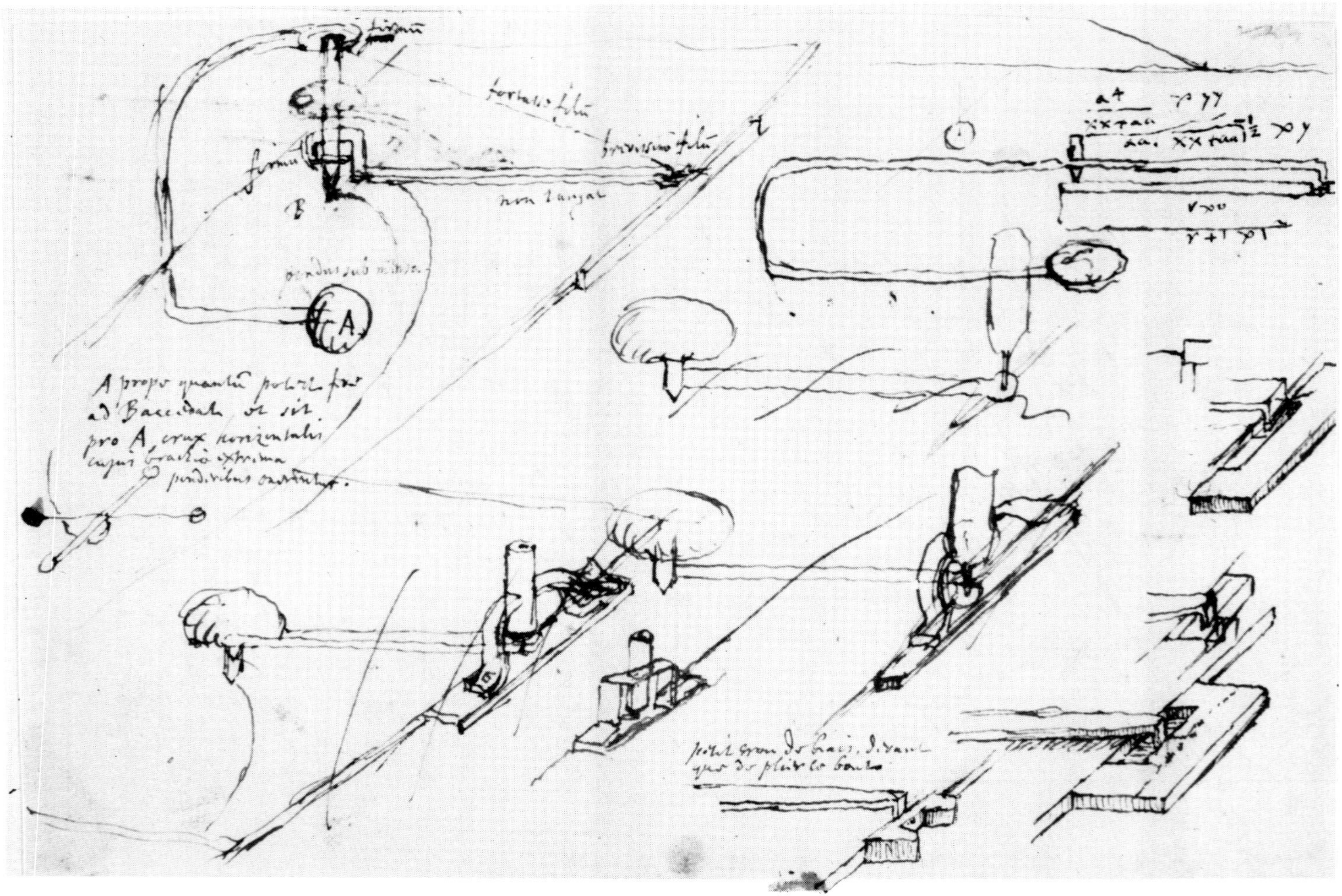

Fig. 24 Sketches for a tractrix instrument, from Huygens' manuscripts (University Library Leyden, Ms. Hug. 6, fol. 59r).

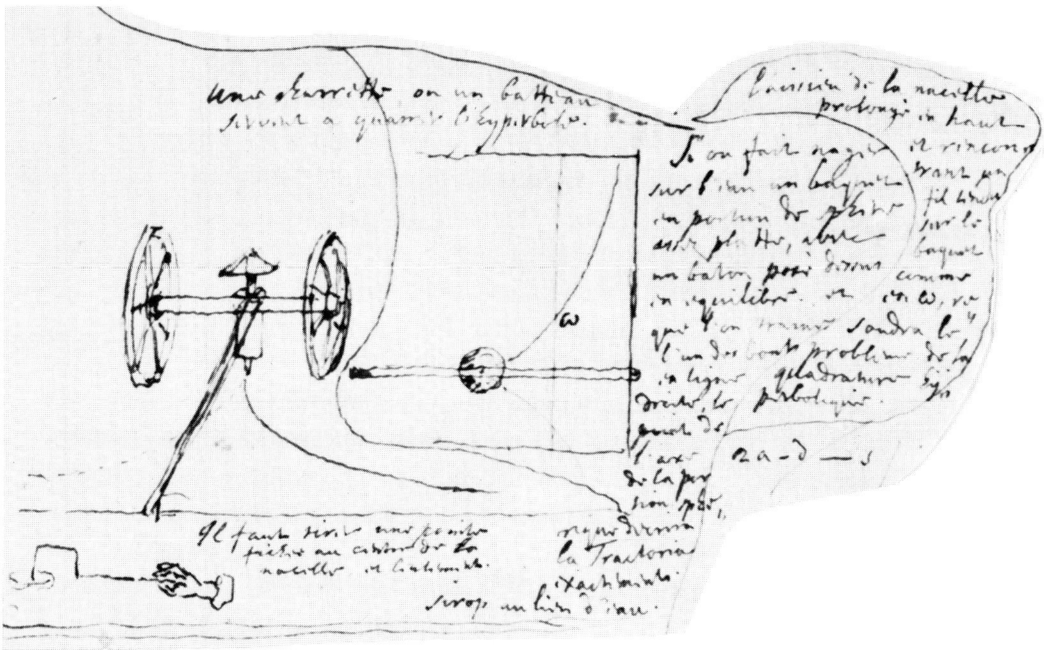

**Fig. 25 Variant designs of tractional instruments, from Huygens' manuscripts (University Library Leyden, Ms. Hug. 6, fol.64r).**

A around which it could turn without friction. A pin F was placed at its other end; it traced the tractrix while B was moved along the rim. A weight E was connected with the beam by a U-shaped lamina in such a way that the centre of gravity of the whole system was below F, thus ensuring sufficient pressure as well as a vertical position of the tracing pin. In fig. 24 we see how Huygens also considered a weight on top of the tracing pin. Figure 25 shows variant designs which Huygens considered as well. One was a two-wheeled cart that was to be drawn over a horizontal surface tracing the curve by a pin between its wheels. The small drawing underneath related to the possibility of tracing by drawing a floating object. Huygens seriously considered this possibility although he finally abandoned it. The fluid would automatically ensure a horizontal position of the surface and friction could be increased by using syrup instead of water. The curve itself would be marked along a horizontal plane above the fluid's surface by a vertical pin fixed to the floating object.

§ 4. Although, as apppears from these figures and from several passages in the manuscripts, Huygens' interest in tractional motion had a decidedly technical bent, his main motive was not technical, not even theoretically mechanical, but abstractly mathematical. His ultimate aim in these studies was to overcome the Cartesian restrictive demarcation of geometry.

According to Descartes, curves belonged to geometry only if they had an algebraic equation, that is, an equation involving no other operations than addition, subtraction, multiplication, division and the extraction of roots. Other curves, for instance those involving logarithmic or trigonometric relations, were merely mechanical and had no place in geometry. I shall not here go into the question why and how Descartes had come to this conviction.[10] Suffice it to say that during the second half of the seventeenth century his work was so influential that all mathematicians were aware of his demarcation of geometry and that many felt compelled to justify their procedures if they found themselves crossing the boundaries which Descartes had prescribed. Also it should be noted that when Descartes first formulated his demarcation of geometry it was not a restriction but rather an extension with respect to earlier conceptions of geometry in which the restriction to straightedge and compass constructions induced limitations much stricter than Descartes'.

Now the tractrix does not have an algebraic equation, and so, from the Cartesian point of view, it did not qualify as an acceptable geometrical curve. But Huygens wanted to accept the curve, he wanted to study it and to use it as a "quadratrix" of the hyperbola, that is, a curve by means of which the quadrature of the hyperbola could be solved. The quadrature of the hyperbola was a problem much discussed since the middle of the seventeenth century. It is equivalent to determining to logarithmic function or the logarithmic curve.[11] This curve has no algebraic equation either and therefore it was not an acceptable curve within the Cartesian view of geometry. Huygens had found that, once one had the tractrix, the logarithmic curve could easily be constructed. Hence if the tractrix could be legitimated, the problem of the quadrature of the hyperbola would be solved.

What Huygens was doing in his manuscripts on tractional motion was to explore arguments to justify the tractrix contra Descartes; in other words he

10. See [Bos 1988] pp. 15-19; compare [Bos 1981] and [Bos 1989].

11. Quadrature of a curve meant the determination of the area bounded by the curve, the axis, a fixed ordinate and a variable ordinate. Taking $y = 1/x$ as the equation of the hyperbola, the quadrature z (fixed ordinate at $x = 1$) can be expressed as

$$\int_1^x dx/x = \log x.$$

Thus the quadrature is given by the equation $z = \log x$, or equivalently $x = e^z$. The curve expressed by these (non algebraic) equations is the logarithmic curve.

searched for a legitimation of the tractrix as a fully acceptable geometrical curve. This motivation is clear in many passages of the manuscripts. For instance:

> It should be admitted that, if my curve is supposed or given, one has the quadrature of the hyperbola. So if I find some means to draw it as exactly as a circle is drawn by an ordinary compass, would I not have found that quadrature? (...) It is true that I need the parallelism of a plane to the horizon; but that is possible, not in utmost precision, but like the straightness of a ruler. For the rest I draw my curve almost as easily as a circle and the machine I use comes very close to thc simplicity of the compass.[12]

And:

> Descartes was wrong when he dismissed from his geometry those curves whose nature he could not express by an equation. It would have been better if he had acknowledged that his geometry was defective in so far as it did not extend to the treatment of these curves; for he was well aware that the properties and uses of such curves can also be investigated by geometrical methods.[13]

In an article on the tractrix which Huygens published in 1693 he wrote about the tracing of the curve:

> If this description, which by the laws of mechanics must be exact, could pass for geometrical, in the same way as those of the conic sections, performed by instruments, one would thereby have the quadrature of the hyperbola, and together with that the perfect construction of all the problems that can be reduced to that quadrature such as among others the determination of points on the *Catenaria* or Catenary, and the logarithms.[14]

12. "On doit avouer que ma courbe estant supposée ou donnée, on a la quadrature de l'Hyperbole. Si je trouve donc quelque moien de la decrire aussi exactement qu'avec un compas ordinaire on decrit un cercle, n'auray je pas trouvé quadrature? (...) Il est vray que j'ay besoin du parallelisme d'un plan a l'horizon; mais cela est possible, non pas dans la derniere justesse, mais comme la droiture d'une regle. Pour le reste je decris ma courbe presque aussi facilement qu'un cercle et la machine que j'emploie approche fort la simplicité du compas." Huygens [ms 1692] fol. $62^r$, [Oeuvres] vol. 10, p. 412, note. (Here, as throughout the present article, the translations are mine.)

13. "Non bene Cartesius e geometria sua rejiciebat curvas quarum naturam aequatione exprimere non poterat. Melius agnovisset geometriam suam hac parte mancam quod ad earum tractationem non attingeret non enim nesciebat talium quoque curvarum proprietates atque usus geometricis rationibus investigari." [mss 1692] fol. $60^v$.

14. "Si cette description, qui par les loix de la Mechanique doit être exacte, pouvoit passer pour Geometrique, de même que celles des sections de Cone qui se font par les instrumens l'on auroit par elle, avec la quadrature de l'Hyperbole, la construction parfaite des Problemes qui se reduisent à cette quadrature; comme sont entre autres la determination des points de la *Catenaria*, ou Chainette, et les Logarithmes." [Huygens 1693a] pp. 409-412. The catenary is the curve formed by a chain or cord suspended from two points; its mathematical description was a famous problem at the time.

From the quotations it is clear how Huygens tried to legitimize the tractrix: he wanted to show that instruments could be devised, and actually built, that traced the tractrix as precisely as compasses or some other then familiar instruments traced circles or conic sections. Thus his legitimatory criterion was practical feasibility. Still in a deeper sense his argument concerned pure abstract geometry rather than practical precision. Indeed if mere practical precision was at stake, Huygens had better means at his disposal: he could use logarithm tables, which were then available in precision of ten digits, to calculate coordinates of points of the tractrix and he could then trace the tractrix through those points with a much more precise result than any instrument could give. Significantly, he did not do so; he relied on the tracing instrument for the legitimation. Thus we can characterise his approach to the justification question as idealization of practical geometrical precision with emphasis on idealization.

So at the beginning of its journey through pure mathematics, tractional motion was invoked and used with a special purpose: legitimation. Tractional motion was precise, it traced the tractrix, therefore the tractrix was a legitimate geometrical curve. Soon others took up the theme and added a new but related purpose, namely construction.

§ 5. Huygens published some results of his studies on tractional motion in an article that appeared in 1693.[15] He explained the mathematical properties of the tractrix and he expounded the idea that if the tractrix could be traced by acceptable motion, then the quadrature of the hyperbola could be solved in a legitimately geometrical manner by means of the tractrix. He did not, however, explain the instrument he had himself designed for tracing that curve.

Huygens' idea was immediately taken up by Leibniz, Johann and Jakob Bernoulli, and l'Hôpital. They explored the following argument.[16] If tractional motion, by its precision in principle, provides valid argument for justifying and legitimating the tractrix, then in fact a new means of construction is introduced in the geometry of curves. That is, we may conceive curves as known when we can indicate a tractional motion process by which they are traced.

Leibniz, the Bernoulli's and l'Hôpital were at that time engaged in developing the recently created Leibnizian version of the differential and integral calculus. By means of that calculus processes of motion could be

15. [Huygens 1693a].
16. For a more detailed discussion of this episode see [Bos 1988] pp. 32-52.

described mathematically by differential equations. These differential equations were still conceived as primarily geometrical problems, their solution therefore required a geometrical construction of the solution curve. Now if tractional motion was an acceptable means of construction, one could study several classes of such motions, derive the pertaining differential equations and classify these as constructible by tractional motion. If then, later, in dealing with some mechanical or other problem, one would encounter such a differential equation one would know that its solution could be constructed by tractional motion.

Several articles[17] with this motivation appeared in the *Acta Eruditorum* in 1693 and 1694. They centred on a problem proposed by the Bernoulli's (it became known as the "Bernoulli problem"). It concerned curves traceable by one particular kind of tractional motion, and the pertaining differential equation was one of the most complicated that were studied at all in the early period of the calculus.[18]

The interest in the subject offered Leibniz the opportunity to testify to an earlier origin of tractional curves; in an article of 1693 he told how in the early 1670's Perrault used to ask mathematicians in Parisian salons what curve was traced by his watch if he moved the end of its chain along the rim of the table and let the watch on the table follow that movement.[19] Leibniz wrote that at the time he had solved the problem (which meant that he had found a construction for the tractrix curve, because that was, obviously, the curve traced by Perrault's watch) but that he had not published it. Nevertheless I have taken Huygens' studies of 1692 rather than Perrault's question of c. 1672 as the origin of the theme of tractional motion, because I take it as essential for the theme that tractional motion was actually accepted as a means of construction. There are no signs that this was the case with Perrault's question nor with Leibniz' solution of it. Figure 26 shows a drawing from a later source on tractional motion;[20] the drawing is clearly a reference to Leibniz' story about Perrault.

§ 6. Thus legitimation and geometrical construction were the motives behind the earliest appearances of our theme. These motives were appropriate

17. [Jakob Bernoulli 1693], [Jakob Bernoulli 1696], [Johann Bernoulli 1693], [l'Hôpital 1693a], [l'Hôpital 1693b], [l'Hôpital 1694], [Huygens 1693b], [Huygens 1694a], [Huygens 1694b], [Leibniz 1693a], [Leibniz 1693b], [Leibniz 1693c], [Leibniz 1694].

18. Namely $yds = p(xdy—ydx)$; cf. my article [1988], pp. 32-43.

19. [Leibniz 1693b], pp. 296-297; This was Claude Perrault (1613-1688); his brother Charles Perrault (1628-1703) became famous as the author of the Mother Goose fairy tales.

20. [Poleni 1729], Plate BB, Figure 3.

Fig. 26 Tracing a tractrix by means of a chain watch, from [Poleni 1729].

and understandable around 1700. At that time, the Cartesian view of geometry was still so strong that the use of non-algebraic curves needed a legitimation. Also the conception of solution of problems was still predominantly geometrical, so a problem like the quadrature of the hyperbola required a construction for its solution as a matter of course.

In the period after 1700 several processes occurred within mathematics that made the use of tractional motion for the purposes above obsolete — so obsolete indeed that it is something of a surprise to find them still as late as 1767 in the writings of Riccati. These processes fall under the general heading of the emancipation of analysis from geometry. Analytical methods, in the sense of calculation and manipulation with formulas, were introduced and elaborated in the seventeenth century in connection with the study of geometrical problems. But they gradually acquired independent meaning and mathematicians tended more and more to disregard their geometrical interpretation or motivation. At the same time a gradual habituation to new material occurred; mathematicians became used to the new curves such as the tractrix, and thereby they no longer felt so strongly the need for a separate legitimation of these curves. One important factor in this habituation process was that new notations were introduced that enabled mathematicians to write

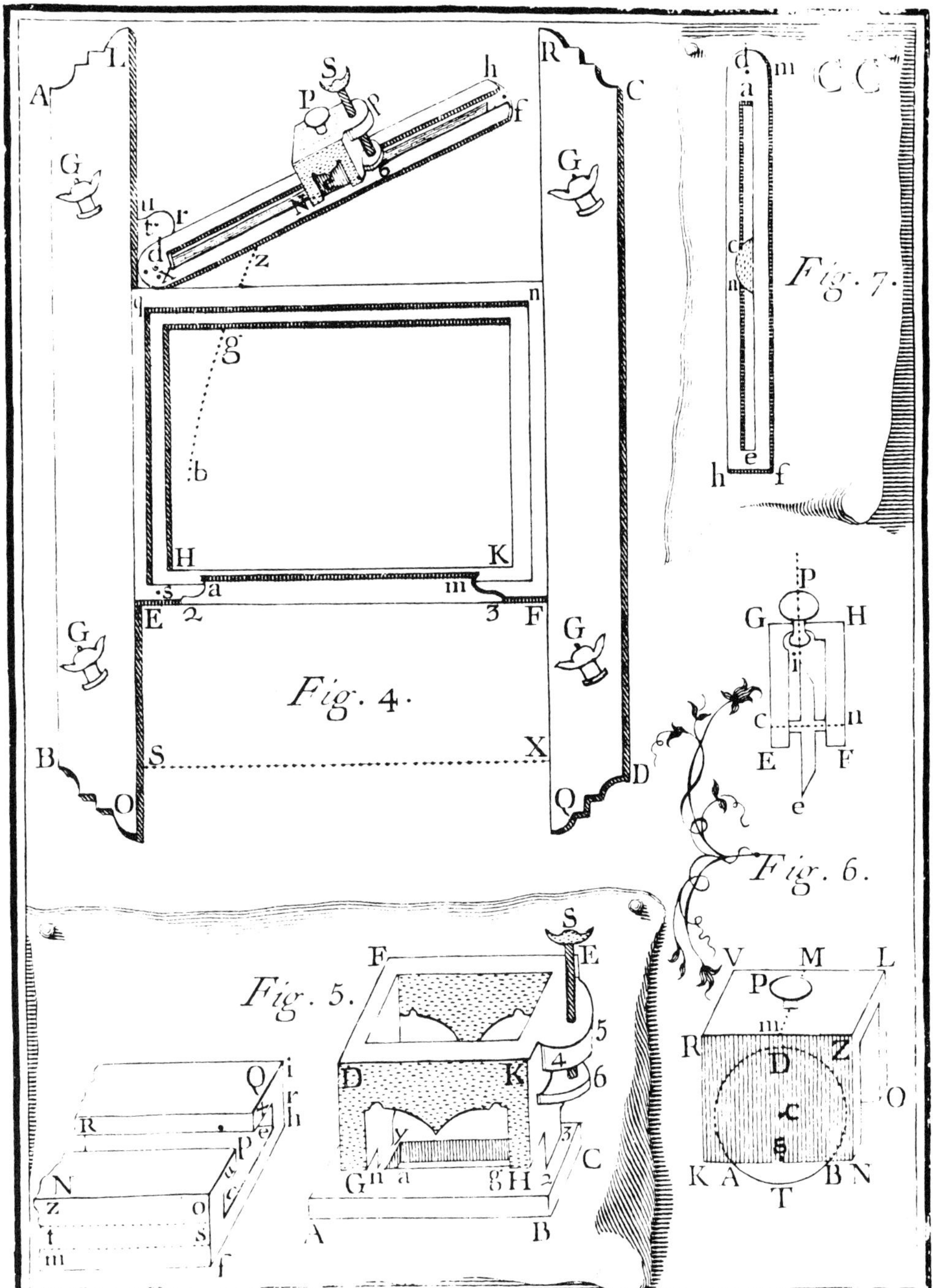


**Fig. 27 Poleni's instrument for tracing the tractrix, from [Poleni 1729].**

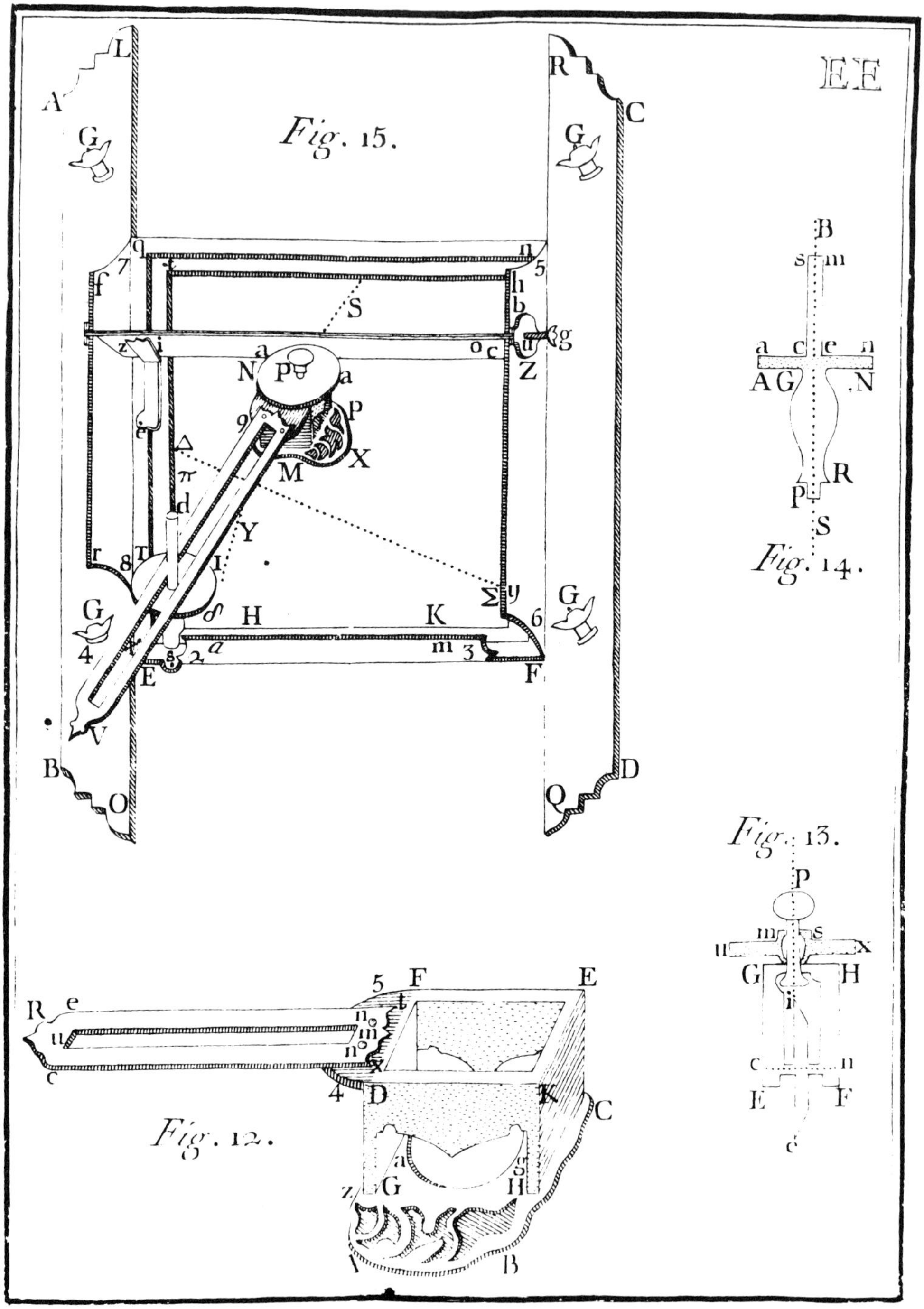

Fig. 28 Poleni's instrument for tracing the logarithmic curve, from [Poleni 1729].

down equations for these new curves. Under the influence of these processes the earlier purposes of the use of tractional motion within pure mathematics became obsolete. Why, then, did the theme of tractional motion stay alive so long? and how, having reached Germany, Switzerland and France, did the theme continue its journey from Holland to Italy amidst a general decline of its interest? In the sequel I shall briefly sketch the answers to these two questions.

§ 7. In the period 1710-1730 our theme made three appearances in mathematical literature. Three authors from different countries independently discussed tractional motion. In each case the motivation was legitimation. These writers were: a Monsieur Bomie in France, who published an article on tractional motion in 1714,[21] John Perks in England with an article in 1715,[22] and G. Poleni in Italy in 1729.[23] The first two authors did little more than to explain a design of an instrument that could trace a tractrix or related curves and to explain the pertaining mathematics.

The case of Poleni is somewhat more interesting. Inspired by Huygens' article (apparently he was not aware of Bomie's and Perks' work) he set out to construct precision instruments to trace the tractrix and the logarithmic curve. Figures 27 and 28 show the plates of the two instruments in Poleni's publication.[24] Before publication, however, he sent copies of the machines to three famous Italian mathematicians, namely Antonio Conti, Gabriele Manfredi and Jacopo Riccati (the father of the Vincenzo whom we have already met). In his covering letter he claimed, in the same way as Huygens, that by making his precision instrument he had finally geometrically legitimized the tractrix, the logarithmic curve, and thereby all other curves that could be constructed from these by classical means. He asked his cor-

21. [Bomie 1714] cf. [Fontenelle 1714]; I have not found any biographical information about Bomie.

22. [Perks 1715]; on Perks see [Pedersen 1963].

23. [Poleni 1729].

24. [Poleni 1729] plates CC and EE. In the tractrix instrument (fig. 27) the frame EFnq is moved between the rulers LO and RQ in the direction qE. The lamina dxfh, which can turn freely around a pin at d, follows the frame. At an adjustable distance dP from d, a wheel can be fixed between the parts of the lamina. The wheel rests on the paper and traces the curve, which has the property that at each of its points the lamina dP is along its tangent — so it is the tractrix (with tangent length dP and axis ds). The design of the second instrument (fig. 28) is based on a property of the logarithmic curve, namely that at each of its points the *subtangent* has the same length. The subtangent is the segment along the axis from the ordinate to the intersection of the tangent. The instrument employs again a frame moving between rulers. Furthermore the lamina VM and the casket PM housing the tracing wheel are now designed in such a way that during the motion the direction of the wheel is towards d, so that the tangent of the curve is along Md, whereas the ruler zZ, which pushes the casket, ensures the subtangent zd to be constant.

respondents to test the instruments and to report their findings. They did so, and Poleni incorporated their answers in his publication in 1729. That publication consisted of a letter to Jakob Hermann, in which Poleni described the instruments and their use, the letters of Conti, Manfredi and Riccati, and a treatise on the mathematical properties of the tractrix.

Conti and Manfredi were generally positive in their reaction, but non-committal as regards the legitimatory purpose of Poleni's instrument. Riccati took the trouble to go into the question of legitimation more deeply. He complimented Poleni on the accuracy of the instrument but rejected the claim that practical accuracy was in any way an argument in support of the geometrical legitimacy of the curve. He also rejected Descartes' restriction of geometry to algebraic curves and claimed that, because pure geometry concerned contemplation rather than practice, legitimation could not be based on the precision of instruments. Rather, criteria for accepting or rejecting geometrical entities (as curves) or procedures ( as constructions) should relate to pure contemplation, not to practice. He mentioned in particular abstract existence and non-contradiction as criteria — a remarkably modern point of view.[25]

Thus we find in the period 1710-1730 three mathematicians, Bomie, Perks and Poleni, still interested in legitimating tractional curves in defiance of the Cartesian demarcation of geometry. By the time of publication this interest was dated and it is therefore not surprising that our three authors were of decidedly less mathematical calibre that those that were interested in tractional motion around 1700. In fact, Jacopo Riccati voiced the opinion of those who were in contact with up to date mathematical thinking: the problem of legitimation was no longer urgent.

§ 8. Although the question of legitimation had lost its urgency, the theme of tractional motion did recur in mathematics, introduced this time by one whose rank as mathematician was unquestionable: Leonhard Euler. In 1741 Euler published an article called

25. Cf. "Munus est Geometrae investigare et expendere Curvarum naturam; ideam illarum exprimere distinctam; idque palam facere, qua ratione progignantur; denique ex earum origine commonstrare, eas non contradicentium numero, sed ejusmodi rerum, quae fieri possunt, contineri. Ac quamdiu proprietatibus illarum utitur, ut propositis quaestionibus difficilioribus satisfaciat, res versatur ipsa in contemplatione sola: cum vero hinc ad usum operamque rem traducit, Mechanica saepissime opus est, et iis instrumentis, quae ad manum pertinent." [Poleni 1729] letter of Jacopo Riccati.

> "On the construction of equations by means of tractional motion and other matters concerning the inverse method of tangents."[26]

Euler was the giant of eighteenth-century analysis, the mathematical science that had emerged through the emancipation from geometry of such seventeenth century methods as analytic geometry and the early versions of the differential and integral calculus. Euler's mathematics, to put it briefly, was about formulas, not about figures. And so in dealing with differential equations his approach was to try and find the formulas for the functions that were the solutions of the differential equations, or the equations of the solution curves. The usual way to do so involved a technique called "separation of the variables". Until c. 1740 this was practically the only way to find the equations of the curves or the formulas for the functions. In his article Euler dealt with a class of differential equations for which separation of variables failed. And for all the primacy of analytical methods in Euler's style of mathematics, when analysis failed, when no formulas could be found, Euler reached back to earlier geometrical canons of solution, namely construction. In the article he dealt with a special class of differential equations[27] which did not admit separation of variables. Therefore he wrote:

> its construction by means of tractional motion merits our attention. Moreover, as that construction is rather simple and easy, it will be worthwhile to reduce the construction of such a difficult equation to tractional motion.[28]

Again, the format of this article does not allow the presentation of the technical mathematical details of Euler's study. Suffice it to say that for a considerable class of differential equations he described a procedure, involving tractional motion, by which the solution curves could be traced.[29] Euler did not present his use of tractional motion as feasible in practice, nor

26. [Euler 1741].

27. They were the differential equations of the form $ds + ssdz = Z(z)dz$. The so-called Riccati differential equation (named after Jacopo Riccati who had published about it in 1724 — not in connection with tractional motion) belongs to this class. About that equation it was known that only in special cases could the variables be separated.

28. "eius constructio ope tractorii motus attentionem meretur. Quae constructio cum sit praeterea admodum simplex et facilis, operae pretium erit aequationis tam difficilis constructionem ad motum tractorium reduxisse" [Euler 1741, p. 84].

29. Euler's construction of the solutions of the differential equation $ds = Z(z)dz - s^2dz$ was as follows (see the Figure): Draw a curve BN with coordinates (expressed in terms of the parameter z) $u = QN = \frac{1}{2} b \log Z$ and $t = AQ = 2b/\sqrt{(Z(z))}dz$, where b is an arbitrary constant (and the logarithm and the integral are assumed to be constructable!). With this curve as *base* apply tractional motion (cord length b) to trace a new curve AM. Call $w(z) = \angle MNQ$ and $q(z) = \tan\frac{1}{2} w(z)$. Then $s = q(z)\sqrt{(Z(z))}$ is the solution of the differential equation.

did he attach any legitimatory force to it — for him it was exclusively a means of abstract geometrical construction, to be invoked when analytical methods of representing the solutions failed.

§ 9. In its open recourse to, by then rather dated, geometrical methods, Euler's article stood outside the main stream of mathematics. Still, others acknowledged and used it. For instance, when Clairaut in an article published in 1745 dealt with a problem in mechanics and encountered a differential equation of the type Euler had studied, he explained that it was

> ... famous among geometers who have applied themselves to the integral calculus, and nobody has yet been able to separate its indeterminates except in a small number of very special cases. Mr Euler is the only one, as far as I know, who has given a general construction of the equation, and although that construction does not have all the advantages of those that are based on the separation of the indeterminates, it nevertheless is worthy of its learned author. As the construction is very little known among geometers I shall give it here.[30]

And he spelled out Euler's construction.

Clairaut did not provide the proof that the solution gained by Euler's construction procedure using tractional motion actually satisfied the differential equation, and this omission may have caused the final move in the journey of our theme. For, as Vincenzo Riccati explained in the preface of his

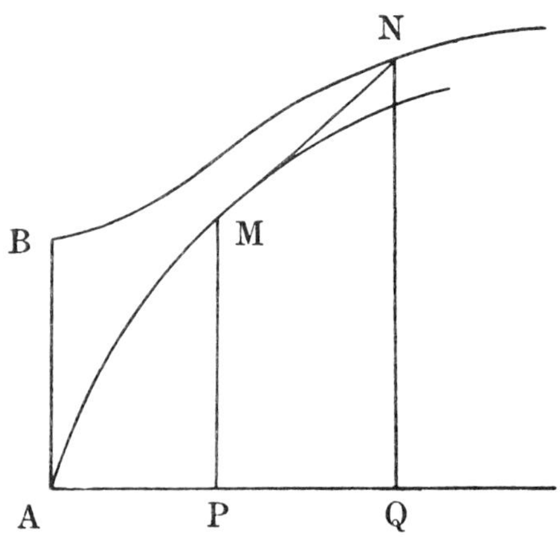


30. "fameuse parmi les Géomètres qui se sont appliquez au Calcul Intégral, et dont personne n'a pu séparer les indéterminées, que dans un petit nombre de cas très-particuliers. M. Euler est le seul, que je sçache qui en ait donné une construction générale; et quoique cette construction n'ait pas tout l'avantage de celles qui sont fondées sur la séparation des indéterminées, elle est cependent digne de son sçavant Auteur. Comme elle est très-peu connue des Géomètres je le mettrai ici." [Clairaut 1745] p. 9.

1752 monograph, he had hit upon Clairaut's article and, because no proof was given, he tried to find the proof, and succeeded (apparently Riccati had no access to Euler's original article to which Clairaut did refer). The exercise spawned a great enthusiasm for the method and Vincenzo Riccati started to generalize and elaborate. His results in that enterprise are the subject of the publications I mentioned at the beginning of this article; indeed Riccati's contributions are essentially no more than generalizations and clarifications of Euler's procedures.

§ 10. This brings us back to the endpoint of our theme's journey through mathematics, from Huygens' perhaps somewhat overconscientious attempts to legitimate the tractrix to Riccati's overdiligent elaboration of Euler's construction. It is not one of the grand themes in the history of mathematics. But it is, in my view, a charming, curious and revealing theme. Its journey testifies to the persistence of geometrical ideas even amidst the general emancipation of mathematics from geometrical ties; such ideas still emerged naturally in cases where the new methods of analysis failed. In this way the story of tractional motion documents a way of thinking that remained present for a long time and that occasionally had an influence on the direction of mathematical research.

The theme also points to a more general question which has always been present in the development of mathematics. This is the question: When is a problem solved? When do we know a mathematical object? Tractional motion for some time served as an adequate solution of a certain type of problem. Giving a construction by tractional motion meant that the problem was solved and that henceforth the curve in question was known. But the issue was dubious and ultimately tractional motion lost its conviction as a means of solution. Similar processes of the interpretation and re-interpretation of what constitutes a solution and what constitutes proper knowledge of mathematical objects have played — and still play — an important role in mathematics, and I find these processes very instructive objects of historical study.

Finally, I think the theme has a certain charm in its seemingly naive attitude to the mechanical intricacies of instruments and the mathematical complexities of differential equations. To me, this charm adds to the pleasure I have in following the theme's journey, in noting the variations and echoes of general ideas and in finding links between mathematical activities of individuals in Holland and Italy.

BERNOULLI, Jakob
[Opera] *Opera,* Geneva, 1744 (reprint Brussels 1967).

[1693] "Solutio problematis fraterni ante octiduum Lipsiam transmissi" *Acta Eruditorum* 1693 (June) pp. 255bis-sqq, in [Opera] pp. 574-576.

[1696] "Constructio generalis omnium curvarum transcendentium ope simplicioris Tractoriae Logarithmicae", *Acta Eruditorum* 1696 (June), pp. 261sqq, in [Opera] vol. 2, pp. 725-728.

BERNOULLI, Johann
[Opera] *Opera Omnia*, (4 vols) Lausanne/Geneva 1742 (reprint Hildesheim 1968).

[Briefw.] *Der Briefwechsel von Johann Bernoulli*, vol. 1 (correspondence with Jakob Bernoulli and with l'Hôpital, ed. O. Spiess), Basel 1955.

[1693] "Solutio problematis Cartesio propositi a Dn de Beaune ..." *Acta Eruditorum* 1693 (May) pp. 234sqq, in [Opera], 1 pp. 65-66.

BOMIE, ?
[1712] "Propriétés de la Tractrice", *Mémoires de l'Académie Royale des Sciences* (1712) Paris 1714, pp. 215-225.

BOS, H.J.M.
[1974] "The lemniscate of Bernoulli", *For Dirk Struik* (eds R.S. Cohen e.a., Dordrecht, 1974) pp. 3-14.

[1981] "On the representation of curves in Descartes' Géométrie", *Archive for History of Exact Science*, 4, (1981), pp. 295-338.

[1984] "Arguments on motivation in the rise and decline of a mathematical theory; the "construction of equations", 1637-c. 1750", *Archive for History of Exact Science*, 30, (1984), pp. 331-380.

[1988] "Tractional motion and the legitimation of transcendental curves", *Centaurus*, 31 (1988), pp. 9-62.

[1989] "The structure of Descartes' *Géométrie*, to be published in *Atti del Convegno Internazionale "Descartes: il Discorso sul Metodo e i Saggi di questo Metodo" (Lecce, 21-24 Ottobre 1987).*

BREGER, H.
[1986] "Leibniz' Einführung des Transzendenten", *Studia Leibnitiana* (Sonderheft 14, 300 Jahre "Nova Methodus" von G.W. Leibniz (1684-1984), Wiesbaden, 1986) pp. 119-132.

CLAIRAUT, A.-C.

[1745] "Sur quelques principes qui donnent la solution d'un grand nombre de problèmes de dynamique", *Mémoires de l'Académie Royale des Sciences* (1742) Paris 1745, pp. 1-52.

DESCARTES, R.

[Oeuvres] *Oeuvres de Descartes* (ed. Ch. Adam & Paul Tannery) 12 vols, Paris 1897-1910.

[1637] *La geometrie* (one of the essays appended to) *Discours de la méthode*, Leiden 1637 (pp. 297-413); facsimile and translation in: D.E. SMITH and M.L. LATHAM (eds), *The geometry of René Descartes*, New York (Dover) 1954.

[1649] *Geometria* (Latin translation by F. van Schooten with notes and additional treatises by van Schooten and others), Leiden 1649.

[1659] *Geometria* (enlarged edition of [1649]) (2 vols), Amsterdam 1659-1661.

EULER, L.

[Opera] *Opera Omnia* (3 Series, many vols) Leipzig etc. 1911- .

[1741] "De constructione aequationum ope motus tractorii aliisque ad methodum tangentium inversam pertinentibus" *Commentarii Academiae Scientiarum Petropolitanae* vol. 8 (1736/1741) pp. 66-85; in [Opera] Ser. 1, vol. 22, pp. 83-107. (References to the [Opera] edition.)

FONTENELLE, B. de

[1714] "Sur la Tractrice" (Report on Bomie [1714]), *Hist. Acad. Roy. Sci.* 1711 Paris 1714, pp. 58-62.

L'HÔPITAL, G.F. de

[1693a] "Problematis à Joh. Bernoullio in hisce actis mense Majo pag 235 propositi solution, a Dn Marchione Hospitalio in literis ad Dn Bernoullium d. 27 Junii exhibita. Conferentur Acta Erud. mensis Jun. Pag 255", *Acta Eruditorum* 1693 (September 1) pp. 398-399.

[1693b] "Solution d' un problème de géometrie que l'on a proposé depuis peu dans le Journal de Leipsic. Par M. le Marquis de l'Hospital", *Mém. Math. Phys.* 1693 (June 30) pp. 97-101 (but published c. December 1693).

[1694] "Dn Marchionis Hospitalii solutio problematis geometrici nuper in Actis Eruditorum (Anno 1693 pag 235) quae Lipsiae eduntur, propositi. Translata ex Commentariis Mathemat. Physic. Parisiensibus. Anno 1693, 30 Jun." *Acta Eruditorum* 1694 (May 1) pp. 193-196.

HUYGENS, Christiaan
[Oeuvres] *Oeuvres Complètes* (22 vols), The Hague, 1888-1950.

[ms 1692] Manuscript "Quadratura Hyperbolae per novam quadratricem ejus, quae uno ductu describitur" (29 October-20 November 1692), University Library Leiden, Ms Hug 6 ff. 59r-86r. Summary and some excerpts in [Oeuvres] vol. 10 pp. 409-413, 418-422.

[1693a] "Lettre à l'auteur de l'Histoire des Ouvrages des Sçavans", *Histoire des Ouvrages des Sçavans* 1693 (February) pp. 244-257. [Oeuvres] vol. 10 pp. 407-417 (References to the [Oeuvres] edition).

[1693b] "De problemate Bernoulliano ...", *Acta Eruditorum* 1693 (October) p. 475, [Oeuvres] vol. 10, pp. 512-515.

[1694a] "Constructio universalis problematis a clarissimo viro Joh. Bernoullio, ... propositi." *Acta Eruditorum* 1694 (September) pp. 338bis-339bis, [Oeuvres] vol. 10, pp. 673-674.

[1694b] "Excerpta ex epistola C.H.Z. ad G.G.L.", *Acta Eruditorum* 1694 (September), pp. 339-341, in Leibniz [M. Schr.] vol. 5, pp. 292-293.

LEIBNIZ, G.W.
[M. Schr.] *Mathematische Schriften* (7 vols, ed. C.I. Gerhardt), Berlin and Halle, 1849-1863 (reprint Hildesheim 1961-1962).

[Schr. Br.] *Sämtliche Schriften und Briefe*, Berlin 1954-. .

[1693a] "Ad problema Majo nupero in his Actis p. 235 propositum", *Acta Eruditorum* 1693 (July), [M. Schr.] vol. 5 p. 288.

[1693b] "Supplementum geometriae dimensoriae seu generalissima omnium tetragonismorum effectio per motum: similiterque multiplex constructio lineae ex data tangentium conditione" *Acta Eruditorum* 1693 (September) [M. Schr.] vol. 5 pp. 294-301.

[1693c] "Excerpta ex epistola G.G.L. cui precedens meditatio fuit inclusa" *Acta Eruditorum* 1693 (October), [M. Schr.] vol. 5, pp. 290-291, [Huygens, Oeuvres] vol. 10, pp. 516-518.

[1694] "G.G.L. Additio" (to Huygens [1694b]) *Acta Eruditorum* 1694 (September), pp. 339-341, [M. Schr.] vol. 5, pp. 293-294.

PEDERSEN, O.
[1963] "Master John Perks and his Mechanical Curves", *Centaurus*, 8 (1963), pp. 1-18.

PERKS, J.
[1715] "An easy mechanical way to divide the nautical meridian line in Mercator's

projection; with an account of the relation of the same meridian line to the curve catenaria. By J. Perks, M.A.", *Philosophical Transactions*, 29 (1714-1716), pp. 411-418 (1715).

POLENI, G.
[1729] *Epistolarum mathematicarum fasciculus*, Padua, 1729. The volume is not consecutively paginated; I refer to the seventh item (which is not paginated at all): "Ioannis Poleni ad virum celeberrimum Iacobum Hermannum (...) Epistola in qua agitur de organica curvarum Tractoriae, atque logarithmicae constructione. Accedunt Problematum ac theorematum de curva tractoria, a celeberrimis Geometris propositum, demonstrationes." (The letter is dated September 1728, the item also contains the answers from Antonio Conti, Gabriele Manfredi and Jacopo Riccati.)

RICCATI, Vincenzo
[1752] *De usu motus tractorii in costructione aequationum differentialium commentarius*, Bologna 1752.

[1753] "Lettera (...) alla signora D. Gaetana Maria Agnesi intorno alla costruzione di alcune formule senza la separazione delle indeterminate", *Symbolae Litterariae Opuscula Varia*, Vol. 10, Florence 1753 (The letter is dated November 3, 1750).

RICCATI, Vincenzo and SALADINI, Girolamo
[1765] *Institutiones analyticae* (2 parts, 3 vols), Bologna 1765-1767.

*Acknowledgements*

This article was partly written during my stay at the Institute for Advanced Study at Princeton as a member of the School of Mathematics. I am very grateful to have been able to take advantage of the excellent opportunities for research offered by the Institute. The Library of the University of Leyden has kindly permitted the reproduction of details from Huygens' manuscripts.

# THE TEXTBOOKS OF 'S GRAVESANDE AND VAN MUSSCHENBROEK IN ITALY

*C. de Pater*

In 1745 the Neapolitan professor Antonio Genovesi wrote a *Disputatio physico-historica de rerum corporearum origine et constitutione* in which he resolutely took sides against Descartes and with Newton. According to him, "[Newton] wanted to philosophize, not with vague hypotheses and conjectures, but with conclusive experiments and solid reasoning, confirmed by experiments (...) As a result it came about that in all famous European academies physics moved away from a «Romanensi» base [i.e., Cartesian] and returned to a firmer and more solid one."[1]

We find Genovesi's treatise in a textbook of van Musschenbroek which was edited by Genovesi, together with a part of a textbook of 's Gravesande.[2] Therefore in using the afore-mentioned phrase 'famous European academies' he must also have been thinking about the University of Leyden. There Willem Jacob 's Gravesande (1688-1742), as the first university professor on the continent, lectured in experimental Newtonian physics from 1717 to 1742. Through his lectures as well through his textbooks and orations he was a very important promoter of Newtonian philosophy.

Again and again the Leyden professor pleaded for the Newtonian method of investigation in physics. This method is characterized by three fundamental aspects: first, only experiments which are carried out carefully should form the basis of physics; second, out of observations and experiments natural laws must be inferred by means of induction and with the help of logic and mathematics; third, all natural phenomena should then be explained with these — mathematically formulated — natural laws.

1. A. Genovesi, *Disputatio physico-historica de rerum corporearum origine et constitutione*, in P. van Musschenbroek, *Elementa physicae conscripta in usus academicos*, 2 vols (Neapoli, 1745), vol. I, 69.

2. Van Musschenbroek, *Elementa* (o.c. note 1), vol. II 'De rebus coelestibus tractatus'; cf. W.J. 's Gravesande, *Physices elementa mathematica, experimentis confirmata. Sive introductio ad philosophiam Newtonianam*, 3d ed., 2 vols. (Lugduni Batavorum, 1742), vol. II, liber VI: 'De mundi systemate' (without the *Scholia* and with some minor deletions).

An important aspect of the presentation of Newtonian philosophy is the polemic with the Cartesians. Descartes was first and foremost interested in deducing the observed phenomena from basic principles evident to him. In doing so he showed hardly any scruples in a priori postulating all sorts of ad hoc hypotheses without testing their validity by means of experiments. Just like Newton himself — hypotheses non fingo — and other Newtonians, 's Gravesande repeatedly made a stand against the hypothetical-deductive method of the French philosopher.[3]

This anti-Cartesianism also emerges in the afore-mentioned quotation from Genovesi's 'Disputatio physico-historica'. A few pages further he pithily summed up his views once again: "Descartes taught nothing but vague and uncertain hypotheses, Newton on the other hand assumed nothing unless it could be proved either by experiment or reasoning. Hence Descartes' hypotheses have been put aside, so that Europe could be called back from uncertain hypotheses to a better founded physics."[4]

There can hardly have been any other Newtonian fulminating as vehemently against hypotheses à la Descartes as 's Gravesande's colleague Petrus van Musschenbroek. The latter continually criticized 'meditation in the study', contriving 'chimeras' and 'trifles'. As he saw it, fancied hypotheses were nothing less than a malady for true physics.

What 's Gravesande did in Leyden — to reform the study of physics in Newton's spirit — van Musschenbroek did in Utrecht (1723). He can be regarded as a pupil of 's Gravesande, to the extent that he briefly attended the master's lectures, but he was by no means his slavish follower. Van Musschenbroek detached himself further from the substance of the *Principia* and the *Opticks* than 's Gravesande. He was, for example, strongly interested in magnetism, electricity, capillarity, meteorology and the theory of the strength of materials, subjects only summarily dealt with by 's Gravesande. Besides, from the method he employs we can infer that van Musschenbroek's Newtonianism has a strong Baconian tendency.[5]

Despite the two physicists' profound admiration for Newton, they had understood him too well not to maintain their independence from him. Not

3. For example: 's Gravesande, *Phys. El.* (o.c. note 2), praefatio I-III; cf. G. Gori, *La fondazione dell'esperienza in 's Gravesande* (Firenze, 1972), 162-163 and C. de Pater ed., *Willem Jacob 's Gravesande Welzijn, wijsbegeerte en wetenschap*, Geschiedenis van de wijsbegeerte in Nederland, M.J. Petry and J. Sperna Weiland eds., vol. 13 (Baarn, 1988), 16-19, 76-79.

4. Genovesi, *Disputatio* (o.c. note 1), 74.

5. Cf. C. de Pater, *Petrus van Musschenbroek (1692-1761), een newtoniaans natuuronderzoeker* (diss. Utrecht, 1979), 58.

Newton, but the experiment, not Newton's words, but his method had the final say.[6]

Both 's Gravesande and van Musschenbroek carried out physical investigations, especially van Musschenbroek. He extensively applied himself to, among other things, magnetism and the strength of materials. His treatises in these fields were quite well-known in Italy, too. When Pope Benedict XIV requested advice in connection with the restoration of the St. Peter's Dome in 1743, the results of van Musschenbroek's investigations in the tensile strength of iron cables (staves) were also taken into account. This was no doubt due to the fact that the Pope's advisers, Giovanni Poleni, Ruggero Giuseppe Boscovich and others were well-informed about van Musschenbroek's work.[7]

The merits of the two Dutch professors lie mainly in their propagation of Newtonianism, both through their lectures, which attracted many people from abroad, and through their textbooks. Both men were seen as experts 'par excellence', in physics generally and in Newtonianism in particular. When Voltaire was looking for someone to counsel him in Newtonian physics in connection with the publication of his *Elemens de la philosophie de Neuton* (1737) he chose 's Gravesande. The then heir apparent Frederic of Prussia thought this was a sensible choice: 's Gravesande was supposed to have completely mastered Newtonian physics. Elsewhere Voltaire complains that his country does not have any physicist of 's Gravesande's or van Musschenbroek's stature.[8]

A few years after they had taken their professorial chairs, both physicists wrote a textbook based on the substance of their lectures. 's Gravesande's *Physices elementa mathematica experimentis confirmata* appeared in 1720-21. The subtitle of this work is probably even more revealing: *Introductio ad philosophiam Newtonianam.* In 1726 van Musschenbroek published his *Epitome elementorum physico-mathematicorum.* Both professors kept a close eye on developments in the field of physics and made sure that their work remained up to date. It was continually enlarged with the results of new

6. W.J. 's Gravesande, *Phys. El.* (o.c. note 2), 2d ed. (Lugd. Bat., 1725), 'Monitum de hac secunda editione', [3]; also in the 3d ed. vol. I, XIII-XIV; De Pater, *'s Gravesande* (o.c. note 3), 53.

7. In 1755 van Musschenbroek was 'extraordinarily' associated to the 'Accademia delle Scienze' of Bologna thanks to the suggestion of Benedict XIV, who had commisioned the Dutch scientist many scientific instruments that were to be donated by the Pope in 1746 to the Institute of Bologna. See: W. Tega ed., *Anatomie accademiche*, I, *I commentari dell'Accademia delle Scienze di Bologna*, Bologna, 1986, 33, 259. Cf. also J. Witteveen, *Lessen uit de geschiedenis van de toegepaste mechanica* (Delft, 1976), 11, 21.

8. Th. Besterman, *Voltaire's correspondence*, vol. VI (Genève 1954), 1, 25.

investigations, partly also their own, in the successive reprints appearing at home. In 1742 the third edition of 's Gravesande's book was double the size of the first, while van Musschenbroek's 1762 edition of his textbook, the *Introductio ad philosophiam naturalem*, had swollen to four times the size of the 1726 *Epitome elementorum physico-mathematicorum*.[9]

The various editions of both physicists' works survived until well into the 19th century, including van Musschenbroek's 1741 edition of his textbook, the *Elementa physicae*, as we shall see after this. They found their way throughout Europe. This was true not only of the authentic, Latin version but also of its various translations. 's Gravesande's physics textbook was translated into English (7 editions), French and Dutch. The philosophy textbook he wrote in 1736, the *Introductio ad philosophiam, metaphysicam et logicam continens* was translated into French, German, Dutch and even Greek. Van Musschenbroek's physics handbook was, in some version, translated into English, French, German and Swedish, and he prepared a Dutch edition himself, *Beginselen der Natuurkunde, beschreven ten dienste der landgenooten* (1736).

Also in Italy there was a strong interest in the works of both Dutchmen. It is therefore quite remarkable that no translations were made. Giambattista Gori is of the opinion — in his book *La fondazione dell'esperienza in 's Gravesande*, (1972) — that the social and economic regression did not warrant the publication of an Italian translation. This does not explain, however, why other works did get translated.[10]

It is possible to give another explanation. The Italian historian Paolo Casini argues in an article (1978) that in Italy Newtonianism hardly surfaced during 1710 to 1740, owing to the strong influence of the Inquisition. Whoever dared accept the Newtonian worldmachine, accepted at the same time Copernicanism, which had been condemned at the Galileo-trial in 1633. The situation changed when the Archbishop of Bologna became Pope Benedict XIV in 1740. The new Pope stimulated natural philosophy and the climate started changing in favour of Newtonianism. It is true, there were some hitches. Take for instance Francesco Algarotti's *Il Newtonianismo per le Dame* (1737), which had been put on the Index and remained on it, despite the modifications in the later editions from 1739 to 1752.[11] Incidentally, this extremely popular work

9. Cf. the bibliographies in De Pater, *'s Gravesande* (o.c. note 3), 152-160 and De Pater, *Musschenbroek* (o.c. note 5), 349-360.

10. Gori, *La fondazione* (o.c. note 3), 114.

11. P. Casini, *Newton e la coscienza europea* (Bologna, 1983), 173-177; P. Casini, 'Les débuts du newtonianisme en Italie, 1700-1740', in: *Dix-Huitième Siècle* 10 (1978), 85-100, in part. 86, 91, 98-99.

was translated into French, English and also into Dutch (*De Newtoniaansche Wysbegeerte voor de Vrouwen*, 1767).

Thus, according to Casini, Newtonianism in Italy was mainly the concern of the élite, for three reasons: first, because of the attitude of the Inquisition; second, the political fragmentation and third, the intellectual stagnation.[12] And, to give my conclusion, because the intellectual élite had the Latin edition of the textbooks of 's Gravesande and van Musschenbroek, there was hardly any need for an Italian translation.

The need for a native Latin student edition was so strong, however, that after 1740 (1745 to 1794 to be precise) various editions of those textbooks appeared, notably van Musschenbroek's. In various Italian cities it was apparently believed that the two Netherlanders were among the best representatives of Newtonian philosophy.

In his book *Newton e la coscienza europea* (1983), in a chapter 'Newton in Italia', Casini mentions father Celestino Galiani (1681-1753) the 'eminenza grigia' of Newtonianism in his native land.[13] As early as 1706 Galiani knew and valued Newton's 'Opticks', which had appeared two years before. His reputation as a scholar was due to his lectures in Rome, where the first edition of the *Principia* must have been available, just as in Florence and Venice. It was also caused by his educational reform in his capacity as minister in the Naples' government.

Galiani kept up a correspondence with 's Gravesande in the years 1713-1717. The Leyden physicist was one of the main editors of the *Journal Litéraire*, which he and others had founded and to which various foreign correspondents also contributed. In this context 's Gravesande had asked Galiani in 1713 to send him the appropriate information from Italy from time to time. In 1715 the printer of the *Journal Litéraire*, T. Johnson, sent him Newton's *Principia* — probably the second edition — as well as the *Praelectiones astronomicae* by William Whiston (1667-1752) and the *Introductio ad veram physicam* by John Keill (1671-1721).

In 1717 Galiani asked 's Gravesande for advice on the possibility of having the river Reno flow into the Po. In the same year 's Gravesande wrote him that he had passed the letter to Newton and the Royal Society for an expert opinion. In the same letter 's Gravesande informed Galiani that he would have his inaugural address carried to him by means of a 'monsieur anglais'.[14]

12. Casini, 'Les débuts' (o.c. note 11), 85-86.
13. Casini, *Newton* (o.c. note 11), 192.
14. F. Nicolini, *Un grande educatore italiano, Celestino Galiani*, Storia e pensiero n. 2 (Napoli, 1951), 162, 163; Casini, *Newton* (o.c. note 11), 192; Casini, 'Les débuts' (o.c. 11), 92-93; Gori, *La fondazione* (o.c. note 3), 158-159.

At the time of this correspondence there were not yet any textbooks available by 's Gravesande and van Musschenbroek. Whenever they were available (1720-21, respectively 1726), they soon turned out to be familiar. Paolo Frisi (1728-1784), who taught philosophy and mathematics in Milan and Pisa, criticized some English authors (Cotes, Keill) on account of their essentialism in the field of gravitation. In 1754 he expressly accused Keill and Cotes of swearing to Newton's words, a phrase which might have come straight from the foreword in the second edition of 's Gravesande's *Physices Elementa mathematica*.[15] Poleni in Padua as well as Frisi in Milan, Giambatista Beccaria in Turin and Genovesi in Naples quoted or paraphrased or directly used in their lessons 's Gravesande's or van Musschenbroek's textbooks.[16] Apparently these textbooks were so frequently used, that there was a need for locally published editions. This included 's Gravesande's philosophy textbook. The second Latin edition of it came out in 1737, a year after the first one, and in the same year a Latin edition appeared in Venice. As late as 1792 a fourth edition was published. Although this work cannot be called Newtonian as a matter of course — it occupied a middle position in between Descartes and Locke — it did contain a few Newtonian elements.[17]

As for physics, in 1723 's Gravesande had, for his students, published a shortened version of his original textbook, namely the *Philosophiae Newtonianae institutiones*. Three years after its third Leyden edition in 1746, a reprint appeared in Bassano and Venice (1749). This Bassano/Venice edition came out only once. There was another edition in Vienna, which may also have been used in Italy.

There appeared to be a preference for van Musschenbroek's physics handbooks, at least for educational purposes. In 1741 van Musschenbroek had published a second enlarged edition of his *Elementa physicae*, dating from 1734. A reprint of this new edition appeared in Venice in 1745. In the same year a similar edition came out in Naples. This had been edited by the aforementioned Antonio Genovesi, professor of metaphysics, ethics and political economics and by Giuseppe Orlandi, professor of mathematics and experimental physics and known as an able experimenter. Genovesi had his *Disputatio physico-historica*, from which I have quoted at the beginning of this paper, incorporated within this treatise. Moreover, many annotations and additions were worked into footnotes, largely by Orlandi.[18]

15. Gori, *La fondazione* (o.c. note 3), 158, 158n17.

16. For Frisi's, Beccaria's and Genovesi's concern for 's Gravesande textbook, see e.g. Gori, *La fondazione* (o.c. note 3), 114, 114n88.

17. Cf. De Pater, *'s Gravesande* (o.c. note 3), 31-37: 'Tussen Descartes en Locke'.

18. E. Garin, 'Antonio Genovesi e la sua introduzione storica agli *Elementa physicae* di Pietro van Musschenbroek', in: *Physis* 11 (1969), 211-222.

Both in Naples and in Venice this Genovesi-Orlandi-edition was reprinted several times. As late as 1794 a fifth edition was issued in Venice. It is curious indeed that van Musschenbroek's *Elementa physicae* remained in print for so long, instead of adopting the extended *Institutiones physicae* of 1748, or — later —the greatly enlarged and completely updated *Introductio ad philosophiam naturalem* of 1762. However, there was also an Italian edition of this latter publication, in Padua in 1768.

An interesting question is: why did people prefer van Musschenbroek's physics manual to that of 's Gravesande? Probably Genovesi and Orlandi believed that van Musschenbroek's work was more readily accessible to students, just as the Dutch professor Petrus Camper did. He counselled students who had little mathematical knowledge to make use of van Musschenbroek's textbooks, although he based his lectures at Franeker (1747-1755) on 's Gravesande's *Philosophiae Newtonianae institutiones.*[19] Further, the fact that van Musschenbroek discussed a greater variety of physical subjects than 's Gravesande, may have played a role as well.

In his request for the imprimatur in favour of the Genovesi-edition the Neapolitan professor Jacobus Martorellius stated that no single 'elementae physicae' was more highly appreciated than van Musschenbroek's. This reason was given: in the Dutch physicist's textbook everything discovered in more recent physics, under the guidance of reason and experience, had been discussed.[20] In connection with a second request for the imprimatur Nicola de Martino remarked in a letter to Galiani that van Musschenbroek had won so much fame, that further recommendations were superfluous.[21]

In the preface of their edition the editors themselves wrote that it would be quite useful for the Neapolitan students, enthusiastically engaged in the study of natural philosophy.[22] In one respect, however, van Musschenbroek's text was supposed to need a supplement. From early on van Musschenbroek had failed to incorporate an explanation of the world system, whereas 's Gravesande had. Genovesi and Orlandi apparently thought that a treatment of the world system had to be part of any physics textbook. In any case, they decided to include the relevant part of 's Gravesande's *Physices elementa*

19. P. Camper, *Prolegomena in philosophiam* (without date, ca. 1750), manuscr. Library of Leyden University (BPL 247/12), 13; cf. R.P.W. Visser, *The zoological work of Petrus Camper (1722-1789)*, (Amsterdam, 1985), 16-22.

20. Van Musschenbroek, *Elementa* (o.c. note 1), vol. I, [408].

21. Van Musschenbroek, *Elementa* (o.c. note 1), vol. I, [409].

22. Van Musschenbroek, *Elementa* (o.c. note 1), vol. I 'De nova hac Neapolitana editione Ad Lectorem'.

*mathematica* (third edition, 1742) in their edition, following van Musschenbroek's text. It went under the title *De rebus coelestibus tractatus*.[23]

They had it preceded by a word of caution — because to say Newton, is to say Copernicus as well. Pointing out that they had chosen 's Gravesande's explication for his lucid treatment of Newton's discoveries, and that they had omitted the mathematical 'scholia' out of consideration for their students, they went on to posit that the arguments for the movement of the earth, put forward by 's Gravesande's and other Copernicans, 'possess no small measure of probability'. But they stopped short of accepting that the issue had been settled beyond dispute, as they believed Newton, Huygens, Gregory and others plainly stated. To bear this out they referred to *l'Existence de Dieu*, the French translation of *Het regt gebruik der Werelt beschouwingen* (ch. 7, 1, 3) by Bernard Nieuwentyt, who published his book in 1715. It could then be argued that there was no compelling reason to deviate from a literal interpretation of the Scriptures, which affirm an immovable earth. But they hastened to add that the Copernican system could very well be used as a suitable hypothesis to explain certain phenomena.[24] It was precisely this prudent attitude which cardinal Bellarmine had once recommended to Galileo, who refused to go along with it.[25]

With regard to this particular edition of the *Elementa physicae* we are facing the remarkable fact that both Dutch Newtonians, often bracketed together —and rightly so — as advocates of Newtonianism in Europe, cropped up in Italy in one volume which had as many as five reprints. But, there is more to say about the interaction between Italy and the Netherlands. It was actually an instance of mutual love. When 's Gravesande was criticized because he hardly mentioned any source in his physics textbook, he took the decision to list the subjects treated, together with the relevant authors consulted, in the preface of the third edition. In addition to Newton quite a number were mentioned, including several Italians: Galileo, obviously, but also Torricelli, the Accademia del Cimento, Giovanni Alfonso Borelli, Giandomenico Cassini, Paolo Casati, Domenico Guglielmini, Guido Grandi and Giovanni Poleni.[26]

In particular Giovanni Poleni (1683-1761), who was professor in Padua since 1709, deserves special attention. He was regarded as a Newtonian, but,

23. See note 2.

24. Van Musschenbroek, *Elementa* (o.c. note 1), vol. II, 'De rebus coelestibus tractatus': 'De hac rerum coelestium tractatione ad lectorem', [III-IV]; cf. vol. I, 60n1: "Observetur extra Ecclesiam nostram ex Copernicana hypothesi Auctorum [= Van Musschenbroek, d.P.] loqui".

25. S. Drake ed., *Discoveries and opinions of Galileo* (New York, 1957), 162-164.

26. 's Gravesande, *Phys. El.* (o.c. note 2), vol. I, p. XVII, XX-XXIII, XXVII, XXIX, XXXI-XXXII.

according to Casini, his *De vorticibus coelestibus dialogus* (1712) or his *De mathesis in rebus physicis utilitate praelectio* (1720) did not warrant this. Curiously, Casini failed to mention Poleni's *De physices in rebus mathematicis utilitate praelectio* (1716), in which he briefly commented on Newton and his principle of gravitation, in positive terms.[27] An unfinished, unpublished physics textbook of Poleni bore the same title as that of 's Gravesande. Some instruments of his *Teatro di filosofia sperimentale*, founded in 1740, were analogous with those described by 's Gravesande in his manual, too.[28]

An issue occupying many 18th-century philosophers and physicists for decades was the so-called vis viva-question: should the effect of a force be measured by mv or by $mv^2$? In a 1722 article, *Essai d'une nouvelle théorie du choc des corps*, 's Gravesande had rejected his original mv-standpoint, and joined the ranks of the $mv^2$-adherents. This change of mind was based on the results of his experiments with balls of different masses dropped into clay from various heights. Shortly after, he read Poleni's *De castellis (...). Quo etiam continentur nova experimenta (...) et ad percussionis vires pertinentia*, published in 1718, in which the author had come to the same conclusion, with similar experiments. Still in 1722 's Gravesande wrote an additional article to his *Essai*, mentioning this.[29] In the third edition of his *Physices elementa mathematica* he inserted Poleni's experiment, having carried it out himself with an improved device. Conversely, Poleni appreciated 's Gravesande's work in this field. Again, in connection with the treatment of liquids he

27. Casini, *Newton* (o.c. note 11), 179-181; G. Poleni, *De physices in rebus mathematicis utilitate* (Patavii, 1716), 24-25.

28. G. Poleni, *Physices elementa mathematica experimentis confirmata* (ms. ca. 1740); M. Pancino, 'La didattica di Giovanni Poleni', in: *Il Teatro di filosofia sperimentale di Giovanni Poleni*, Mostra di strumenti scientifici (Padova, 1986), XV-XXIV, in part. XX-XXI. Poleni had a correspondence with Petrus van Musschenbroek and some of the apparatus of his Theatre of experimental philosophy were made by Jan van Musschenbroek in Leyden. The extant letters of this correspondence at the Marciana Library in Venice are to be published in M.A. Pancino-G.A. Salandin, 'Relazioni scientifiche fra Leida e Padova: Pieter van Musschenbroek e Giovanni Poleni', *Incontri. Rivista di studi italo-nederlandesi,* n.s., IV, 1-2 (1989).

29. W.J. 's Gravesande. 'Essai d'une nouvelle théorie sur le choc (du choc) des corps', in: *Journal litéraire*, t. XII (1722), I, 1-54; sep. La Haye, 1722; Sup(p)lément à l'essai sur le choc des corps, t.XII (1722), II, 190-197; both art. also in J.N.S. Allamand ed., *Oeuvres philosophiques et mathématiques de Mr. G.J. 's Gravesande*, 2 vols. (Amsterdam, 1774), vol. I, 217-251; P. Costabel, ''s Gravesande et les forces vives ou des vicissitudes d'une expérience soi-disant cruciale' in: I.B. Cohen and R. Taton eds., *L'aventure de la science*, (Paris, 1964), 117-134, in part. 123; G. Poleni, *De castellis (...). Quo etiam continentur nova experimenta (...) et ad percussionis vires pertinentia.* (Patavii, 1718), nrs. 315-318; cf. T.L. Hankins, *Science and the Enlightenment* (Cambridge, 1985), 30-35.

admitted that he had derived a number of things from Poleni.[30]

Van Musschenbroek also mentions various Italian physicists in his books. In the auction catalogue of his library a large number of Italian authors is listed.[31] In his *Introductio ad cohaerentiam corporum firmorum*, containing his experiments regarding the strength of materials, he points to Galileo who discussed the subject in his *Discorsi* (1638).[32]

Galileo is also extensively commented on in van Musschenbroek's preface to his Latin translation of the *Saggi di naturali esperienze fatte nell' Accademia del Cimento* (1667), the *Tentamina experimentorum naturalium captorum in Academia del Cimento* (1731). According to van Musschenbroek Accademia del Cimento's fame had been spread all over Europe so that he felt a Latin translation of the *Saggi* to be of great importancc.[33] Actually, he added countless 'addidamenta' with experiments of his own, including a description of his well-known pyrometer.[34]

Accademia del Cimento's Baconian approach is bound to have appealed to our compatriot. In the preface of his translation he names Bacon and Galileo in one breath as men who broke with the word-mongering of Aristotelianism and who took a completely new course. It was they who started experimenting; and it was Galileo who successfully employed mathematics in the physical domain. Van Musschenbroek strongly fulminated against the whole bunch of ignorant Scholastics who in despicable narrow-mindedness had tried to demonstrate that Galileo's Copernican sympathies were contrary to the Holy Scriptures.[35]

Even in this work van Musschenbroek pays tribute to Newton: after the Florentines' experiments, physics was not a dead-end affair. Newton "applied a sublime mathematics to finely tuned experiments (...) and taught us how to penetrate into the profoundest secrets of nature and to acquire a true and reliable science (...). He taught us that hypotheses should be banned, only that which is demonstrated should be accepted."[36]

30. 's Gravesande, *Phys. El.* (o.c. note 2), vol. I, XXVII and 239 (par. 834); XXIX; liber III, pars II (cap. VII-XI); cf. Allamand, *Oeuvres* (o.c. note 29), vol. I, XIV-XIX.

31. *Bibliotheca Musschenbroekiana, sive catalogus librorum viri celeberrimi Petri van Musschenbroek* (Lugdini Batavorum, 1762).

32. P. van Musschenbroek, *Introductio ad cohaerentiam corporum firmorum* in: *Physicae experimentales, et geometricae (...) dissertationes* (Lugduni Batavorum, 1729), 421-672, in part. 425; G. Galilei, *Discorsi e dimostrazioni matematiche, intorno à due nuove scienze* (Leida, 1638), 'giornata prima' and 'giornata seconda'.

33. P. van Musschenbroek, *Tentamina experimentorum naturalium captorum in Academia del Cimento*, etc., 2 partes (Lugduni Batavorum, 1731), praefatio [4, 5].

34. Van Musschenbroek, *Tentamina* (o.c. note 33), pars II, *'additamentum'*, 12-57.

35. Van Musschenbroek, *Tentamina* (o.c. note 33), praefatio [2].

36. Van Musschenbroek, *Tentamina* (o.c. note 33), praefatio [6].

This Newtonianism, enthusiastically commended and worked out by 's Gravesande and van Musschenbroek in their textbooks, was what many students all over Europe became familiar with through these books, not least the students at various Italian universities. The influence of the two Dutch physicists on the spreading of Newtonianism in Italy discussed in this paper may well turn out to have been a very interesting problem, and deserves further investigation.

# ITALIAN HYDRAULICS AND EXPERIMENTAL PHYSICS IN EIGHTEENTH-CENTURY HOLLAND. FROM POLENI TO VOLTA

*Cesare S. Maffioli*

## I. *Introduction*

In the introductory remarks to the studies in the historiography of eighteenth-century science edited in 1980 by G.S. Rousseau and Roy Porter it has been emphasized that the whole picture of eighteenth-century science lacks structure and many of its elements do not yet seem to be well focalized. In a word, the period itself represents a historical and historiographical problem.[1] For a better understanding of its distinctive features, the significance and the usefulness of a comparative approach between different 'national' or 'regional' traditions become evident. Italy and the Low Countries are particularly suitable cases since they have been relatively less studied, in comparison with England and France, in spite of the persistence in both areas of strong seventeenth-century autonomous traditions originated by men like Galileo, or Stevin and Huygens. On the other hand, confining to eighteenth-century physical sciences, the emphasis laid on Newtonian science and its diffusion in Europe has brought in past years the undesired consequence of largely neglecting other streams of scientific thought. Besides studying neglected traditions and alternative attitudes in the key subjects, in order to put forward a more balanced picture it is also essential to enlarge the usual coverage of eighteenth-century physical sciences so as to include a great deal of the magmatic field between science and technology as well as a good deal of the so-called 'mixed mathematics'. Hydraulics, from this point of view, deserves much more attention if we wish our reconstructions of eighteenth-century physical sciences to be in harmony with the richness and complexity of the subject-matter. This is particularly evident in our comparative case. Indeed, both in Italy and in the Low Countries hydraulic science and technology played a crucial role from the late Renaissance until the Enlightenment and we

1. *The ferment of knowledge. Studies in the Historiography of Eighteenth-Century Science*, ed. by G.S. Rousseau and R. Porter, Cambridge, Cambridge UP (1980), pp. 1-7, in part. pp. 2-4.

shall see that significant exchanges of learning between the two areas took place in these fields.

In this paper I shall illustrate just some instances of the complex web of relationships and mutual influences between eighteenth-century Italy and Holland in the fields of hydraulics and experimental physics. Although I concentrate on the stream from Italy to Holland, the width of the subject is such as to render premature any general survey and interpretation in the actual state of knowledge. However, I hope that, even in the presence of a historical framework calling for integration and refinement, situations (§§ II-III) and open questions (§ IV) put forward may stimulate discussion and further research.

## II. *Hydraulics*

A curious draft of Newton, written in Latin and in English in reply to a letter of 's Gravesande of 1717, is evidence that the controversy about the river Reno which had placed the Italian cities of Bologna and Ferrara in opposition at least from the sixteenth century onwards, assumed in the early eighteenth century an international dimension.[2] The move to ask the judgement of the Royal Society in the controversial matter came from Celestino Galiani (1681-1753), a key figure in intellectual life of early eighteenth-century Rome, who had supported during an official visit to the Reno in 1716 the reasons of Bologna.[3] It was not an individual attempt but the consequence of a precise strategy to involve the foreign academies: e.g. in 1719 an anonymous pamphlet appeared written in French by Eustachio Manfredi (1674-1739), whose aim was to gain the support of the Paris Academy of Sciences.[4]

2. *The Correspondence of Isaac Newton*, VI, ed. by A.R. Hall and L. Tilling, Cambridge, Cambridge UP (1976), pp. 420-421, letter 1266.

3. In a letter to Galiani written on 17 August 1717 from The Hague, 's Gravesande informed his Italian correspondent of having received his letter as well as the writings pertaining to the Reno controversy. He was to send a specimen of these papers to Newton to have the opinion of the Royal Society, and in order to give that of the Academy of Leyden he promised he would try and discuss the matter with his colleague J. Bernard (*SNSP,* Ms. XXXI. A.2., ff. 105r.-106r.). 's Gravesande's correspondence with Galiani had started on 14 November 1713 (and not December as reported in F. Nicolini, 'Su taluni rapporti di cultura tra l'Italia, l'Olanda e l'Inghilterra al principio del Settecento', *Atti della Reale Accademia di Scienze Morali e Politiche di Napoli,* LIII, 1 (1930), pp. 153-171, in part. p. 158) thanks to William Burnet. Galiani acted for a few years as Italian correspondent of the *Journal Literaire,* printed at The Hague by T. Johnson since 1713, which had 's Gravesande among its main collaborators.

4. [E. Manfredi], *Memoire Qui contient les raisons pour la jonction du Reno avec la Riviere du Po* [n.p., n.d., but 1719]. It should be remembered that, in addition to astronomical work, Manfredi spent much of his time as a superintendent to the waters of Bologna.

At the beginning of the seventeenth century the Reno was a tributary of a silted branch of the river Po which bordered on the town of Ferrara. In 1604, however, a brief of Clement VIII (Ferrara had been part of the Papal States since 1598) ordered the provisional diversion of the Reno with the main aim of digging the channels of the Po of Ferrara and of the Primaro, the Southern branch into which the Po was divided, in order to restore the navigation of Ferrara and to by-pass the Po Grande controlled by the Venetians. Another aim of the project was to reclaim the marshy valleys on the right side of the Primaro with the help of the silt deposited there by the Reno. The ambitious project largely failed: Ferrara did not recover its navigation and the Reno transformed the different valleys and formerly reclaimed soil into an almost uninterrupted marshy area. The damage particularly affected the lower side of the Bolognese plain which threatened to be periodically submerged. The remedy looked for was to give a new alveus to the Reno, and many projects were to be devised to this purpose during the seventeenth century. The Bolognese side and particularly Domenico Guglielmini (1655-1710), on the occasion of an official visit in 1692-1693 when he further developed and improved former projects, considered that the real solution could only be that of bringing the Reno into the Po Grande. E. Manfredi had the same view but, given the strong opposition of Ferrara and other cities as well as of the Venetians, opinion in Bologna gradually turned in favour of the idea of canalyzing the Reno towards the Po of Primaro, a project fully realized only in the second half of the eighteenth century (in fig. 29 a map is reproduced from Frisi's *Del modo di regolare i fiumi e i torrenti* (1762), where many of the proposed lines converging towards the Primaro are marked out).

The hydraulic debate around the Reno involved some of the best Italian scientists of the time[5] and raised many scientific questions, the most significant being that of measuring the discharge of a river. To face the problem of determining the speed of a river two approaches were possible: the first one consisted of studying (by way of mathematical hypotheses or of physical analogies) the relations between speed, slope, depth, ... so as to obtain some theoretical rule; the second one of actually measuring the speed (or at least of estimating the ratios of different speeds) by means of experimental devices. An example of the first line of research is represented by a well-known letter of

5. In addition to Guglielmini, Manfredi and Frisi we at least have to mention Gian Domenico Cassini and, above all, Benedetto Castelli (c. 1577-1643), the close friend and collaborator of Galileo who shares with Guglielmini the honour of having founded the Italian scientific school of hydraulics. For further information, I refer to my forthcoming *New Science in Italy after Galileo. The «Scienza del Moto delle Acque» 1690-1718* (Amsterdam, exp. 1991).

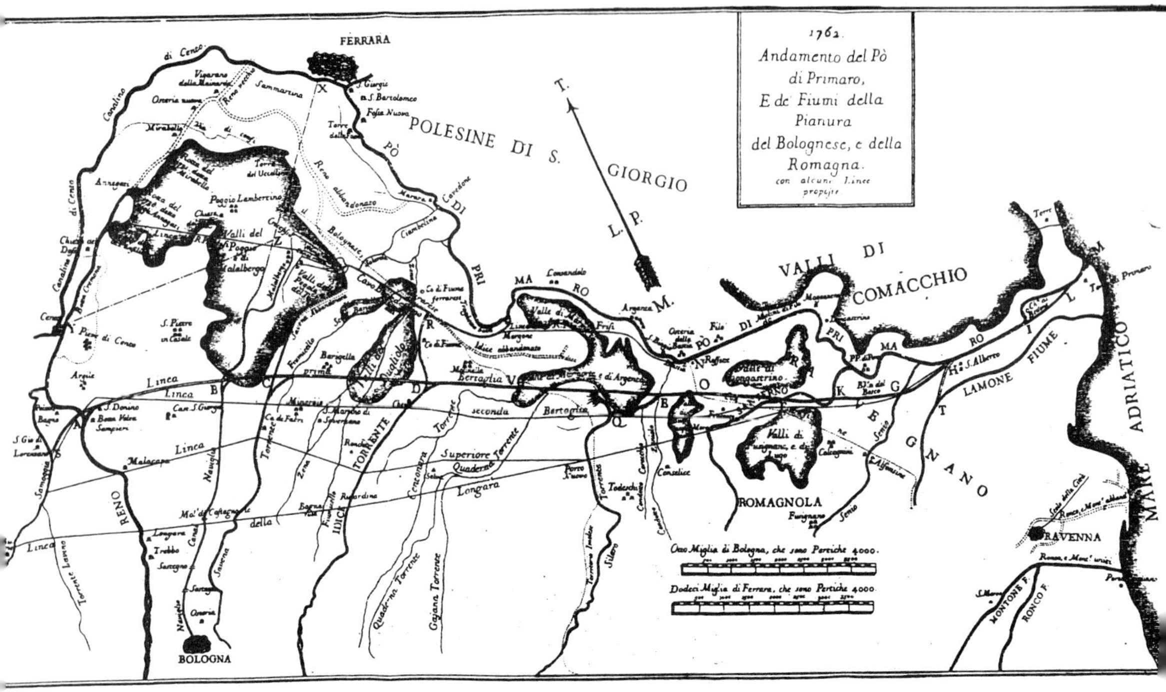


**Fig. 29 'Map of the Po of Primaro, and of the Bolognese and Romagna rivers. With some of the proposed lines' (From: P. Frisi, *Del modo di regolare i fiumi, e i torrenti, principalmente del bolognese, e della Romagna*, Lucca, V. Giuntini, 1762).**

Galileo on the river Bisenzio (dated 16.01.1631), in which he analyzed the flow of water along an inclined plane with the same conceptual tools he was to publish some years later in the *Discorsi*.[6] An instance on the other side of a coupling of both methods, although with a bias towards the building of physico-mathematical hypotheses, is represented by the *Aquarum fluentium mensura* (1690-1691) of Domenico Guglielmini. In this work the Bolognese scientist tried to produce a coherent theory of the motion of running waters along open channels by a generalization of Torricelli's law of efflux from small orifices in the wall of a vessel full of water to the cases of inclined or horizontal channels.[7]

6. *Le Opere di Galileo Galilei. Nuova ristampa della edizione nazionale,* VI, Firenze, G. Barbèra (1968), pp. 627-647.

7. An outline of Guglielmini's hypothesis in C.S. Maffioli, 'Guglielmini vs. Papin (1691-1697). Science in Bologna at the end of the XVIIth century through a debate on hydraulics', *Janus*, LXXI, 1-4 (1984), pp. 63-105, in part. pp. 71-73.

In the early eighteenth century Guglielmini's approach to river hydraulics, epitomized by his *Della natura de'fiumi* (1697), soon became a reference work for the practitioners in the field.[8] Also his theoretical mistakes, such as the parabolic profile of the speeds along a given perpendicular in the cross section of an ideal frictionless river, were largely followed. Although Guglielmini cannot be blamed for the lack of the concept of internal pressure in dynamic cases, it is evident that the Galilean tradition in the mechanics of fluids (characterized by the lack of the concept of hydrostatic pressure in its initiator) did not help Guglielmini in this instance. The Torricellian generalization presupposes indeed that the flow of water in such a perpendicular is 'free', as in the case of the outflow from a narrow and long opening along the vertical of a vessel always kept full of water while, on the contrary, each cross-section layer of a river interacts with the adjacent ones and thus the flow cannot be considered 'free' except for the points on the upper surface at atmospheric pressure. Only very slowly did this idea gain ground among the hydraulicians of the eighteenth century, both in Italy and outside. The first important step forward in the right direction was taken by the Venetian scientist Giovanni Poleni (1683-1761), professor at the University of Padua since 1709, in a work devoted to the study of the motion of water in the Venetian lagoons.

Poleni's *De Motu Aquae Mixto* (1717) is a complex work which defies any simple labelling, at least within our modern disciplinary context.[9] Its aim was a theoretical one, namely to extend the hydraulic theory to the case of the 'mixed motion' of waters. Poleni called 'aqua mortua' (dead water) that in which the upper surface is equidistant from the centre of all heavy bodies, and 'aqua viva' (living water) that receiving its motion from the pressure of superincumbent

8. An account of the diffusion of Guglielmini's hydrometry and river hydraulics in the first half of the eighteenth century in C.S. Maffioli, 'Domenico Guglielmini, Geminiano Rondelli e la nuova cattedra d'Idrometria nello Studio di Bologna (1694)', *Studi e memorie per la storia dell'Università di Bologna*, n.s., VI (1987), pp. 81-124, in part. pp. 115-124.

9. G. Poleni, *De Motu Aquae Mixto Libri Duo. Quibus multa nova pertinentia ad Aestuaria, ad Portus, atque ad Flumina continentur*, Patavii, Typis I. Comini (1717). A mere description of the contents shows its irreducibility to our disciplinary context. In the first book after a few definitions (§§ 2-14), a short examination of the nature of water (§§ 17-22), an account of the theory of the 'simplex motion' (§§ 23-35), Poleni dealt with the 'mixed motion' of water by way of some very accurate experiments (§§ 36-55). Afterwards many problems and theorems were proposed (§§ 56-101) and the deepest features of the 'mixed motion' studied with the help of differential calculus and three-dimensional analytic geometry (§§ 102-140). In this respect it represented one of the first Italian books dealing with physical subjects by way of the new mathematics. In the second book, without forgetting the mathematical and experimental approach followed in the first, the scene was constituted by the Venetian lagoons which were analyzed taking account of the action of men (§§ 142-153), sea (§§ 154-202) and rivers (§§ 203-217).

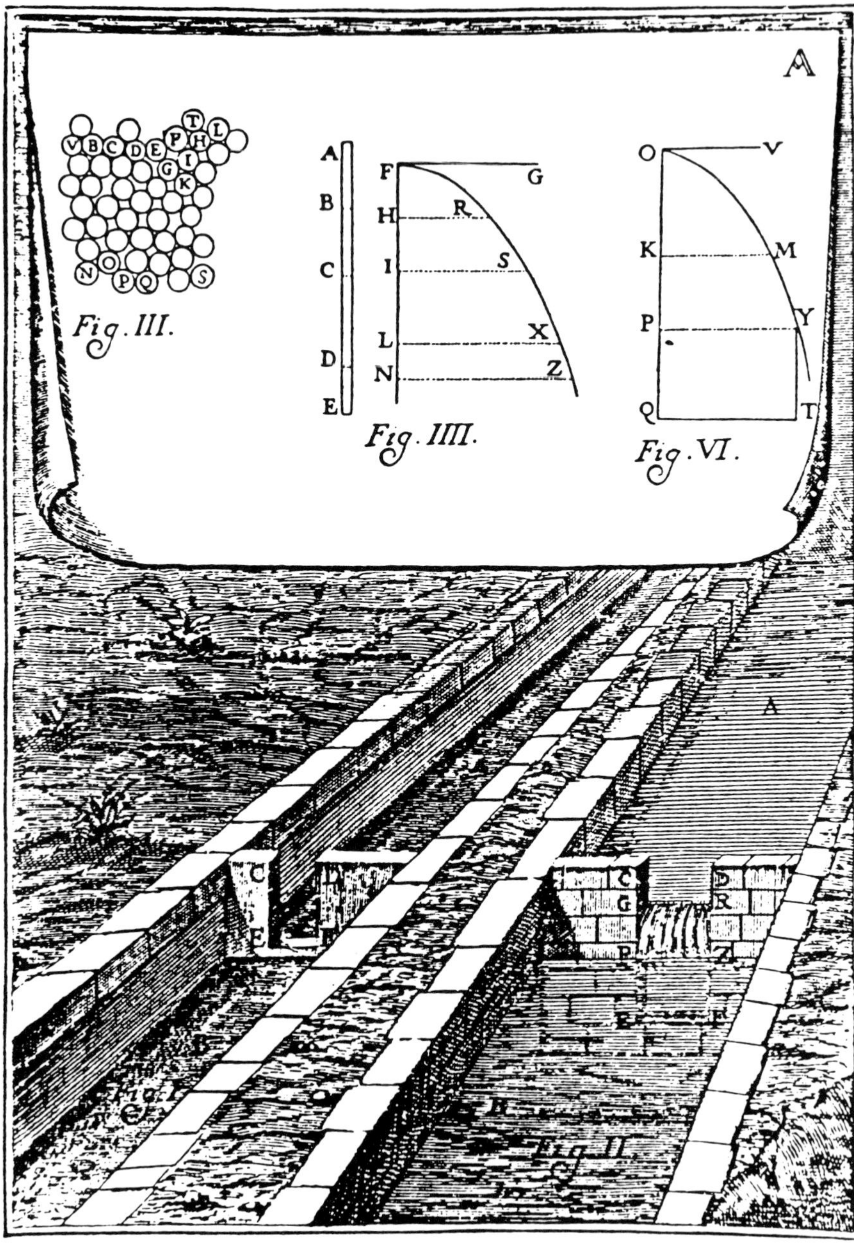


Fig. 30 From: G. Poleni, *De Motu Aquae Mixto Libri Duo. Quibus multa nova pertinentia ad Aestuaria, ad Portus, atque ad Flumina continentur*, Patavii, I. Comini, 1717: *Tabula A*, p. 3.

water and meeting no opposition from other water. The motion of living water was called 'Motus Simplex' by Poleni and according to him it was the only one studied until then, although the dead water too could be put in motion by the living water above it. To distinguish compound motion in which both living and dead water were carried on from simplex motion, Poleni devised for the former the name of 'Motus Mixtus'. In fig. II of *tabula A* of *De Motu Aquae Mixto* (see here fig. 30, where Poleni's original plate is reproduced) was pointed out an example of mixed motion: the upper level of water 'A' was higher than the upper level of 'B', so that 'GPZR' ran through the rectangular opening 'CDEF' as living water. Also 'PEFZ', although dead water, ran through the opening because of the pressure of living water and so all water 'GEFR' flowed by mixed motion.[10] While the profile giving the velocity distribution along a given perpendicular 'FN' of simplex motion was a parabolic one as shown by Torricelli and Guglielmini (*tabula A*, fig. IIII), the profile of the distribution along a given perpendicular 'OQ' of mixed motion (*tabula A*, fig. VI) registered two important variations. The upper part 'OMY' was still a parabola, even with a reduced parameter to take into account the diminution of velocities (confirmed by the experimental data collected by Poleni) caused by the amount of dead water to be put in motion. The second change is, however, more interesting: the lower part 'PYTQ' was a rectangular one, that is all the dead water flowed at constant speed regardless of the depth, given that running water through the portion 'PQ' (representing the 'altitudo mortua') was subject to the same 'pressure' of living water as the lowest point 'P' of the living portion 'OP' of the same perpendicular.[11] To find the relations between the three variables constituted by the ratio of the parameters of mixed and simplex motion, the dead height and the living height, Poleni performed many accurate experiments with a highly original apparatus (see *tabula B* of *De Motu Aquae Mixto* reproduced in fig. 31), coupling experimental physics and new mathematics in a creative way.[12]

According to the Paduan professor the case was fairly general: water flowing in and out of the lagoons through the 'Porti' (opening in the sea-shores) during high and low tide, water streams in a harbour or at the mouth of a river, canal-water entering a basin, all these were instances of mixed motion. And in the second part of his work he specified that many of the things related

10. *Ibidem*, pp. 2-3, §§ 2, 4, 6, 10.

11. *Ibidem*, pp. 27-30, §§ 57-59.

12. *Ibidem*, pp. 20ff. An account of Poleni's experimental research in A. Ghetti, 'Giovanni Poleni idraulico teorico', in *Giovanni Poleni (1683-1761) nel bicentario della morte*, Padova, Accademia Patavina di Scienze, Lettere ed Arti (1963), pp. 19-42, in part. pp. 21-25.

Fig. 31 From: G. Poleni, *De Motu Aquae Mixto Libri Duo. Quibus multa nova pertinentia ad Aestuaria, ad Portus, atque ad Flumina continentur*, Patavii, I. Comini, 1717: *Tabula B*, p. 21.

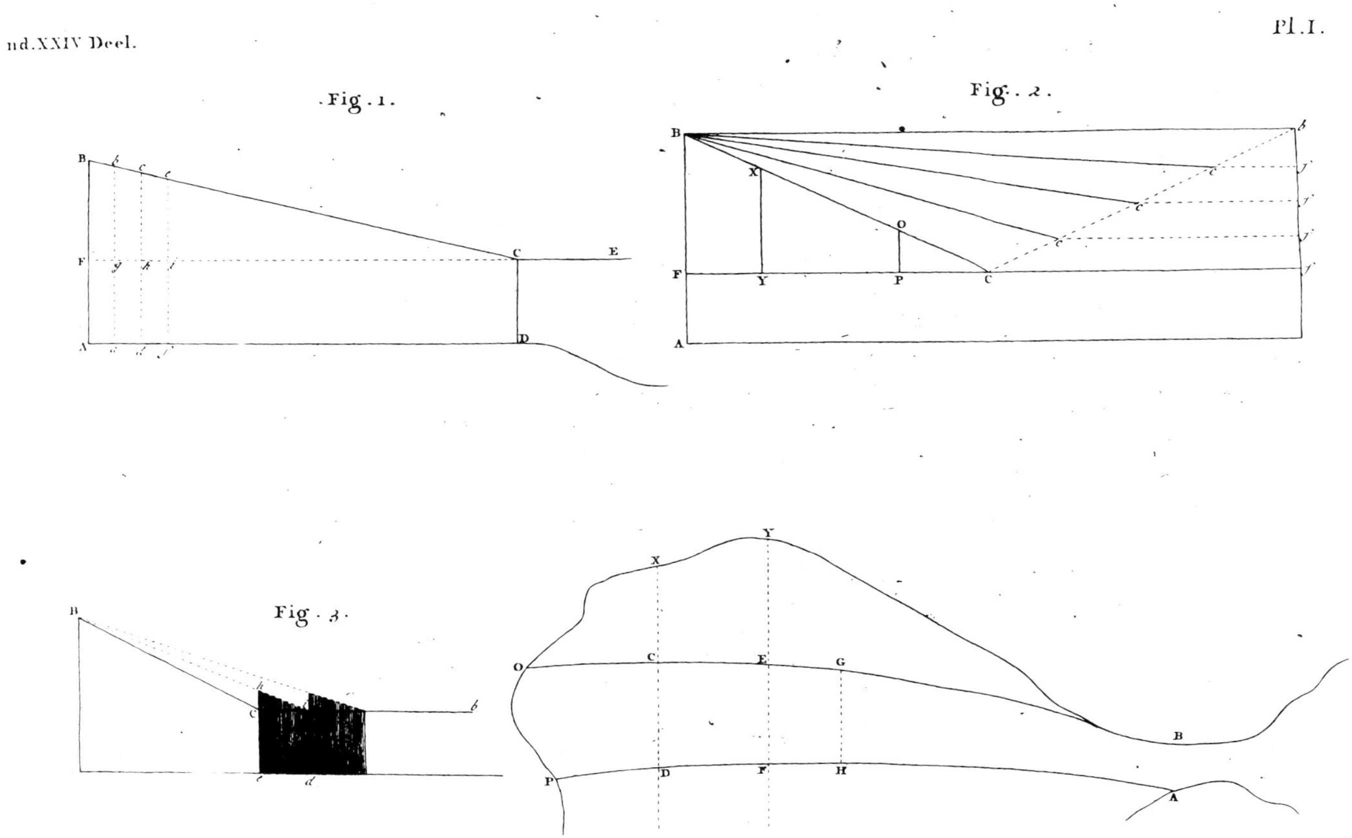


Fig. 32 From: C. Brunings, 'Antwoord op de vraag: Is de algemeene Grondregel der Hydrometrie, ...', *Verhandelingen uitg. door de Hollandsche Maatschappye der Weetenschappen*, XXIV, 1787; *Plaat I*, p. 58.

to mixed motion applied perfectly to some "Batavorum locis."[13] Thus it is not merely by accident that we are able to trace a positive influence of Poleni's work on Dutch hydraulicians of the second half of the eighteenth century such as Christiaan Brunings (1736-1805), who in 1769 succeeded the Leyden professor Johan Lulofs (1711-1768) as general superintendent to the rivers of Holland and West-Friesland and who is considered one of the fathers of the modern Dutch organization devoted to water control. The occasion was a prize essay written by Brunings to answer a question proposed in 1782 by the Haarlem Hollandsche Maatschappij der Wetenschappen, an essay rewarded in 1784 and published in 1787.[14] The formulation of the question showed a strong practical concern, in this case for the improvement of the navigability of the Ye (or IJ according to the modern spelling) between Amsterdam and Spaarndam, which was typical of the enlightened scientific societies and academies of the time. However, its beginning ("Is de algemeene Grondregel der Hydrometrie, ...") recalled unequivocally Italian terminology and theoretical attitude.[15] And, more to the point, the theoretical part of Brunings' answer was largely founded on Poleni's theory of mixed motion. The Dutch hydraulician proposed the case of a horizontal conduit (see Fig. 1 of Brunings' original plate, here reproduced in fig. 32) ascribing the cause of motion to the slope of the free surface, i.e. to the small differences respectively between the "levendige hoogte" 'BF' and 'bg', 'bg' and 'ch', .... The difference of pressure between the different columns of water then put in motion the "doode water" 'FAag', 'gadh', ... too. Afterwards Brunings considered the case of the ebb and flow ( see Fig. 2 of Brunings' original plate, here reproduced in fig. 32) and concluded that between the two cases there was no essential difference, both being instances of "gemengde beweeging" ("motus mixus").[16]

Brunings' paper represents significant evidence of the influence of Poleni's theory in Holland, but it also put forward additional problems if considered in the view of the transmission of scientific ideas between the two areas. Why e.g. was there such an intolerable delay of more than half a century before a theory

13. *Ibidem*, respectively on p. 50, § 101 and pp. 99-100, § 175.

14. C. Brunings, 'Antwoord op de vraag, (...): Is de algemeene Grondregel der Hydrometrie, dat de grootste diepten steeds op de naauwste plaatsen der Rivieren bevonden worden, insgelyks toepasselyk op de Zeeboezems, gelyk het Ye, alwaar de stroom door de Getyen veroorzaakt word? (...)', *Verh. HMWH*, XXIV (1787), pp. 1-58.

15. The term 'hydrometry' to indicate in a concise way the science of the measuring of running waters was introduced by Guglielmini and by other Italian scientists (see C.S. Maffioli, 'Domenico Guglielmini ...', op.cit. in note 8, in part. pp. 89-90), and often used in eighteenth-century Italy as an alternative to the word 'hydraulics'.

16. C. Brunings, 'Antwoord op de vraag, ...', op. cit. in note 14, in part. pp. 11-24, §§ 6-11. Poleni's *De Motu Aquae Mixto* is quoted on p. 37.

which was so useful for the Low Countries was transmitted from Italy and actually adopted in Holland? We cannot explain the delay by a break in information, considering the relatively prompt notice of the issue of *De Motu Aquae Mixto* in the *Journal Literaire* of The Hague[17] and the fact that the Venetian scientist was fairly well-known in Holland. Also the paucity of direct relationships or the difficulty of a regular book trade between Italy and Holland seem inadequate explications at least in this instance. To find a better one, it is necessary to recall that Poleni's work was not a treatise on river hydraulics. Poleni seemed to consider subject to the laws of mixed motion just the stretches of rivers close to their mouths, while for upper parts with an inclined bed he apparently still relied upon the theory of simplex motion.[18] In eighteenth-century Holland, where the hydraulic scene was dominated by the fear of being submerged by the floods of the main rivers (cf. 'infra'), Guglielmini's magisterial treatise of river hydraulics certainly appealed more than mathematical and experimental finesses in the mechanics of fluids. However, this was largely the case also in Italy where Poleni's work did not have the influence it deserved. Its theoretical features, partially alternative to Guglielmini's (i.e. on the question of the relation velocity/ pressure), and its new approach founded upon experiment and calculus were not easily accepted either by people devoted to practical hydraulics or by mathematicians still tied up with classical methods. Last but not least, the opposition of the Venetians to the Bolognese project of bringing the Reno into the Po Grande probably contributed to the limitation of the diffusion of Poleni's ideas.[19]

*

In the first edition of the *Physices Elementa Mathematica* (1720) by Willem Jacob 's Gravesande (1688-1742), professor of mathematics and astronomy at the University of Leyden since 1717, a short chapter is devoted to river hydraulics.[20] Although the Dutch scientist did not quote any of his sources, the theory summarized there was clearly Guglielmini's as 's Gravesande was to acknowledge in the preface to the third edition (1742) of his work.[21] However,

17. *Journal Literaire de l'année 1717*, IX, 2 (1718), pp. 469-470.

18. G. Poleni, *De Motu Aquae Mixto* ..., op. cit. in note 9, in part. p. 76, § 146 and p. 97, § 169.

19. A concise account in C.S. Maffioli, 'Domenico Guglielmini ...', op.cit. in note 8, in part. pp. 119-123.

20. W.J. 's Gravesande, *Physices Elementa Mathematica, Experimentis Confirmata. Sive Introductio ad Philosophiam Newtonianam. (...), Tomus Primus*, Lugduni Batavorum, Apud P. Vander Aa & B. Janssonium Vander Aa (1720), pp. 143-147.

21. W.J. 's Gravesande, *Physices Elementa Mathematica, Experimentis Confirmata. Sive*

the Leyden professor, like his colleague Poleni at Padua, was also requested to apply himself to practical hydraulics. The States of Holland decided e.g. in 1729 on an inspection of the Merwede from Gorinchem downwards with the aim of finding the means to free Gorinchem, the Alblasser Waard and the Land van Arkel from the danger of floods .[22] In the inspection 's Gravesande and his Leyden colleague J. Wittichius participated, as well as the land-surveyor N.S. Cruquius. It resulted in a very detailed report accompanied by a map of the river provided with depth curves.[23] After 's Gravesande's death, Johan Lulofs succeeded him not only at the University of Leyden but also as technical adviser in hydraulic matters. In 1750 he published a geographical treatise in which, in the chapter devoted to rivers, he adopted as his "beroemden Leermeester 's Gravesande" Guglielmini's parabolic profile of the speeds.[24] In 1754 Lulofs was appointed 'Inspecteur Generaal van de Rivieren' of the province of Holland, a new function created to give the Leyden professor a formal title in a period of great concern about the situation of the main rivers.[25] In a map drawn by the land-surveyor C. Velsen (here reproduced in fig. 33), we see the different branches of the rivers Rhine and Meuse (the most important were the Waal, the Lower Rhine, the IJssel, the Lek and the Merwede) in about 1750. At the end of the seventeenth century the Rhine, just before entering the territory of Gelderland, split into the Lower Rhine and the Waal but the latter carried out more than 90% of all the water. To give more outflow to the northern branches, an essential task for the improvement of internal navigation and the defence of the Republic, a new mouth for the Lower Rhine and a channel were dug out in 1705-1707. But the partition of the water was all but stable and the Lower Rhine began to capture more and more water thus provoking, particularly in winter time when ice-dykes hindered the outflow, overflowing and breaching of the banks.[26] The most feared disaster was the breaking of the northern-banks of the river Lek, an event when,

---

*Introductio ad Philosophiam Newtonianam. (...), Tomus Primus, Editio Tertia duplo auctior*, Leidae, Apud J.A. Langerak, J. et H. Verbeek (1742), p. xxix ("Caput X agit de motu fluminum. Quae de determinandi velocitate aquae habemus, dedit Guglielmini (*Mensura Aquarum fluentium*). Ex quo auctore etiam quaedam alia desumsimus (De Fluminum Natura)").

22. *Resolutien* (1729): 10.05.1729, pp. 368-370.

23. *Resolutien* (1730, I): 12.07.1730, pp. 553-582.

24. J. Lulofs, *Inleiding tot eene Natuur- en Wiskundige Beschouwinge des Aardkloots*, Leiden en Zutphen, By J. en H. Verbeek en A.J. van Hoorn (1750), p. 346.

25. *Resolutien* (1754, I): 08.05.1754, pp. 385-386.

26. While in the sixteenth and seventeenth centuries only 12 great breaches of the banks of the main rivers were recorded, their number increased to 23 in the first half and to 136 in the second half of the eighteenth century (J. Blanken, *Verhandeling over de Algemeene Rivier en Waterstaatkundige Ontwerpen (...)*, Utrecht, bij N. van der Monde (1836), p. 29).

Fig. 33 'Map of the rivers Rhine, Meuse, Waal, Merwede and Lek, through the Provinces of Gelderland, Holland and Utrecht' (From: C. Velsen, ***Rivierkundige Verhandeling***, Amsterdam, I. Tirion, 1749).

according to accounts of the age, the greater part of the Provinces of Utrecht and Holland risked being totally submerged. In 1754 the situation was considered so serious that Lulofs and the land-surveyor M. Bolstra wrote a dramatic appeal to the States of Holland,[27] trying in this way to surmount the internal opposition of the towns of Dordrecht, Rotterdam and Gorinchem to a project of diverting part of the waters of the Lek in case of emergency. A brief outline of this project was given by Lulofs and Bolstra in a report of 15.10.1751,[28] but the complete version was not presented to the States before 09.04.1754.[29] The aim was to divert more than 400 cubic meters per second by means of 10-12 sluice-gates and to discharge this amount of water into the Merwede near Gorinchem (see fig. 33), so as to diminish by about two feet the surface-level of the Lek. The idea of a lateral diversion when a river threatened to overflow onto all the neighbouring land was by no means a new one in Holland: 's Gravesande, Wittichius and Cruquius had e.g. envisaged in their report of 1730 (cf. 'supra', note 24), in the event of ice-barriers stopping the flow of the Merwede, such an extreme remedy. But these kinds of projects met with a lot of opposition, both for their huge cost and the fear of the population concerned. Also the federal constitution of the Republic, with several involved sovereign Provinces which often had opposing interests, certainly did not help their realization and only a side diversion of the Meuse was achieved during the eighteenth century.[30]

However, the project by Lulofs and Bolstra also met opposition of a technical nature, in particular from Cornelis Velsen who in 1749 had already defined the former projects of 'overlaaten' (artificial lowering of a tract of the banks to let part of the stream flow out laterally) a "chimericq en imaginair" remedy against overflowing and ice-barriers.[31] Last but not least, even in Leyden somebody dared to contradict the influential professor and, still worse, through arguments of a scientific nature which appealed to fluvial observations and to experiments performed with a small scale model of an

27. J. Lulofs-M. Bolstra, 'Kortbondig Vertoog van het eminente gevaar, in het welk zig de Provincie van Holland bevindt, wegens de kwaade gesteldheid van de Leck', in *Resolutien* (1755, I): 18.04.1755, pp. 541-547.

28. *ARA, Derde Afd., Gecommitt. Raden van de Staten van Holland and West-Friesland 1621-1795*, 3970, Bijlagen, vol. 6: Nr. 306.

29. *Ibidem*, 3976, Bijlagen, vol. 12: Nr. 566; *Resolutien* (1754, I): 09.04.1754, pp. 322-334.

30. G.P. van de Ven, *Aan de wieg van Rijkswaterstaat. Wordingsgeschiedenis van het Pannerdens Kanaal*, Zutphen, De Walburg Pers (1976), p. 274.

31. C. Velsen, *Rivierkundige Verhandeling, (...)*, Amsterdam, By I. Tirion (1749), p. 203. Some papers in the *ARA* witness the criticism by Velsen about the projected diversion of the Lek (note 29: Nr. 555, 573 and 583; *Ibidem, 3964, Journal der besognes (...) 02.01.1751-31.05.1754*, sub 1754, dates 14.02, 28.02, 06.03 and 03.04).

inclined canal. He was Léopold Genneté (1706-1782), a former student of 's Gravesande who, in the period 1750-1756, settled again in Leyden to build a new kind of water engine with the support of the Hoogheemraadschap van Rijnland and of the States of Holland.[32] On 30.09.1755 Genneté addressed to the burgomaster of Leyden a long letter in which the results of the experiments performed with his artificial river were produced. In Genneté's opinion these experiments clearly showed the uselessness of river diversions. In the letter, by quoting Guglielmini, the importance of keeping all the water together to increase the speed and to realize an efficient scour of the river-bed was further emphasized. The experiments had also to do with the confluence of rivers, showing no apparent increase in the surface-level of the water in a small canal after the confluence of another similar one. In spite of the fact that no exact description of the experimental apparatus was given, Genneté's letter after its publication in Paris in 1760[33] was particularly welcomed in Italy by people supporting the project of joining together the Reno and the other rivers and streams of the region into a sole channel so as to discharge the whole mass of water in the Primaro. One of them was the young Milanese mathematician Paolo Frisi (1728-1784), in those years professor at Pisa and involved in the matter at the request of Clement XIII. In his work *Del modo di regolare i fiumi, e i torrenti* (1762), Frisi did not hesitate to call "aureo libro" the pamphlet of Genneté and to quote the Leyden experiment to support the existence of a simple proportionality between the speed and the amount of water carried, before and after the confluence of two rivers.[34] It originated in a controversy with the Ferrarese mathematician T. Bonati, who was to perform similar experiments to Genneté's ones obtaining a different outcome. Thus when Frisi, after his call to Milan in 1764, started a long journey through Western Europe in the spring of 1766, he probably had Holland in mind above all to exchange information about hydraulic subjects with Dutch colleagues. This is indirectly confirmed by the still unpublished part of Frisi's journal[35] related to

32. A.C.J. Habets, 'Het waterrad van Léopold Genneté', *Spiegel Historiael*, XIII, 10 (1978), pp. 620-625.

33. L. Genneté, *Expériences sur le Cours des Fleuves: ou Lettre a un Magistrat Hollandois; (...)*, Paris, Chez M. Lambert (1760).

34. P. Frisi, *Del modo di regolare i fiumi, e i torrenti, principalmente del bolognese, e della Romagna, Libri Tre*, Lucca, Per V. Giuntini (1762), pp. 135, 139-140, 142-143. Not all the supporters of the Primaro project were, however, enthusiastic about Genneté's experiments (see e.g. in *BAM*, 'Cod. Y 148 Sup', ff. 59r.-v., a letter by E. Zanotti to Frisi of 21.11.1761).

35. A copy of Frisi's journal is kept in *BAM*, 'Cod. Y 163a Sup', ff. 42r.-49v. It has been only partially published by Franco Venturi (*Illuministi Italiani*, III, Milano-Napoli, Ricciardi (1958), pp. 305-314). I am to publish the part related to Holland (ff. 48r.-49v.), together with an index and summary of the contents of the 'Dutch' correspondence of the Milanese mathematician, in

Holland.[36] A more or less direct consequence of Frisi's journey was the beginning of his correspondence with two Utrecht professors, Rijklof Michaël van Goens (1748-1810) and Johan Frederik Hennert (1733-1813), and his association in May 1770 to the Hollandsche Maatschappij der Wetenschappen, a scientific society founded in 1752 in Haarlem.[37] Since the Society requested from its fellows at least one dissertation every seven years to be published in its *Verhandelingen*, Frisi chose as the subject of his paper fluvial hydraulics and precisely the crucial problem of the confluence and division of rivers. Frisi's Latin autograph, having as its title *De divisione, et confluxu Fluminum*, is still in the archive of the Hollandsche Maatschappij.[38] It was received by the Society through van Goens and, at a meeting held on 06.08.1771, it was decided to translate Frisi's dissertation into Dutch and to publish it both in Latin and in Dutch.[39] However, in the end it was only published in Dutch, probably because of its practical character.[40] In this paper Frisi immediately put his finger on the dolent punctum of Genneté's experiments, as well as on the debate which had arisen in Italy about them. He defended Genneté (and himself) observing that the experimental apparatus used in Italy was different from the Leyden one, but he also pointed out that the experiments performed with small models of canals pertained more to physics than to the theory of rivers. However, he emphasized that the Italian hydraulicians were in general against diversion as a means of preventing floods, and he used the opposite case of the confluence of the rivers Samoggia

---

my forthcoming work 'Paolo Frisi (1728-1784) e l'Olanda. Relazioni scientifiche, il vaggio del 1766, la corrispondenza 'olandese' ', *Incontri. Rivista di studi italo-nederlandesi*, n.s., IV, 1-2 (1989).

36. Frisi sejourned in Holland from the 7th until the 19th of September 1766. The 9th he met Lulofs while on the 14th he met W. Bentinck van Rhoon and J.N.S. Allamand, who had both been present at Genneté's experiments and gave him further details on them.

37. *Verh. HMWH*, XII (1770), *Voorbericht*, p. VI. Frisi was elected together with R.J. Boscovich, "beiden Professoren in de Mathesis te Milaan". They were the first Italian fellows of the *HMWH*, the oldest society of sciences in The Netherlands (Giovan Francesco Salvemini da Castiglione might be considered the first Italian, but he was elected in 1760 in his quality of professor of mathematics at Utrecht and the standard history of the *HMWH* by J.A. Bierens de Haan classifies him correctly among the "binnenlandse leden"). For further information, see my 'Paolo Frisi e l'Olanda', cit. in note 35 (in press).

38. *Paulli Frisii/Mediolanensis Mathematum Professoris/De divisione, et confluxu Fluminum/ad Societatem Batavo-Harlemensem*, in *Arch. HMWH*, map 1772.

39. *Arch. HMWH*, 'Notulen 1767-1781': sub 06.08.1771, f. 648.

40. P. Frisi, 'Berigt (...) nopens de Verdeeling en Zamenloop der Rivieren', *Verh. HMWH*, XIV (1773), pp. 112-130. Another dissertation of Frisi, to which in 1785/7 was awarded the prize on the irregularities ('ongelykheeden') of the motion of the satellites of Jupiter, however, had the honour of being published both in Dutch and in Latin (*Verh. HMWH*, XXV (1788), pp. 225-361).

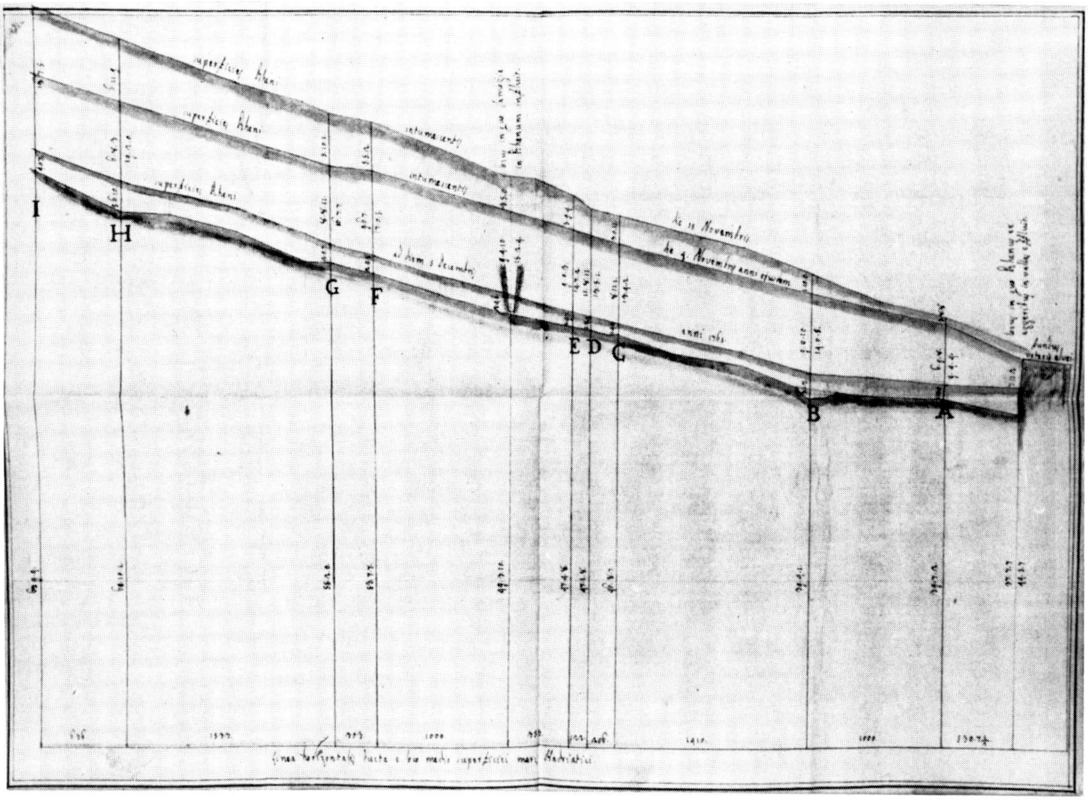


**Fig. 34 Longitudinal profile of the Reno of Bologna, in *Paulli Frisii ... De divisione, et confluxu Fluminum* (Archive of the Hollandsche Maatschappij der Wetenschappen, Haarlem: map 1772, original cm 51.5 x 35.5).**

and Reno to show that the area of the liquid cross-section after the confluence was much less than the sum of the areas of the two liquid cross-sections before (see in fig. 34 the profile of the Reno produced by Frisi). Frisi's point against any kind of diversion was too pretentious if assumed as the general opinion of the Italians on the matter. Christiaan Brunings, in judging in 1776-1777 a prize essay on the usefulness of the derivations to prevent inundations, put forward that the author (C. Zillesen) had refused the derivations just on the basis of the dissertation of Frisi without distinguishing a permanent derivation from a temporaneous one controlled by means of gates. Besides, Brunings quoted a passage from Poleni's *De Castellis* (1718) in which the latter kind of derivation was described as an "artificium utile."[41] In other words the matter, both in Italy and in Holland, was a controversial one.

41. I refer to three letters written by Brunings in 1776-7 in his capacity as a member of the jury

In 1789 a prize essay by Brunings on the speed of running water was published in vol. XXVI of the *Verhandelingen* edited by the Hollandsche Maatschappij, an essay awarded the gold medal three years before.[42] I cannot analyze here this important contribution, in which a new kind of tachometer was proposed. It is, however, worth noticing the great acquaintance with the Italian authors of hydraulics shown by Brunings in this dissertation. Not just works published in the second edition of the *Raccolta d'Autori che trattano del moto dell'acque* ( 9 vols, Florence 1765-1774) and in the *Nuova Raccolta* (...) (7 vols, Parma 1766-1768) were reported there, but also original works of F.D. Michelotti, A. Lecchi, A.M. Lorgna, L. Ximenes, T. Bonati and others, edited in the period 1765-1784 and not included in both collections. Brunings showed a predilection for authors oriented towards the empirical side of hydraulics, and he valued very much the works of the Italian Jesuits Lecchi and Ximenes. On the other side, it is also worth noticing the reaction to Brunings' approach by two members of the jury. Both declared that Brunings' dissertation deserved the prize although they were, in quite different tones, not fully in harmony with the empirical approach chosen. Pybo Steenstra, who lectured in mathematics, navigation and astronomy in Amsterdam, was the most critical: he thought that Brunings' drastic refusal of Guglielmini's mathematical theory was a frontal attack against any kind of mathematical generalization in hydrometry and fluvial hydraulics, and he defined it "een geheele omverre-werping van die Weetenschappen."[43] The second one, the Utrecht professor of mathematics, astronomy and philosophy Johan Hennert, was more positive. In a letter to the Society's secretary C.C.H. van der Aa, written on 19 April 1786, he pointed out that in his "publique Lessen over den Loop des Rivieren" he had proposed a new theory to determine the average speed of a river in a given cross-section. Besides his (largely favourable) observations upon the dissertation, he had included a sketch of his theory to be used by the author of

---

of a prize proposed in 1771 by the *HMWH* (see *Arch. HMWH*, map 1771, I, prijsvraag 20). The original Latin passage of Poleni is as follows: "Neque pluribus artificium utile persequar, quo efficitur, ut per diversionum Fossas tantum divertere aquarum quis possit, quantum superfluere necesse esset eo tempore, quo Flumina immodico illarum proventu turgescunt." (G. Poleni, *De Castellis per quae derivantur Fluviorum Aquae habentibus latera convergentia Liber. Quo etiam continentur nova Experimenta ad Aquas Fluentes, & ad Percussionis Vires pertinentia*, Patavii, Typis I. Comini (1718), p. 10, § 16).

42. C. Brunings, 'Verhandeling over de Snelheid van Stroomend Water, en de middelen, om dezelve op allerleïe diepten te bepaalen: (...)', *Verh. HMWH*, XXVI (1789), pp. 1-187 (with two appendixes, pp. 188-233).

43. *Arch. HMWH*, map 1781, II, prijsvraag 43: Steenstra's judgement is signed by the motto "Quaestioni Responsionem adaptatam esse oportet, non autem Responsioni Quaestionem", 16 pp. n.n., quot. on p. 11.

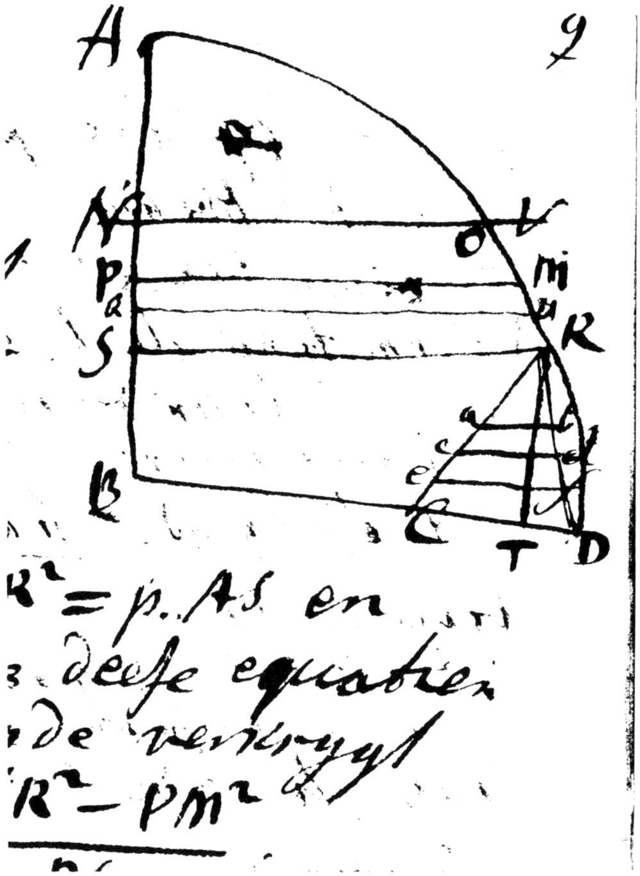

**Fig. 35 The profile (NORCB) of the speeds along a given perpendicular (NB) of a river proposed by J.F. Hennert (From: *Schets van eene Theorie, nopens het bepaalen der middelbaaren snelheid van het stroomend waater*, in Archive of the Hollandsche Maatschappij der Wetenschappen, Haarlem: map 1781, II, prijsvraag 43).**

the same dissertation (Brunings) or by the Society.[44] Hennert indeed thought that a theory (in our case some hypotheses) was necessary to compare experimental results in different ways so as to produce a coherent body of knowledge. However, it is a symptom of the difficulty in joining the rational mechanics of fluids with practical hydraulics that even a pupil of Euler such as Hennert did not find anything better than to propose a modified version of Guglielmini's parabolic profile of speeds (see fig. 35).

44. *Ibidem*: J.F. Hennert, *Schets van eene Theorie, nopens het bepaalen der middelbaaren snelheid van het stroomend waater.*

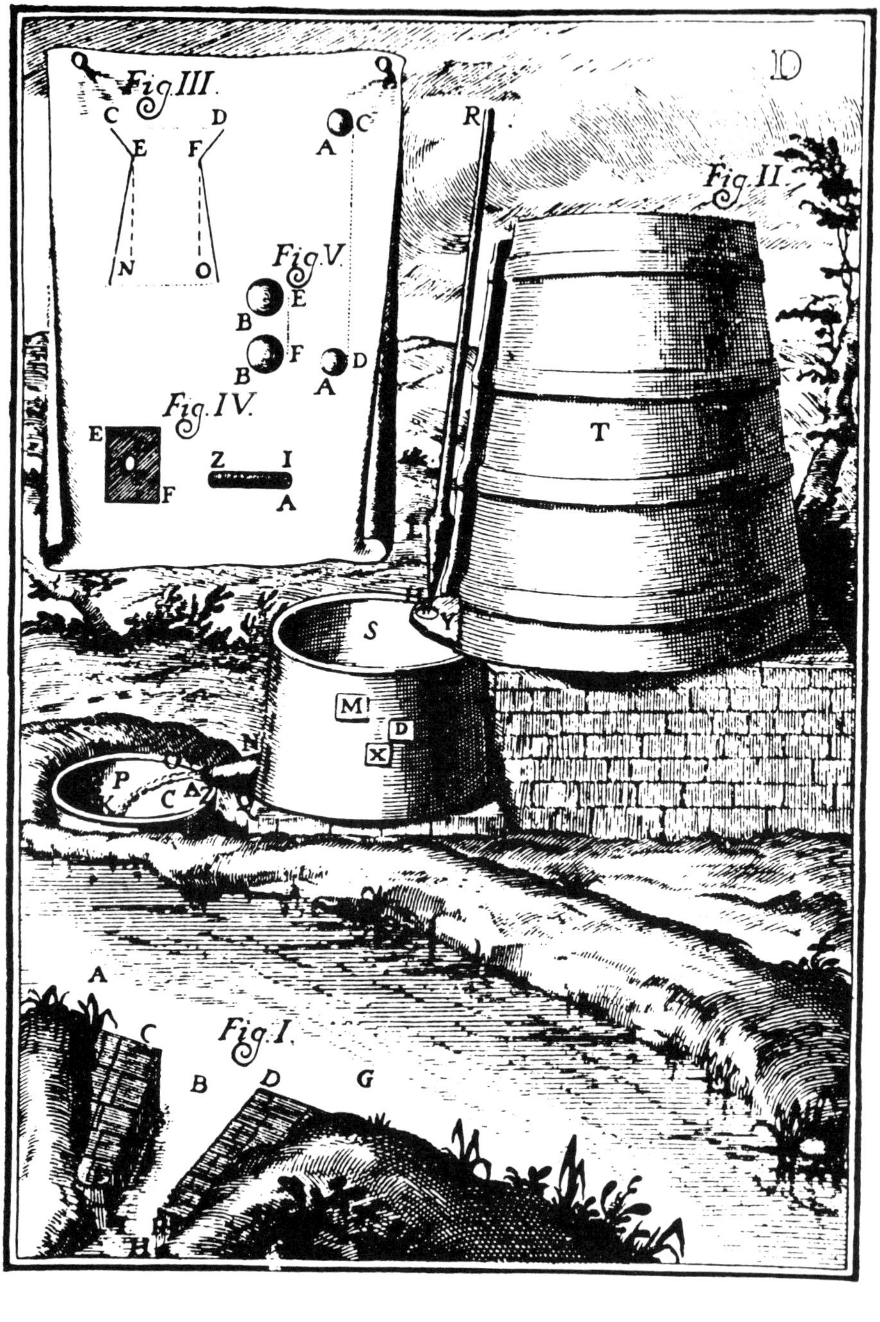


Fig. 36 From: G. Poleni, *De Castellis per quae derivantur Fluviorum Aquae habentibus latera convergentia Liber*, Patavii, I. Comini, 1718: *Tabula D*, p. 11.

## III. *Experimental Physics*

Hydraulics and experimental physics were much more bound together in the seventeenth and eighteenth centuries than nowadays and many hydraulic problems were interpreted in terms of experimental physics. A significant instance is to be found in the way in which the problem of side diversions of rivers in the event of floods was dealt with in Poleni's *De Castellis* (1718). According to Poleni, there were two main points in planning such gates or 'castella': they had to derive a definite 'measure' of water and they had to resist the 'force' of water running and pressing against their sides. Poleni proposed investigating the first point in the case of 'castella' having converging sides and performed many accurate experiments of efflux from orifices furnished with additional mouthpieces (see Fig. II and IV of *Tabula D* of Poleni's *De Castellis,* here reproduced in fig. 36). He concluded by way of analogy that the outflow from the 'castellum' CDEF (*Tabula D*, Fig. I), the other factors being unchanged, was primarily determined by the magnitude of the exit mouth EF and not by that of the entrance mouth CD.[45] For the right assessment of the second point, he thought it essential to know first of all the 'force' of running water. He performed to this end the well-known experiment of the two spherical balls of equal radius falling on a vessel filled with tallow (or with clay or wax) from heights in inverse ratio to their weights, so that the product '(weight) x (height of fall)' was the same in both cases, and he observed that the two small pits made in the tallow were precisely the same. By the equality of the effects Poleni inferred that of the causes, namely of the 'living forces', and he concluded that the 'living force' of a falling body was proportional to the square of its speed.[46] It was the first real experiment performed to decide the 'vis viva' controversy and it is interesting to see that the experiment took place in a technological (but also theoretical) context, in spite of the later description of the controversy as a purely metaphysical one or as a pure question of words.[47]

A few years after the issue of Poleni's *De Castellis*, 's Gravesande became 'converted' to the Leibnizian measure of the 'force' and he described in his *Essai d'une nouvelle theorie sur le choc des corps* (1722) an experiment similar to Poleni's.[48] The Leyden professor, however, arrived independently at the

45. G. Poleni, *De Castellis per quae derivantur Fluviorum Aquae ...*, op.cit. in note 41, p. 36, § 73.

46. *Ibidem*, pp. 56-57, § 118.

47. J. le Rond d'Alembert, *Traité de Dynamique, (...). Nouvelle Edition (...)*, Paris, Chez David (1758), (Johnson Reprint Corp., New York and London, 1968), p. xxij.

48. W.J. 's Gravesande, 'Essai d'une nouvelle theorie sur le choc des corps', *Journal Literaire de l'Année 1722*, XII, 1 ( 1723), pp. 1-54, in part. pp. 21-23, §37 (the essay is dated: 'A Leide ce 10 d'Avril 1722').

experiment: in the *Journal Literaire* no news appeared of *De Castellis*, and it is quite possible that 's Gravesande had also missed the review of Poleni's book in the *Acta Eruditorum* of 1720. The Dutch scientist, in a *Supplement* to his theory (1723), stated that only after the printing of his essay had he seen Poleni's tract and recognized the priority of the Paduan professor in performing the experiment.[49] Thus here we do not see the result of a direct influence but of the independent emergence of the same theme in Padua and Leyden. This emergence, as we shall see later, if considered in the long perspective of scientific traditions, is evidence of the persistence in both areas of the doctrines respectively of Galileo/Torricelli and Huygens as well as of the close relations between them. However, if we look at local factors, the influence of Leibniz and his followers was evident both in Padua where J. Hermann and Nicolaus I Bernoulli taught mathematics respectively in 1707-1713 and 1716-1719, and in Leyden where Johann Bernoulli (in 1695-1705 at Groningen) had succeeded in 1700 in convincing B. de Volder of the new measure of the force.[50] Not by chance thus did notice of Poleni's experiment come to 's Gravesande from Johann Bernoulli in a letter of 31.10.1722, while on the other side Nicolaus I Bernoulli informed Poleni on 30.04.1723 that the Leyden professor in the answer to his uncle Johann had apologized by saying that before the publication of the *Essai* he had not heard of the Paduan experiment and assured that at the first opportunity he would give Poleni his due.[51]

's Gravesande's 'conversion' put him in serious trouble with the English Newtonians, but he could justify the title *Philosophiae Newtonianae Institutiones* given in 1723 to his compendium by saying that although it did not propose Newton's opinion on the measure of forces the book however followed both Newton's results in other branches of physics and, above all,

49. W.J. 's Gravesande, 'Suplement à la Nouvelle Theorie du Choc, inserée dans ce Journal pag. 1 & suiv.', *Ibidem*, pp. 190-197, in part. p. 191.

50. G.W. Leibniz, *Die philosophischen Schriften*, 2, hrsg. von C.J. Gerhardt, Hildesheim-New York, G. Olms Verlag (1978) (Nachdruck der Ausgabe Berlin 1897), the correspondence Leibniz-de Volder, in part. the letter by de Volder of 05.04.1700, pp. 207-210. See also G. Gori, *La fondazione dell'esperienza in 's Gravesande*, Firenze, La Nuova Italia Editrice (1972), pp. 32-34.

51. Johann Bernoulli's letter was published by J.N.S. Allamand in his *Histoire de la vie et des ouvrages de M.r 's Gravesande* (*Oeuvres Philosophiques et Mathématiques de M.r G.J. 's Gravesande, (...), Premiere Partie*, Amsterdam, Chez M.M. Rey (1774), pp. ix-lix) in note U (pp. xxxvi-xlv). Notice of Poleni's experiment on p. xxxix and of the 'conversion' of de Volder (cf. 'supra') on pp. xl-xli. The minute of Nicolaus I Bernoulli's missive is in *OBUB*, 'Hs. L Ia 21/2', ff. 289r.-290r.

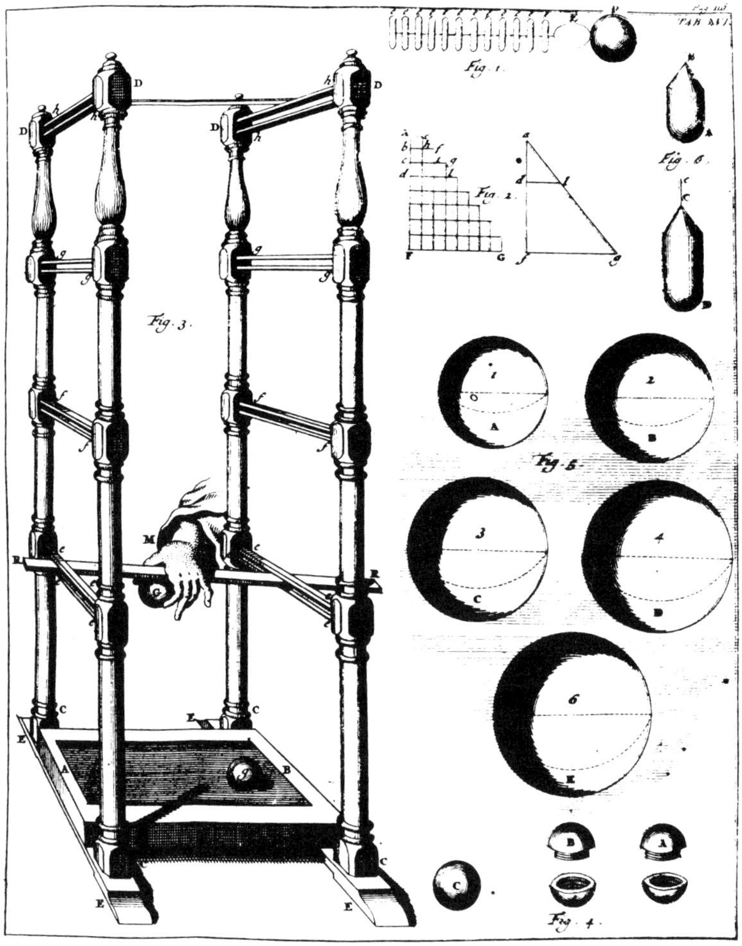

**Fig. 37 's Gravesande's apparatus to perform Poleni's experiment (From: W.J. 's Gravesande, *Physices Elementa Mathematica, Experimentis Confirmata. Sive Introductio ad Philosophiam Newtonianam, Tomus Primus, Editio Secunda*, Lugduni Batavorum, P. Vander Aa & B. Janssonium Vander Aa, 1725: *Tab. XVI*, p. 116).**

Newton's philosophical method.[52] He adopted the same strategy in 1725 in the second edition of his *Physices Elementa Mathematica (...) Sive Introductio ad Philosophiam Newtonianam*. Only in the preface to the third edition of the latter work (1742), where he had finally decided to quote all his sources, was it possible for the reader to perceive the extent of 's Gravesande's debt to continental physics. The move was a late answer to the criticism of men like

52. W.J. 's Gravesande, *Philosophiae Newtonianae Institutiones, In Usus Academicos*, Lugduni Batavorum, Apud P. Vander Aa (1723): *Ad Lectorem*.

Johann Bernoulli, and represented a symptom of a more oecumenical attitude. Among the Italians quoted by 's Gravesande, Poleni had a distinguished position and the reader was referred to an improved apparatus especially made to perform Poleni's experiment (see fig. 37). Of course in the preface he also referred to the Latin translation of the *Saggi di Naturali Esperienze* of the Accademia del Cimento, a translation ("cum commentario pulcherrimo & multis experimentis novis", said 's Gravesande[53]) published in 1731 by that other great Dutch Newtonian, Petrus van Musschenbroek (1692-1761).

*

If we turn from Poleni's hydraulics and mechanics to other sections of the physical sciences, no doubt Volta's electrical researches command major attention in order to assess Italian influence in eighteenth-century Holland. Alessandro Volta (1745-1827), professor of experimental physics at the University of Pavia since 1778, had indeed close relations with Holland through Martinus van Marum (1750-1837), director at Haarlem both of the cabinet of natural history of the Hollandsche Maatschappij (since 1777) and of Teyler's physical cabinet (since 1784) as well as secretary (since 1794) of the Maatschappij. While their scientific relations have already been studied, less care has been devoted to analyze more generally the influence of Volta in Holland.

Volta became famous throughout Europe thanks to three experimental devices: the 'elettroforo perpetuo' (1775), an apparently indeficient supplier of electricity; the 'condensatore' (1782), an instrument to reveal the weakest electricity; and, last but not least, the 'pila' (1800), which opened new and vast fields to nineteenth-century electrical research. On each of these discoveries there was a significant printed output in Holland:

1) The electrophore with its extraordinary inexhaustible electricity was very popular. A Dutch translation of two letters on the electrophore by the abbot Jacquet at Vienna and by Volta had appeared already in 1777.[54] Experiments

53. W.J. 's Gravesande, *Physices Elementa Mathematica, ..., Editio Tertia ...*, op.cit. in note 21, *Praefatio Hujus tertiae Editionis*, pp. xv-xxxviii, in part. p. xxvii and p. xxxi. 's Gravesande referred to the *Tentamina Experimentorum Naturalium captorum in Academia del Cimento (...) Ex Italico in Latinum Sermonem conversa. Quibus commentarios, nova experimenta, et orationem de methodo instituendi experimenta physica addidit Petrus van Musschenbroek*, Lugduni Batavorum, Apud J. et H. Verbeek (1731). The relationship of van Musschenbroek and Poleni are the subject of a forthcoming article of M.A. Pancino and G.A. Salandin, which is to appear in *Incontri. Rivista di studi italo-nederlandesi*, n.s., IV, 1-2 (1989).

54. *Vier natuurkundige verhandelingen, behelsende verscheiden opmerkens-waardige en*

with the electrophore made by J.R. Deiman (1743-1808) were published in 1780 in the *Algemeene Vaderlandsche Letteroefeningen Mengelwerk*,[55] while in 1781 the Bataafsch Genootschap der Proefondervindelijke Wijsbegeerte, a society devoted to experimental philosophy founded in 1769 in Rotterdam, proposed a prize in order to produce the best electrophore, to explain the particular phenomena observed and to show the contribution the study of the new apparatus could give to electrical theory. The prize was awarded in 1782 to an essay of van Marum and A. Paets van Troostwijk (1752-1837), in which they followed Franklin's theory with Aepinus' modifications.[56]

2) The 'condensatore' was a modified electrophore coupled by an electrometer. The finding was based on Volta's electrostatic law $Q = C \cdot T$, where C was the capacity of a conductor, T its tension and Q the quantity of electric fluid localized on it. Volta had shown that by approaching one conductor to another one the capacity of the latter increased while on the contrary such capacity decreased by removing the first conductor. In this second case, with a given Q, a diminution of C was accompanied by a parallel increase of T, so that electrical signs became detectable on the electrometer. In spite of the novelty of Volta's theoretical ideas and of the difficulty in grasping them, the apparatus was a source of new experiments particularly to detect weak atmospheric electricity. And precisely along this empirical path a prize on the subject of the 'condensatore' was proposed in 1783 by the Hollandsche Maatschappij and awarded in 1786 to J. van Breda.[57]

3) The pile was the result of the debate on galvanic electricity started by Volta in April 1792. Only half a year later we notice the first printed record in Holland: on 12, 19 and 26 October a partial Dutch translation of a letter on

---

*nieuwe ontdekkingen in het Ryk der Natuur*, Leyden, By J. van Tiffelen en L. Herdingh (1777), pp. 31-46 (Jacquet) and pp. 47-52 (Volta). I have found a copy of this work in *ARA*, *Archief J.H. van Swinden* (2.21.187), Nr. 7.

55. Information from H.A.M. Snelders, *Het Gezelschap der Hollandsche Scheikundigen*, Amsterdam, Rodopi (1980), p. 42 and note 7. We may add that, in the same year, a Dutch translation by J. van Breda of Jan Ingen-Housz' Bakerian lecture on the electrophore, read on 04.06.1778 at the Royal Society, appeared (*Proeve over den Electrophorus, door Johan Ingen-Housz, ... Uit het Engelsch vertaald door J. van Breda*, Te Delf, By E. van der Smout en J. de Groot, 1780). On Ingen-Housz' Italian relationships, see the paper by R. Mazzolini in this volume.

56. *Verhandelingen van het Bataafsch Genootschap der Proefondervindelyke Wysbegeerte te Rotterdam*, VII (1783), pp. iv-v (awarding on 12.08.1782) and pp. 195-244 (the dissertation).

57. *Verh. HMWH*, XXIII (1786), p. xi (awarding on 10.07.1786); *Verh. HMWH*, XXV (1788), pp. 363-456 (the dissertation). In the text of the question reference was made to Volta's article in the *Philosophical Transactions* of 1782 (vol. XXI, part I, pp. 237-280 and pp. vii-xxxiii).

**Fig. 38** From the autograph of the first letter on animal electricity of Volta to van Marum, dated "à Côme ce 30me Août 1792" (Archive van Marum, Hollandsche Maatschappij der Wetenschappen, Haarlem).

animal electricity by Volta to van Marum (dated 30.08.1792, see fig. 38) was indeed published in the weekly journal *Algemene Konst- en Letter-Bode* of Haarlem.[58] In the issue of 28 December of the same journal a letter was published to van Marum (dated 14.12.1792) written by C.P. Schacht, professor at Harderwijk, who had immediately repeated the experiments and confirmed Volta's observations.[59] In the issue of 4 January 1793 an abstract of the second letter on animal electricity by Volta to van Marum (dated 11.10.1792) was given,[60] but it was unfortunately so short that it omitted the most interesting part, namely the description of what has been called the first making of "un élément de la pile voltaique."[61] After Volta's letter of 20.03.1800 to Sir Joseph Banks announcing the discovery of the pile, new research started in Holland (cf. 'infra') and in 1801 a prize was proposed by the Hollandsche Maatschappij on the question of whether it was possible to explain the working of the pile by the known laws of electricity or whether the pile revealed the existence of a new fluid, different from the electric one.

Apart from scientific prizes and from records in journals of the time, another way to assess Volta's influence is to look at the contents of the lectures given in scientific societies. I only mention here the case of the department of physics of the Maatschappij van Verdiensten Felix Meritis at Amsterdam, where in the period 1787-1801 at least eight lectures were given by C.R.T. Krayenhoff (1758-1840) and W. van Barneveld (1747-1826), which had directly to do with Volta's discoveries.[62] In 1801 van Barneveld, a chemist of Amsterdam, published two articles in which, after describing his experiments (partially proposed in his Felix Meritis lecture of 30.12.1800, see note 62), stated that he saw no reason to consider the electricity of the pile to be of a different nature from that of an ordinary triboelectric generator and he tried to explain the working of the pile in terms of chemical reactions.[63]

58. *AKL*, IX (1792): Nr. 224 (12.10.1792), 115a-116b; Nr. 225 (19.10.1792), 121b-123a; Nr. 226 (26.10.1792), 131a-131b.

59. *Ibidem*: Nr. 235 (28.12.1792), 201b-202a.

60. *AKL*, X (1793): Nr. 236 (04.01.1793): 2b-3b.

61. *Bosscha*, p. 79.

62. *GAA*, 'Particulier archief 59' (Felix Meritis), 'Notulenboeken': inv. Nr. 141 (Krayenhoff lectured on 21.12.1787 on the electrophore; on 28.12.1787 on the 'condensateur'); inv. Nr. 142 (van Barneveld lectured on 24.01 and 31.01.1797 on the electrophore, on 13.02. and 20.02.1798 on the 'condensateur der electrique stoffe', on 30.12.1800 and 19.05.1801 on the galvanic fluid). (Information by courtesy of drs. Marijn van Hoorn who, for the preparation of his doctoral thesis on J. van Swinden, has made a detailed research in Felix Meritis archive).

63. W. van Barneveld, 'Waarnemingen en Proeven, ..., door middel van een colom van twee verschillende metaalen', *AKL voor het Jaar 1801*, I, pp. 290-297. An enlarged version appeared in *Vaderlandsch Magazyn, van wetenschap, kunst en smaak*, 1 (1801), II, pp. 661-688. I have not

Finally we have to consider direct scientific relationships and joint research. During a long trip through Western Europe (September 1781-October 1782), Volta visited Holland for about three weeks. He arrived at Rotterdam on 21.11.1781, stayed for three days at The Hague and on the 26th he was in Leyden. The 28th he met van Marum in Haarlem; afterwards he spent two weeks at Amsterdam. Finally during his return journey he passed through Utrecht.[64] The following year on the 21st of May he was elected fellow of the Hollandsche Maatschappij at the suggestion of van Marum.[65] Since 1778, according to the will of the silk manufacturer P. Teyler van der Hulst, another scientific society, called Teylers Tweede Genootschap, had existed in Haarlem. Through it van Marum was in a position to have at his disposal by the end of 1784 a large electrical machine. In a description of this "ongemeen groote Electrizeer-Machine" published in 1785 both in Dutch and in French, he asked the physicists of the whole of Europe to suggest new experiments to him.[66] Volta was among those who reacted positively to van Marum's call. However, he immediately put forward a theoretical element: by referring to his printed letter to de Saussure on the capacity of conductors, he proposed increasing the prime conductor of the generator in order to get "sinon une étincelle plus longue, une qui sera plus large, et plus foudroyante."[67] In June 1786 van Marum tried to follow Volta's advice, adding to the prime conductor another one "de 4 pouces de diametre et 30 pieds de longueur", but the machine did not gain any force.[68] The following correspondence related to the question shows clearly their different scientific styles, namely Volta's attitude towards the quantification of electrical concepts and van Marum's practical approach. Volta maintained that in general with a larger prime conductor the sparks should have had a greater effect, because "les effets produits par un torrent électrique doivent être non seulement en raison de la *vitesse* ou force avec laquelle le fluide est lancé, mais aussi en raison de la *quantité* de fluide, qui est lancé avec cette vitesse, quantité qui repond à la *capacité* du conducteur qu'on

---

seen this second article, I quote from H.A.M. Snelders, 'Het departement Natuurkunde van de Maatschappij van Verdiensten Felix Meritis in het eerste kwart van zijn bestaan', *Documentatie-Blad Werkgroep Achttiende Eeuw*, 15 (1983), pp. 197-214, in part. 206-207 and note 59 on p. 213.

64. *Ep. AV*, II: letters 458 (pp. 67-69) and 462 (pp. 71-73). Volta visited on 28.11.1781 the cabinet of natural history of the Hollandsche Maatschappij (*Arch. HMWH*, 'Album bezoekers', 'Gabinet van Naturalien', 1772-1797).

65. *Arch. HMWH*, 'Notulen 1781-1794', f. 1169 (meeting of 05.03.1782).

66. *Verh. TTG*, III (1785), pp. xviii-xix.

67. Volta to van Marum, 08.03.1786: *Bosscha*, pp. 11-12; *AV*, IV, p. 65.

68. Van Marum to Volta, 07.05.1787: *Bosscha*, pp. 20-21; *Ep. AV*, II, p. 387.

decharge."[69] Large capacities were indeed the reason for the great effects obtained from Leyden jars. But the length, added Volta, of a cylindrical conductor of about one inch in diameter should have been of at least 800 feet to reach the capacity of a jar with a coating of 1 square foot. In his answer van Marum emphasized that in the experiments made with an enlarged prime conductor only when the air was extraordinarily dry had he noticed an increase in "la force de l'Etincelle". Although he agreed with Volta's theory, it was in his opinion of little use in the damp air of Teyler's hall. In fact by increasing "le volume des conducteurs on augmente ici la dispersion" of the electric fluid.[70]

Between Volta and van Marum there was, however, not just a difference in style and method, but also in the programme and contents of research. The Dutch scientist believed in gigantism: "l'histoire de l'électricité — he wrote in 1785 — nous apprend, qu'on a graduellement avancé dans cette science, à proportion qu'on a aggrandi les machines, et qu'on a acquis par là une plus grande force électrique."[71] It was a position largely shared, particularly in England,[72] but it was not Volta's. This may seem in contradiction with Volta's proposal of further enlarging the prime conductor of Teyler's machine, but such contradiction is more apparent than real. Indeed it was the same van Marum who, in his description of the machine, declared that "il s'excite par le frottement de ces plateaux beaucoup plus de matière électrique, qu'on en peut conserver dans le conducteur."[73] Volta, in his letter of 08.03.1786 (note 67), simply resumed the point ("Voulez-vous voir en effet, — he wrote — que votre conducteur est trop petit vû la quantité étonnante de fluide que fournissent les deux plateaux?") and took the opportunity to refer to his electrostatic theory. The Italian scientist drew in reality a neat distinction between his line of research and van Marum's. In 1789 e.g. he wrote to his Dutch correspondent: "Pendant que vous vous occupez, Monsieur, des expériences d'une électricité très-forte, puisque vous en avez les moyens, je m'occupe de celles de l'électricité très-foible, qui ne sont pas moins instructives; ...."[74] Here it was the

69. Volta to van Marum, 15.06.1787: *Bosscha,* pp. 27-28; *AV,* IV, p. 72 (the words in italics are underlined in the original in *Archief van Marum, HMWH*).

70. Van Marum to Volta, 31.08.1788: *Bosscha*, pp. 37-38; *Ep. AV*, III, pp. 5-6.

71. *Verh. TTG*, III (1785), p. x.

72. W.D. Hackmann, 'The design of the triboelectric generators of Martinus van Marum, F.R.S.. A case history of the interaction between England and Holland in the field of instrument design in the eighteenth century', *Notes and Records of the Royal Society of London*, XXVI (1971), pp. 163-181, in part. p. 167.

73. *Verh. TTG*, III (1785), p. 58.

74. Volta to van Marum, 23.07.1789: *Bosscha*, p. 49; *AV*, V, p. 233.

case of his works on electric meteorology, while seven years later the occasion was his research on contact electricity: "Pendant que vous travaillez, mon cher ami, — wrote Volta — a perfectionner toûjours plus les Machines électriques, à en rendre l'action plus vigoureuse, (...), je continue à mon tour de m'occuper d'une Electricité extremement foible, qui échappe aux plus sensibles Electroscopes, qui par-là n'a été aperçue que dans ces dernieres années, ...."[75] In the meantime, however, van Marum had turned towards chemical research since in the physics of electricity he could not see, as he wrote in the second continuation of experiments made with Teyler's machine, "aucune nouvelle carrière de recherches, dont il y a lieu d'espérer des résultats bien intéressantes."[76] In fact, from about 1791 Teyler's large generator was scarcely used for active research, in part however because van Marum had at his disposal a new handy and powerful plate machine. Also when Volta suggested van Marum to try to obtain with his large generator comparable effects to those got with a small pile,[77] the proposal was first realized in November 1801 by using the smaller machine since the damp season precluded the use of the larger one.[78] Only in the spring of 1802 was it possible for the Dutch scientist to compare the speed of the decomposition of water produced respectively by the large machine and by a pile of 200 pairs.[79] An account of the experiments realized at Haarlem in November 1801 by van Marum with Volta's 'messenger' C. Pfaff was published, in the form of a letter addressed to Volta dated the 29th of November 1801, in the most important scientific journals of the time (see note 78). A Dutch translation of this letter was also published in the *Algemene Konst- en Letter-Bode*.[80] Unfortunately van Marum did not give the Haarlem journalists a copy of Volta's answer, in which a qualitative statement of what was to become known as the first Ohm's law of electric circuits was produced,[81] and the letter remained unpublished until 1905.

## IV. *Conclusions*

I would like to emphasize, by way of conclusion, just a few open questions put within a more general historiographic perspective.

75. Volta to van Marum, 08.10.1796: *Bosscha*, pp. 108-109; *Ep. AV*, III, p. 314.
76. *Verh. TTG*, IX (1795), p. xii.
77. Volta to van Marum, 29.08.1801: *Bosscha*, p. 142; *AV*, IV, p. 211.
78. Van Marum to Volta, 29.11.1801: *Bosscha*, p. 153; *Ep. AV*, p. 101.
79. Van Marum to Volta, September (?) 1802: *Bosscha*, pp. 200-202; *Ep. AV*, IV, pp. 224-225.
80. *AKL voor het jaar 1801*, II, pp. 370-378 and 388-397.
81. Volta to van Marum, 22.06.1802: *Bosscha*, pp. 184-185; *AV*, IV, p. 226.

1) The use of all covering labels such as 'Newtonianism' or 'Newtonian age' to characterize the eighteenth-century scientific movement in the physical sciences has already been criticized in the historiography. Our case supports this new trend. As we have seen, not only Poleni but also 's Gravesande, the great propagandist of Newton in Continental Europe, shared a scientific tradition not reducible to Newton's one. When 's Gravesande in his *Suplement à la Nouvelle Theorie du Choc* (1723) added "un mot touchant l'Auteur de la découverte" of the new theory concerning the force and pointed out that Christiaan Huygens had an important part in it,[82] he was not just claiming a historical priority for the great Dutch scientist whose *Opera varia* (1724) and *Opera reliqua* (1728) he was to edit. 's Gravesande wished to emphasize a Dutch connection with Leibniz' theory, namely the existence of a continental movement autonomous from England. It is worth recalling in the context of a congress devoted to commemorate the 350th anniversary of the Leyden editio princeps of the *Discorsi e dimostrazioni matematiche intorno a due nuove scienze* (1638), that precisely in Galileo's masterpiece and in Torricelli's *De motu* (1644) resided the most important sources of Huygens' famous axiom of the centre of gravity proposed in the *Horologium oscillatorium* (1673). It was the science of energy in its cradle, a tradition that on the contrary was largely ignored in Newton's *Principia* (1687).

2) In our analysis we have met many instances of a remarkable difference in approach between Italian and Dutch scientists, some more theoretical and the others more practical. In the eighteenth century the enlightened ideal of usefulness permeated scientific life in both areas, but it was interpreted in different ways according to different contexts and traditions. Pure mathematics was, for example, not exceptionally flourishing in eighteenth-century Italy, but it was certainly more lively than in Holland. Also in two men so akin as 's Gravesande and Poleni, who from 1739-1740 also taught experimental physics, we perceive a greater interest in mathematical and theoretical researches in the Venetian who gained from the stimulus represented by that great mathematician Jacopo Riccati (1676-1754), who was living in the neighbourhood of Padua. The fact that in Italy river hydraulics was not left to practical engineers, but became an autonomous university subject in Bologna at the end of the seventeenth century, is another example of the difference between the two cultural areas. However, generalization from a few cases is not always the best historical guide. Alternative interpretations are possible.

82. W.J. 's Gravesande, 'Suplement à la Nouvelle Theorie du Choc, ...', op.cit. in note 49, pp. 196-197.

For example, the small contribution made by Dutch scientists to river hydraulics in the seventeenth century may be partially attributed to a fluvial situation not so dramatic as in the following century. The Dutch were famous for their practical skill in drainage works and, when L.F. Marsili (1658-1730) visited Holland in 1722-1723, he did not fail to praise Dutch expertise and to make a comparison with the 'erudite' hydraulics of Guglielmini, Manfredi and Grandi, which was certainly not favourable to the Italians. In a letter from Amsterdam to the Papal Legate at Bologna, dated the 26th of March 1722, Marsili proposed to imitate the Netherlanders to solve the problem of the Reno of Bologna by using windmills and dredgers. He wished to produce a sort of anatomy of the "Terra d'Olanda" to apply it to the "Paesi bassi di Bologna".[83] Marsili was a supporter of a more practical trend which flourished also in Italy, and not just among the heirs of the Renaissance tradition in engineering. I may refer e.g. to the case of the Jesuit mathematicians, although they were not the sole exponents of this trend, who contributed notably to practical and experimental hydraulics and were, as we have seen, particularly appreciated by Christiaan Brunings. On the other hand, also in Holland there were mathematicians with a more theoretical attitude towards hydraulics (remember e.g. the case of Steenstra's and Hennert's judgements of Brunings' paper). Hennert and Frisi are comparable figures, they shared for instance in 1783 an astronomical prize of the St Petersburg academy,[84] although Frisi was much more involved in practical hydraulics than his Dutch colleague. In conclusion, in these situations the professional distinction between university teachers and people devoted to water control seems at least as effective as national or local traditions.

3) Not only in the work of men such as Volta and van Marum in the field of electricity, or in that of Poleni and 's Gravesande in the realm of mechanics, but also in eighteenth-century hydraulics in which native traditions counted more, we perceive that political and linguistic barriers were easily crossed. What for reasons of 'bilateral' comparison is restricted to two countries was often a 'multilateral' dialogue among partners spread all over Europe. So what about comparisons between just two cultural areas? Of course it might be answered that bilateral comparisons are only pieces of a more complex mosaic, and that also mutual influences of countries like Italy and Holland are of great interest. However, if I had to reply on the basis of the personal experience gained in preparing this lecture, I would have no doubts in giving a

83. *BUB*, 'Fondo Marsili', Ms. 87 (F), ff. 39r.-44v., in part. ff. 39v.-40r.

84. *Nova Acta Academiae Scientiarum Imperialis Petropolitanae*, I (1787): 'Histoire de l'Académie Impériale des Sciences Année 1783', pp. 23-25.

positive answer. In studying the influence exerted by the Italians in Holland, I have indeed not only learnt a lot about scientific movement in the Low Countries, which is quite obvious for somebody who enters a new field. What I find more interesting is that the general picture I had of eighteenth-century Italian science and technology comes out enriched both in details and general features from my Dutch 'immersion'. It has been a fresh stream in my current research. And this alone is enough in my eyes to justify the effort.

*Acknowledgements:* I cannot mention here the many individuals and institutions, both in Italy and in the Low Countries, who have helped my research. To all of them my respectful thanks. However, I have a particular debt to the Hollandsche Maatschappij der Wetenschappen at Haarlem, for the permission to study and use archive materials freely, as well as for the assistance of Mr J.G. de Bruijn, adjunct-secretary and librarian of the Maatschappij, and of Mr H.J. Pol.

*Abbreviations* (archives and libraries):

| | | |
|---|---|---|
| *ARA* | = | *Algemeen Rijksarchief*, The Hague |
| *Arch. HMWH* | = | *Archief van de Hollandsche Maatschappij der Wetenschappen*, Haarlem |
| *BAM* | = | *Biblioteca Ambrosiana*, Milan |
| *BUB* | = | *Biblioteca Universitaria*, Bologna |
| *GAA* | = | *Gemeente-archief*, Amsterdam |
| *OBUB* | = | *Öffentliche Bibliothek der Universität*, Bâle |
| *SNSP* | = | *Società Napoletana di Storia Patria*, Naples |

*Abbreviations* (periodicals and other printed sources):

| | | |
|---|---|---|
| *AKL* | = | *Algemene Konst- en Letter-Bode* |
| *AV* | = | *Le Opere di Alessandro Volta, Edizione Nazionale (...)*, vols I-VII, Milano, U. Hoepli Editore, 1918-1929 (New York and London, Johnson Reprint Corporation, 1968) |
| *Bosscha* | = | *La correspondance de A. Volta et M. van Marum publiée par J. Bosscha*, Leiden. A.W. Sijthoff (1905) |
| *Ep. AV* | = | *Epistolario di Alessandro Volta, Edizione Nazionale (...)*, vols I-V, Bologna, N. Zanichelli Editore (1949-1955) |
| *Resolutien* | = | *Resolutien van de Heeren Staten van Holland and Westvriesland* |
| *Verh. HMWH* | = | *Verhandelingen uitgegeeven door de Hollandsche Maatschappy der Weetenschappen, te Haarlem* |
| *Verh. TTG* | = | *Verhandelingen, uitgegeeven door Teyler's Tweede Genootschap*, Haarlem |

# THE VISIT TO THE LOW COUNTRIES IN 1786 OF THE ASTRONOMER BARNABA ORIANI OF MILAN*

*Guido Tagliaferri & Pasquale Tucci*

Among the Italian scientists who visited The Netherlands in the 2nd half of the 18th century, two were mathematicians and astronomers from Milan: Paolo Frisi (1728-1784) and Barnaba Oriani (1752-1832). Their visits took place in 1766 and in 1786, respectively. Both visitors recorded the notable events of their journeys in diaries that are extant but partly (Frisi's) or wholly (Oriani's) unpublished. A few years before Frisi, in 1761, another scientist, soon to become a leading personage in astronomical activity in Milan — the Dalmatian-born Jesuit Roger Joseph Boscovich — had crossed The Netherlands on his way from London to Venice; but he left no account of his journey. Our paper will centre on the figure and activities of Oriani, whose diary and related correspondence are kept in the historical archive of the Brera observatory in Milan.[1]

1. Barnaba Oriani was born of destitute parents on July 19th, 1752, in the village of Garegnano, now incorporated into Milan. The locality was notable for being the site of a Carthusian monastery (Certosa di Garegnano) dating from the 14th century. As luck would have it, the monks noticed the brightness of the Oriani boy, and took care of his education, sending him to the reputable Barnabite schools in Milan (known as the "Scuole Arcimbolde"). He completed the required studies there, and became a cleric, but in the meantime developed a strong interest in science and showed particular propensity for applied mathematics. This inclination was to determine the course of his life.

*We would like to thank Dott. Cesare Maffioli for helping us to identify several Dutch *literati*, and for his search in Holland of documents related to Italian visitors of the second half of the 18th century.

1. This archive has recently been reorganized, and a complete inventory of its contents is now available: A. Mandrino, G. Tagliaferri, P. Tucci, *Inventario di Archivio dell'Osservatorio Astronomico di Brera* (Università di Milano, 1987). We are to publish the part related to the Low Countries of Oriani's diary in our forthcoming work 'La visita dei Paesi Bassi nel diario dí viaggio Barnaba Oriani, astronomo milanese del Settecento', *Incontri. Rivista di studi italo-nederlandesi*, n.s., IV, 1-2 (1989).

Oriani's career began with his entrance at age 23 as an apprentice astronomer in the Milanese observatory, erected some ten years before onto the Brera palace by Boscovich. Appointed 3rd astronomer two years later, Oriani remained there — rising in status and sharing the responsibilities of managing the observatory with Boscovich' successors F. Reggio and A. de Cesaris — until his retirement in 1817. Soon after entering the observatory he made his way into the community of European astronomers by working indefatigably as an observer, and acquiring competence in celestial mechanics. Thus when the news of the discovery by Wilhelm Herschel in March 1781 of a strange heavenly body reached Milan, Oriani was in possession of the skills needed to take part in the international effort to ascertain its nature. The problem presented by that body looked tantalizing. Once verified that it could not be a fixed star, it would have been natural to assume that it was a comet, since every astronomer of the 18th century was prepared to accept the appearance of a new comet. But it moved much too slowly, and had no tail. On the other hand, to imagine that it might be a planet conflicted with the deeply-rooted idea that there could not be other planets beyond the ones known since antiquity, because it was hard to believe that if there had been some other planet it could have remained undetected till then.

However, the strange body — the *singularis cometa* as Oriani called it at first — was indeed a planet that later was named Uranus. But to recognize it as such, required first of all to determine its true orbit and its period of revolution. This was an exacting task, to which Oriani contributed observations and calculations that enabled him to publish in 1783 the mean elliptic elements of Uranus, and the first astronomical tables allowing its heliocentric position to be known at any time. It was valid work, that won him widespread appreciation.[2]

The study of the dynamics of the new planet was to be among the main interests of Oriani in the following years, and one in which he attained considerable success. It is not, however, within the scope of this paper to dwell on the various aspects of his long scientific activity. We intended only to give some information about Oriani's background and professional

2. Oriani's paper was entitled *Observationes Novi Planetae a D. Herschel reperti et Tabulae Astronomicae ad eius locum heliocentricum pro quolibet dato tempore supputandum*, and appeared in *Ephemerides astronomicae anni 1785 ad meridianum Mediolanensem* ... (Milano, 1783). As for its reception, we may quote the Parisian journal *Le Journal des Sçavans* which, in its *Nouvelles Litteraires* of June 1784 reported Oriani's elements with the final comment "Ainsi l'on ne peut espérer, quant à present, [rien] de plus exact, pour la théorie de la nouvelle planète".

Fig. 39 Barnaba Oriani (1752-1832) (Archivio dell'Osservatorio Astronomico di Brera, Milano).

standing at the time when he set out to travel to several countries of Western Europe. (A portrait of Oriani is reproduced in fig. 39.)

2. The first archival document we found related to Oriani's journey is the copy of a proposal (AOB Cat. n. 50)[3] addressed sometime in 1785 by the Brera astronomers to count Giuseppe Wilczeck, plenipotentiary minister in Milan in charge of Austrian Lombardy. The aim of the proposal was to obtain approval and financing for the journey of an astronomer to England. It reported in detail the itinerary, that included visits to Brussels, Amsterdam, London, and Paris, the time and duration (about eight months from April 1786), the anticipated expenditure (555 sequins), and the purpose. This last point was stated as follows:

> The main end [...] is to become acquainted with living astronomers and mathematicians, to make arrangements for a continuous correspondence between them and our observatory, to see the most renowned observatories, the best astronomical instruments, and to know the artists who make them. If, moreover, the Royal Government should decide to order in London the mural quadrant of which the observatory is in need, as submitted in a note by the [present] astronomers some time ago, this would be a most favourable opportunity, not only because it would enable to make a good choice of the artist to be entrusted with the task of manufacturing [the instrument], but also because it would be possible to devise with the artist himself whatever improvement might be deemed feasible. The cost of this instrument is 1000 sequins [...].[4]

Undoubtedly, the Brera astronomers were keen to establish or strenghten connections with other European observatories, and to bring up to date their scientific instrumentation. The sum of money requested was substantial, however, in the eyes of the sparing Austrian administrators. Thus the astronomers offered to sell some books and obsolete instruments of the observatory to cover one third of the cost of the quadrant; moreover, they pointed out that the instrument would have the capability of allowing precise determinations of the positions of some fixed stars to be used as astronomical references in the making of a modern map of Lombardy, an undertaking that

3. The copy bears no date nor indication of addressee or sender, but the evidence for the attributions given here is strong. This document is listed in the catalogue *1726-1799: Catalogo della corrispondenza degli Astronomi di Brera*, edited by A. Mandrino, G. Tagliaferri, P. Tucci (Università di Milano, 1986). We identify the items catalogued therein by the abbreviation "AOB Cat." followed by the serial number.

4. AOB, Cat. n. 50. The copy is written in Italian. We are to publish the original text in our forthcoming work quoted in note 1 'supra'. All translations in the present article are ours.

had long since been in the plans of the government.[5] One can surmise that this practical return had weight in inducing the government to approve the astronomers' proposal; at least, in the official letter dated April 18, 1786 (AOB Cat. n. 1304), in which Wilczeck communicated to Oriani the instructions regarding his journey (which by then had evidently been authorized), the map took first place in the list of the plenipotentiary minister's concerns. In fact, in this letter we read:

> Since the Government has granted Your Lordship Very Revered permission to make a literary expedition to Germany, Holland, England, and France, I have given the appropriate orders [...] so that 400 sequins be paid to you [...]. Given also that the plan to provide a geographical map should be executed, I approve the proposed purchase of the quadrant, which shall be of use equally for the most subtle operations of the observatory [...]; and I have no doubt that in procuring the instrument you will be able to combine maximum economy with [...] the perfection of the machine. To this end you may make personal contact with the most renowned astronomers [&c].

Even if the travel allowance was somewhat less than had been asked for, the astronomers' request had been fulfilled, and Oriani could leave; which he did within a few weeks, carrying several letters of recommendation or introduction to prominent people abroad.

The sources of information on the journey of Oriani are his journal (already mentioned above), and the correspondence related to it. In the journal he noted down the day by day occurrences, the remarkable things seen, the people met, the expenses incurred, and so on. The notes he wrote are frequently accompanied by terse comments and personal judgements; but although his chronicle is brightened up by these additions, the journal appears to have been kept by its author primarily for the sake of future recollection. Perhaps Oriani intended to amplify and polish his notes later on, because he wrote some words to that effect (even though facetiously) in a letter from Amsterdam (AOB Cat. n. 1084) to a friend of his, the sculptor Giuseppe Franchi in Milan. But that project, even if ever seriously contemplated, never materialized.

Oriani's journal covers a period of about five months, from his departure from Milan on May 12th until October 6th, 1786, when he was preparing to leave Paris to return home. The notebook in which Oriani wrote ends with the entry for that day, and we have not found the continuation of his diary

5. Cf. N. Paolucci, G. Tagliaferri, P. Tucci: 'Le vicende scientifiche ed extrascientifiche della realizzazione della prima carta della Lombardia con metodi astronomici', in *Atti dell'8° Congresso Nazionale di Storia della Fisica*, Napoli, 1987, pp. 383-409.

that should have covered the last leg of the journey. It is reasonable to assume, that he managed to be back in Milan around the end of October, because as his stay in Paris was drawing to a close, he expressed eagerness to return soon to his work in Brera.

Out of the almost half year that his journey lasted, Oriani spent five weeks — from June 6th to July 12th — in the Low Countries, including eight days in the Austrian (southern) Netherlands, now part of Belgium. In the following section we shall concentrate our attention on the section of the diary concerned with those five weeks. Reference to any specific passage of the notebook will be made by quoting the date under which it appears.

3. Oriani entered the Low Countries by way of Maastricht, about three weeks after leaving Italy. He proceeded to Brussels, where on June 8th, the day after his arrival, he paid a courtesy call on the Minister for the Austrian Low Countries, Count di Belgioioso. During his stay in Brussels Oriani had a busy time, as usual for him. He met some astronomers there, one of whom was the Hungarian-born Baron von Zach, then in the service of the Duke of Sachsen-Coburg-Gotha at the Seeburg observatory. Zach was destined to remain a friend of Oriani and of Italy for years, also taking part in the establishment of observatories near Lucca and at Naples. On this first occasion Zach was convinced by Oriani to remain in Brussels one day longer than planned, in order to help to observe the flight of the aerostatic balloon of Blanchard,[6] whose course over the city Oriani charted. He immediately presented a copy of his sketch (see fig. 40) to Prince Albert of Sachsen-Teschen, consort of the Grand Duchess Christina of Habsburg-Lorraine who, as a sister of the Emperor Joseph II, was Governor-General of the Austrian Low Countries. The Prince was interested in cartography, and showed him the originals of the maps of the region drawn by General Ferrari. Oriani had already seen copies of the maps at Belgioioso's residence, and criticized their lack of references to measured trigonometric points, as well as a discrepancy of between two and three leagues with the surveys of Cassini.[7]

The quality of Ferrari's maps was a matter of concern to Oriani, not only

6. The French aeronaut Jean-Pierre-François Blanchard (1753-1809) achieved notoriety as the first man to make (with the U.S. physician John Jeffries) the aerial crossing of the English Channel, on Jan. 7th 1785.

7. Oriani does not specify whether he refers to Cassini III (César François) who spent most of his life on field measurements for his formidable chart of France in 180 sheets, or to his son Cassini IV (Jacques Dominique) who completed the work after his father's death in 1784.

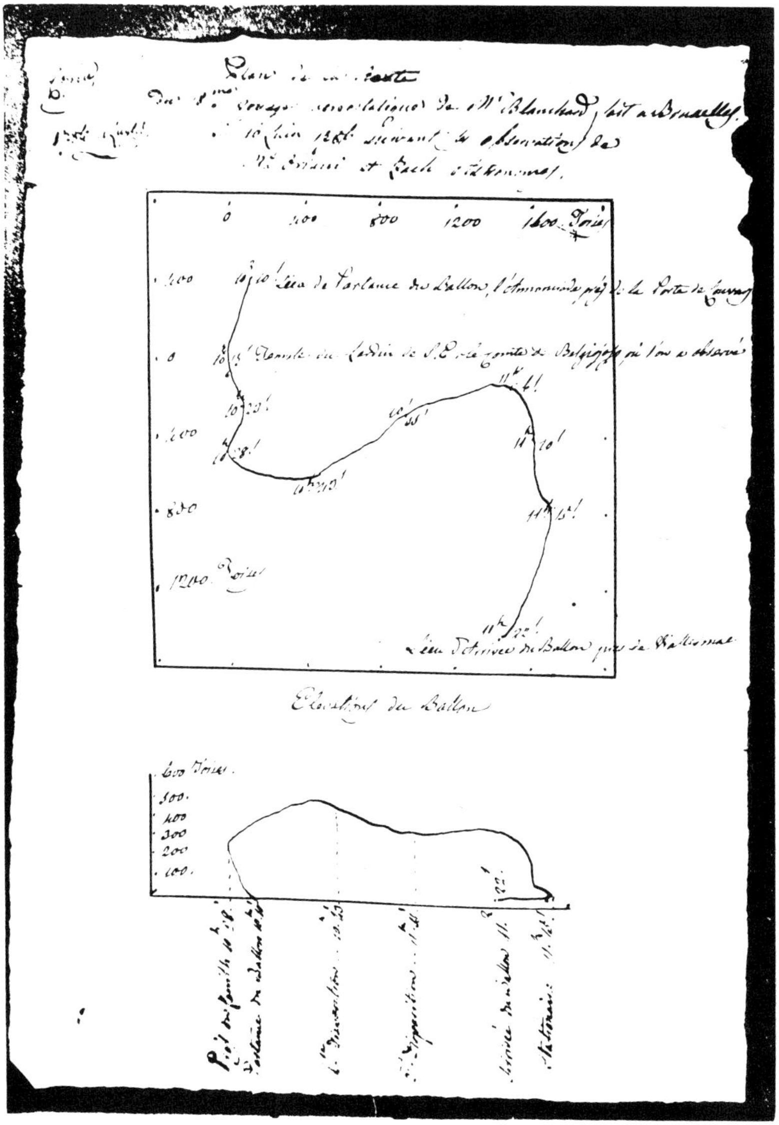


**Fig. 40** Oriani's observation of Blanchard's balloon flight over Brussels. Copy (Archivio dell'Osservatorio Astronomico di Brera, Milano: A.A.V., cartella 9, fasc. 18, n. 129). The original is in Archivio di Stato, Milano.

since knowing the state of the art of map-making was one of the purposes of his journey, but also specifically because the powerful chancellor Kaunitz had reprimanded the Milanese administrators (not the innocent Brera astronomers!) from Vienna for their inability to produce a good chart of Lombardy, extolling the virtues of the one made in the Low Countries.[8] It would seem that Kaunitz, like most centralizers, was not well-informed about all the things he ruled over. Before leaving, anyhow, Oriani measured the latitude of Brussels, obtaining the quite good result of 50°50′40″; and did not omit some sightseeing, noting *i.a.* the magnificent new park, the fountain of the Manneken-Pis, and "two small girls observing him" (June 12th).

On the same day Oriani proceeded to Antwerp, where he admired *The Descent from the Cross* by Rubens. He deplored the ban on navigating the Schelde established by the Dutch with thc treaty of Münster in 1648 (which ended the so-called war of the eighty years); and wrote: "... but now [Antwerp] is a depopulated and miserable city, although it is great, beautiful, and well situated" (June 13th).

From Antwerp Oriani headed directly for Amsterdam, with only an overnight stop in Rotterdam. He was to remain in Amsterdam eight days on this first visit, but intended to pay another visit before leaving The Netherlands. He began his sojourn by calling on some Italian bankers and on the imperial consul Signor Carli; he then went to see Signor P.A. Bolongaro Crevenna. This man was, so to speak, a classical exponent of the enlightened *bourgeoisie*. Italian-born and educated in the Jesuit-college of Brera, he had settled in Amsterdam and amassed considerable riches owing to successful trading. So he could indulge in his passion for books, of which he had collected a great many, including valuable specimens like *incunabula*, manuscripts and first editions. He employed librarians, and in 1783 had even offered the position to Oriani, who had given serious consideration to the proposal but finally refused. At the time of Oriani's visit Crevenna's house was something of an attraction for learned people. Oriani, who was received very cordially and afterwards frequently entertained and shown round Amsterdam by Crevenna, met the librarian in charge there Tommaso de Ocheda, an Italian, and the French bibliophile Barthélemy Mercier, abbot of St. Léger in Soissons. Among the rarest things he could see in that library, one can mention some works by Coster, the alleged Dutch rival of Gutenberg as the inventor of printing.

8. In a letter (kept in the State Archive of Milan, Confini, P.A., Cart. 5) dated Dec. 8th, 1777, Kaunitz wrote "... One works very differently in the Low Coutries, where the topographical map soon to be published [...] is perhaps superior to any other of this kind." For more details, cf. our paper quoted in note 5 'supra'.

In the following days, while not shunning social engagements, and devoting as usual some time to seeing the sights, Oriani contacted two scientists of repute: Pybo Steenstra and Jan Hendrik van Swinden. From the former, lecturer of mathematics and astronomy, owner of a library well stocked with scientific books, he heard opinions on the works of Italian students, especially of hydraulics like Benedetto Castelli and Domenico Guglielmini, and summarized his judgement of him with the words: "Cold but good Dutch" (June 19th). Of the latter, a renowned professor and researcher, best known for his studies of electricity and magnetism, he wrote: "Impressive man who speaks with great fluency, [... he is] very knowledgeable about physics [and is] a good mathematician and metaphysician" (June 19th). He was invited to visit van Swinden's library and scientific cabinet, the next day received a gift of six books, and was asked to dinner. More details of this meeting are reported in the diary; and just to give an idea of how the diary looks, the two pages describing that event are reproduced in fig. 41.

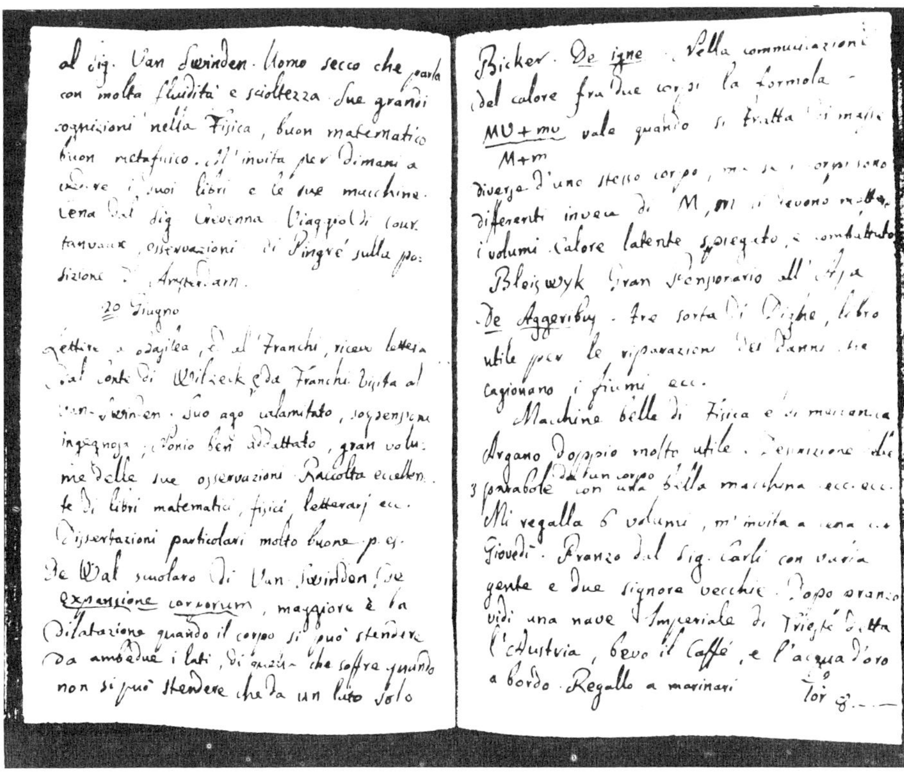

**Fig. 41** Two pages from Oriani's diary (Archivio dell'Osservatorio Astronomico di Brera, Milano: Fondo B. Oriani, cartella 206, 2).

Oriani was to see van Swinden again in July. On this second occasion he was also shown a "simplified" version of Volta's eudiometer (but it is not clear from the sketch in the notebook in what respect the instrument had been simplified).

If Oriani kept his diary for informal personal documentation, in writing about his impressions of Amsterdam he still used in it more restraint than in the letter (already quoted, AOB Cat. n. 1084) he sent Franchi on that same June 19th. Perhaps he was not in his best mood; anyway, here are some excerpts from that letter:

> I have become fish, cow, and goat. Here one does not see anything but water, and what is worse, water that stinks terribly.
>
> In Amsterdam all the streets are full of people and hardly anybody speaks and much less laughs; all the solemn Dutchmen wear round wigs with three or four layers of curls, all are clothed in black [...]. The women [...] wear nightcaps, all of them are blonde or whitish, with small mouths, noses barely protruding, round faces, stiff carriage and lacking agility.
>
> The Milanese living here are all melancholy, the rich ones think always of increasing their wealth, the poor ones of deceiving each other to survive. [...]
>
> Signor Crevenna is a very good man and he lavishes many attentions on me, but he is a perfect Dutchman in his house, and I feel much relieved that luck prevented me from becoming his slave. [...]
>
> I have already met a few *literati*, among whom a square wig [...] by the name of van Swinden, who is a great student of magnetism, and who deigned to come to my antechamber because he hopes that I will sing his praises in the book of my journey. Another one tall, sharp, and Quakerishly encyclopaedic, condenscended to let me see his well-supplied library, in which the best-bound books were just the works by himself, containing very boring elements of geometry and astronomy in Dutch; his name is Steenstra. [...]

On June 23rd Oriani left Amsterdam for Utrecht. The most interesting person he met there was Johan Frederik Hennert, professor of mathematics and astronomy at the University, and successor to the Italian mathematician Giovan Francesco Salvemini da Castiglione. He does not report having scientific discussions with him, but he was made aware of some aspects of the religious and political unrest in Holland. He notes (June 24th):

> Hennert scarcely an Orangeman, rather unhappy about his country, would like to come to Italy. He falters when speaking, and talks quickly; he is short, aged between 55 and 60 years. [In fact he was 53 at the time].

It would appear that Hennert had misunderstood, or perhaps taken too literally, something that Oriani may have said about chairs in Italian Universities. Anyway, on the 12th of July Hennert wrote to Oriani (AOB

Cat. n. 827) remarking on the possibility of settling in Italy:

> "Vous serez peut étre étonné de mon silence, qui ne repond pas aux precieux momens que j'ai passé avec Vous. Les imprimeurs sont comme le reste, lents et phlegmatiques. Il ne falloit q'une fueille presque l'ouvrage, que j'ai l'honneur de Vous presenter, fut achevé, et celle fueille a eté trois semaines à dommage. Jugez de cette ouvrage avec celle affabilité que je Vous connois, et deignez moi de Votre amitié. Je ne pers pas de vue ce que Vous m'avez fait l'honneur de me proposer sur une chaire de mathematiques à Pavie ou à Padoue. Je laisse le tout à Votre decision. Mais, mon cher Monsieur, il y a une difficulté qui s'opposera à nos desseins, c'est que je ne suis pas un Catholique. S'il s'agissait de faire confession publique de la religion Catholique, je ne pourrai pas m'y resoudre [...]."

But Oriani was not in a position to offer chairs in Italian Universities; at the most he might have been able to recommend candidates informally through friends. We have not found, however, any further mention of the matter.[9]

On the night of June 24th to 25th Oriani travelled by boat to The Hague, where he made the usual courtesy call on the Austrian ambassador Baron von Reischach. He was received familiarly: "Of the Dutch [Reischach] has the same opinion as me" noted Oriani (June 26th), who described the ambassador as a man who looked twenty years older than his age of about 50, owing to his libertine life.

In The Hague Oriani met several *literati.* He mentions Jacob Visser ("very phlegmatic man"), historian of the art of book printing; Frans Hemsterhuis, philosopher and writer of aesthetics; Aernout Vosmaer, director of the Cabinet of Natural History of the Stadholder, Prince William V of Orange; and others, perhaps less known, but interesting for their valuable collections of antiquities, curios, old of rare books, &c. Leaving The Hague, he went to Leyden, where he met Laurens van Santen, philologist and poet, for whom he developed a liking evident from these very appreciative words in the diary (June 29th and 30th):

> [...] the young literary man van Santen, full of wit, and talented in Greek and

9. In fact, at the time of Oriani's visit, Hennert was worried about his future. A few months afterwards, he resigned from the University, as we read in a letter of Jan. 9th, 1787, from de Ocheda to Oriani: "On vient de me dire que M. Hennert est passé aux secondes noces, et que il a quitté la Chaire de Mathematiques à Utrecht." H.J.M. Bos, in his article 'Johan Frederik Hennert, wiskundige en filosoof te Utrecht aan het eind der achttiende eeuw', *Tijdschrift voor de geschiedenis der geneeskunde, natuurwetenschappen, wiskunde en techniek* 7 (1984), pp. 19-31, reports that Hennert, accused of atheism and criticized for his teaching methods, left Utrecht in February 1787 and went to Hanau in Germany; but following the overthrow of the political forces governing the city he was formally invited to resume his post and returned to Utrecht in 1788, remaining there until his retirement at the beginning of the 19th century.

> Latin letters [... He] treats us with a cordiality worthy of a true philosopher [...] In general he loves Italians a lot.

The feeling was mutual. On a more professional side, in Leyden Oriani saw Christiaan Damen, professor of mathematics, who showed him the plans of the observatory that the University intended to build. "It is similar to ours", commented the Milanese, who the next day relaxed a little by going to spend a weekend at the country residence of Signor Crevenna near Haarlem. He was not completely idle, however, because he met the secretary of the *Hollandsche Maatschappij der Wetenschappen* (to whose membership Boscovich and Frisi had been elected in 1770) to deliver a letter from Boscovich; and went to see the physical Cabinet of Teyler's Museum, directed by the physicist and botanist Martinus van Marum, credited with the building of the most gigantic electrostatic generator then in existence. Oriani had been informed by Damen that van Marum was away, but went there notwithstanding, and reported (July 1st):

> I went [...] to see the hall were [the machine] is housed, and it is magnificent, oval in shape, full of cupboards, the lower ones to keep the instruments of physics, the upper ones for the books. Over the hall there is a sort of observatory with a anemometer. This hall with its adjoining apartments was made with the legacy of a certain Teyler [, a wealthy merchant].

After spending the Sunday first at Zandvoort ("very beautiful view of the ocean") and afterwards at the outdoor festival ("kermis") at Haarlem, on the 3rd of July Oriani was again in Amsterdam. There, in addition to most of the people he had met during his previous stay, he met P. Nieuwland, A. Paets van Troostwijk, and possibly J.R. Deiman, three scientists who were to gain merit in the years to come for helping to develop Lavoisier's chemistry in Holland[10]; and John Cuthbertson, a famous English maker of electrical instruments then working in Amsterdam.

4. By now Oriani's visit to the Low Countries was nearing the end. He left Amsterdam on July 9th and headed for Rotterdam, stopping *en route* in Leyden to see the houses of Pieter van Musschenbroek, William Jacob 's Gravesande, Herman Boerhaave, and Hugo de Groot, all of them very important men in physics, medicine and law — but dead at the time of Oriani's journey. He called again on van Santen, then proceeded to Delft, where he admired the sepulchral monuments of some prominent Dutchmen:

10. See H.A.M. Snelders, 'The New Chemistry in the Netherlands', *Osiris*, 2nd s., 4 (1988), pp. 121-145.

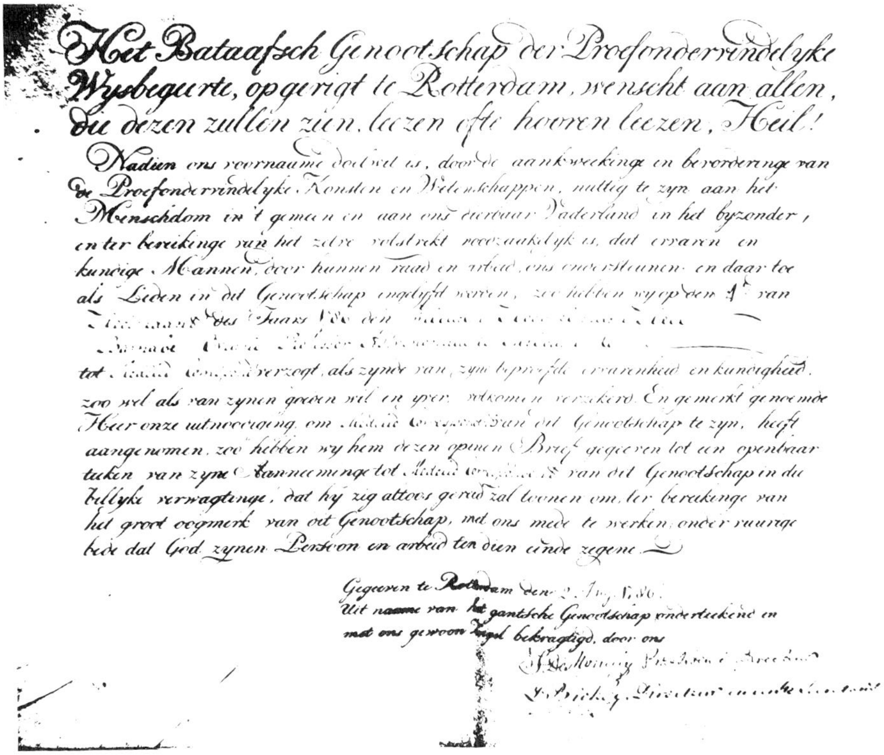

Het Bataafsch Genootschap der Proefondervindelyke Wysbegeerte, opgerigt te Rotterdam, wenscht aan allen, die dezen zullen zien, leezen ofte hooren leezen, Heil!

Nadien ons voornaamste doelwit is, door de aankweeking en bevordering van de Proefondervindelyke Konsten en Wetenschappen, nuttig te zyn aan het Menschdom in 't gemeen en aan ons dierbaar Vaderland in het byzonder, en ter bereikinge van het zelve volstrekt noodzaakelyk is, dat ervaren en kundige Mannen, door hunnen raad en arbeid, ons ondersteunen en daar toe als Leden in dit Genootschap ingelyfd worden, zoo hebben wy op den [illegible]

Gegeeven te Rotterdam den 2 Aug. 1786.
Uit naame van het gantsche Genootschap onderteekend en met ons gewoon Zegel bekragtigd, door ons

**Fig. 42 Oriani's diploma of admission to the Batavian Society of Experimental Philosophy of Rotterdam (Archivio dell'Osservatorio Astronomico di Brera, Milano: Fondo B. Oriani, cartella 206, 1).**

those of William the silent ("superb"), Maarten Tromp ("very beautiful bas-relief of a naval battle"), Piet Hein ("he lies on a double mat of white marble, finely carved"), Antoni van Leeuwenhoek ("with his portrait in marble").

On the night of July 10th Oriani reached Rotterdam. He had a letter of introduction to Dr Salomon de Monchy, a distant relative of van Swinden, and was shown around by him, and even more by his son, who was free with advice and help on practical matters concerning the passage to England. Oriani did not mention that Dr Monchy was President-Director of the *Bataafsch Genootschap der Proefondervindelijke Wijsbegeerte* (Batavian Society of Experimental Philosophy) of Rotterdam, nor that one week before that body had decided to invite him to join as a Member Correspondent. Maybe at the time he did not yet know of the bestowal of such an honour. Only about one month afterwards, while in London, he recorded tersely: "Received a letter from the Secretary of the Philosophical Society of

Rotterdam, in which he tells me that I have been made a member" (August 12th). This letter has not been found, but a diploma attesting to his admission is in the Brera archive, and is reproduced here (fig. 42).

Oriani had a good impression of Rotterdam ("Magnificent view along the Meuse [...] cheerful town [...] well populated"), and commented favourably on the famous statue of Erasmus ("The head is well-made and is expressive") (July 11th). The day after he went to Hellevoetsluis and boarded the packet-boat to Harwich in England.

As far as we know, Oriani never again visited the Low Countries. He had a tempting opportunity in 1795 to settle in Holland, when van Santen wrote to him about the vacancy of a chair at the University of Leyden (AOB Cat. n. 1331):

> [...] il s'offre en ce moment une favorable occasion [...] de voir remply mon ardent desire. La Chaire de professeur en Astronomie est vacant à notre celèbre Université de Leide: l'obstacle de la difference de religion est levé, et surement les curateurs de l'Académie se feront un plaisir d'offrir cette place à un homme de votre merite et renommée [...].

But, even though the proposal looked attractive, in the end Oriani declined the invitation, the proffered reason being the lack of facilities for astronomical observations in Leyden.

In closing our contribution, we might remark — perhaps unnecessarily —that in condensing in a few pages the detailed account that Oriani left of his visit to The Netherlands, we could not render adequately the wealth of information contained in his diary. Page after page, it is a very readable document that attests to the competence, education, perception and wit of its author.

# CONTACTS BETWEEN ITALIAN AND DUTCH SCIENTISTS BETWEEN 1750-1789

*Renato G. Mazzolini*

The aim of this paper is to provide some general information about contacts between Italian and Dutch scientists in the period 1750-1789. As far as I know, this topic has never been the subject of any detailed research and what follows has but a preliminary character. In the first part of the paper I shall make some general remarks on the contacts between the Italians and Dutch from 1750 to 1789, while in the second I shall concentrate on those contacts between Italian and Dutch scientists during the same period. I have given more emphasis to the Italian side of the story with which I am more familiar.

According to the works of Orbaan, Hoogewerff, Luijdjens, van Veen and McCormick it seems that in the second half of the 18th century the number of Dutch visitors and especially of students and artists to Italy declined conspicuously in comparison with the previous period.[1] The *peregrinatio academica* to Italy now belonged to the past and had been replaced by the Grand Tour.[2] On the other hand the number of Italian visitors to the North now increased considerably. A significant number of diplomats and travellers going to Paris, London or Germany, often included The Netherlands in their Grand Tours. We know, for instance, of the journeys of Giuseppe Garampi,[3] of Giovanni Battista Caprara Montecuccoli,[4] of Carlo

1. *Bescheiden in Italië omtrent nederlandsche kunstenaars en geleerden*, 3 vols. 's-Gravenhage: Martinus Nijhoff, 1911-1917 (the first volume was edited by J.A.F. Orbaan, and vols. II and III by G.J. Hoogewerff); and A.H. Luijdjens, "Chronologische lijst van beschrijvingen van Italië en Rome tot 1900 in de Nederlanden geschreven of verschenen", in *Mededeelingen van het Nederlandsch Historisch Instituut te Rome*, 2e reeks, I (1931) 205-229; Henk Th. van Veen and Andrew P. McCormick, *Tuscany and the Low Countries: An Introduction to the Sources and an Inventory of Four Florentine Libraries*, Florence: Centro Di, 1985.

2. Anna Frank-van Westrienen, *De Groote Tour. Tekening van de educatiereis der Nederlanders in de zeventiende eeuw*, Amsterdam: Noord-Hollandsche Uitgeversmaatschappij, 1983.

3. Gregorio Palmieri, *Viaggio in Germania, Baviera, Svizzera, Olanda e Francia compiuto negli anni 1761-1763. Diario del Cardinale Giuseppe Garampi*, Roma: Tipografia Vaticana, 1889.

4. Giovanni Battista Caprara Montecuccoli, "Relazione sull'Olanda, e l'Inghilterra" dated 9 July 1772, in: Archivio Segreto Vaticano, Segreteria di Stato, Colonia 185, ff. 22-29.

Castone della Torre di Rezzonico, of Lorenzo Da Ponte, of Luigi Angiolini, of Giuseppe Gorani,[5] and of Carlo Antonio Pilati.[6]

For Italian travellers of the second half of the 18th century, The Netherlands, together with England of course, was the land of liberty and the land of commerce. This was the most wide-spread image of The Netherlands among Italian scholars, an image shaped in the early 17th century when Traiano Boccalini declared in his *Ragguagli di Parnaso* that it was a "provincia fecondissima, bellissima, amenissima", whose inhabitants were "sopramodo civili, facoltosi e industriosi". After the publication, in 1629, of the *Relatione delle Provincie Unite* by Guido Bentivoglio, Italian intellectuals also showed a considerable interest in the political model developed in the United Provinces with their federal and unitarian character. Some, like Vincenzo d'Andrea, even tried to import this character into Southern Italy.[7] I will not discuss to what extent Italian intellectuals idealized the Dutch reality. It is clear, however, that what motivated them above all was the situation of their own States back in Italy and also that they emphasized contrasts in order to elicit a kind of internal reaction to Italian decline in all fields of civil, commercial and literary life. What they mostly admired was the freedom of the press.

During the second half of the 18th century, however, things were slowly changing in Italy. The general atmosphere was loosening up both within the Church and among Italian rulers who, often assisted by technically oriented and enlightened scholars, started introducing reforms pertaining to all aspects of organised life. That things were changing is clear if we look more closely at the question of the freedom of the press.

The abbé Coyer wrote in his *Voyages d'Italie et de Hollande* of 1775:

> "On imprime en Italie, à Venise, à Milan, à Lucques, à Florence des traductions d'ouvrages français et anglais prohibées dans des pays qui ne sçauraient se piquer d'être plus catholiques, des ouvrages même italiens qui ne manquent pas de hardiesse ... on est tout étonné d'acheter, sans façon, chez les libraires de Rome, des livres qui ne se vendent à Paris que sous le manteau."[8]

For Coyer too Holland was the land of liberty, of good government, of the

5. *Letterati, memorialisti e viaggiatori del Settecento*, a cura di Ettore Bonora, Milano-Napoli: Ricciardi, 1951, pp. 787, 832, 999, 1060.

6. [Carlo Antonio Pilati], *Lettres sur la Hollande*, 2 vols. La Haye: Chez J.H. Munnikhuizen & C. Plaat, 1780.

7. Guido Bentivoglio, *Relatione delle Provincie Unite*, Facsimile dell'edizione "elzeviriana" Brusselles 1632 a cura di S. Mastellone, Firenze: Centro Editoriale Toscano, 1984, p. 24.

8. Gabriel-François Coyer, *Voyages d'Italie et de Hollande*, 2 vols. Paris: Duchesne, 1775, vol. II, p. 190.

useful sciences and of commerce. Commerce, of course, was most important.

> "Elle [l'Hollande] trafique aussi de l'esprit des nations. Sa librairie forme une branche de commerce très florissante et plus certaines nations gênent la presse, plus elle se réjouit."[9]

Freedom of the press was, in The Netherlands, firstly a commercial enterprise, and most of the production of enlightened literature during the second half of the 18th century was intended for the foreign market, not for internal consumption. It is true that freedom of the press made The Netherlands the most important store of political and philosophical ideas and one of the bases for the diffusion of the Enlightenment, but it seems that this diffusion took place mainly abroad and not in The Netherlands itself. It is interesting to note that when in the Seventies many more books of the *philosophes* were being printed in French, a language more likely to be known to the lower classes than Latin, some Dutch authors started complaining about excesses of press freedom.[10] Furthermore, it is also interesting to note that in the second half of the 18th century Italians produced a great amount of enlightened literature, far more than the Dutch. One could, of course, easily argue that Italians also needed this literature far more than the Dutch. Notwithstanding, however, the number and quality of Italian *philosophes*, and the number of their publications freely circulating in Italy, there were still authors whose books could not be published in Italy, and were therefore printed in The Netherlands. One of them, Carlo Antonio Pilati, is of particular interest because he also wrote about The Netherlands. At least six of his works appeared between 1769 and 1782 in The Netherlands.[11] They were all in French, not only in order to guarantee them wider circulation, but especially because Italian had become less and less well-known and therefore it would not have been commercially profitable to publish them in Italian.

The market for Italian books abroad had already declined during the first half of the century because they were regarded as too expensive and because Italian was by now little known. The firm Cramer and Philibert wrote to Giovanni Lami in Florence in 1741: "dans ces pays ultramontains, on entend peu l'italien, le françois est trop à la mode."[12] In 1773 Pilati wrote to one of

9. *Ibidem*, vol. II, p. 255.

10. Franco Venturi, "La rivoluzione delle Province Unite", in *Settecento riformatore*, vol. IV/2 Torino: Einaudi, 1984, 504-614, on pp. 520-521, 538.

11. Luciano Borrelli, "Le edizioni degli scritti pilatiani e la bibliografia critica (1765-1984)", in *Annali dell'Istituto storico italo-germanico in Trento*, XI (1985) 389-488.

12. Françoise Waquet, "Les registres de Giovanni Lami (1742-1760): de l'érudition au commerce du livre dans l'Italie du XVIIIe siècle", in *Critica storica*, XVII (1980) 435-456, on p. 453.

his correspondents that there was no possibility of publishing in Italian in The Netherlands.

> "Dans ces pays vous ne trouveriez pas un libraire qui voulût seulement se charger d'une feuille en langue italienne. En Hollande on ne pourrait s'adresser qu'à Marc Michel Rey d'Amsterdam, mais ce monsieur n'a d'autres correspondance que des Français, et il verse presque tous ses livres en France. Ici [Berlin] on ne veut pas de l'Italien non plus. Je me flatte de réussir en Angleterre. Les anglais aiment assez nos fadaises quand ce ne serait que pour se rappeller la langue d'une nation, dont ils débauchent les femmes, et entendent les operas."[13]

That so many works of Italian physicians and natural philosophers had been published in The Netherlands at the end of the 17th century and beginning of the 18th century is due not only to the careful and intelligent choices made by the Dutch publishers of the time, but also to the fact that they were written in Latin and could therefore have a wide European circulation.

Few Italian scholars of the period lived for a prolonged time in The Netherlands. I know only of two really significant cases: that of Carlo Antonio Pilati and that of the mathematician Giovan Francesco Salvemini known as Castillon or de Castillon. What is striking about both cases is that they show the persistence, well into the second half of the 18th century, of the Swiss-Dutch connection aiding catholic dissenters.

Pilati was a man of the Trentino.[14] He never became a protestant, but was strongly anti-clerical. The book which established his reputation, *Di una riforma d'Italia*, first appeared in 1767 and was at once prohibited by the Holy Office, but was nonetheless reprinted several times. Because of the prohibition he did not feel safe in Trento and left for The Netherlands (1767). There he made the acquaintance of the Swiss natural philosopher Jean Nicolas Allamand, at the time professor at Leyden University. It is interesting to note that during Pilati's exile at Coira, in the Grigioni, where he played an important part in advising the newly founded Tipographical Society, it was Allamand who supplied him with books published in The Netherlands.[15] In Switzerland Pilati had strong ties. Back in The Netherlands

13. From the letter of Carlo Antonio Pilati to Ulysses von Salis-Marschlins, written in Berlin on 8 May 1773: Coira, Staatsarchiv Graubünden, Familienarchiv: von Salis, Ms. D VI./CM, V. D7, 153-154. This letter, translated into German, has been published by Meta von Salis-Marschlins, "Ein genialer Abenteurer", in *Jahresbericht der Historisch-antiquarischen Gesellschaft von Graubünden*, (1939) 124-160, on pp. 158-160.

14. Maria Rigatti, *Un illuminista trentino del secolo XVIII: Carlo Antonio Pilati*, Firenze: Valecchi, 1923.

15. See the letter of Carlo Antonio Pilati to Giovanni Andrea Cristani, written from Coira on 20 March 1768, and preserved in the Biblioteca Comunale of Trento, Ms. 457, 28-34.

in 1771 Pilati was introduced in The Hague to the Russian ambassador, Prince Golitsuin, a very open-minded man and an amateur mineralogist, by the curator of Leyden University, Count Willem Bentinck, and recommended for a job in Russia by Allamand.[16]

The case of Giovan Francesco Salvemini is more interesting.[17] Salvemini was a Tuscan born in Castiglion Fiorentino. His father was a lawyer. We know he studied at Pisa University and had among his teachers the mathematician Guido Grandi. As a young man Salvemini showed no interest in academic work, he much preferred to enjoy life, he loved dancing and music. He managed, however, to get a degree in jurisprudence. He probably became a freemason and an unbeliever. The only thing that seems to be known for sure is that in 1736 he had to leave Tuscany surreptitiously. The reasons why he had to do so are unclear: perhaps because of his unorthodox beliefs. From now on he changed his name from Salvemini to Castillon or de Castillon, and withdrew to Lausanne first as private tutor in the house of a patrician, later he moved to Vevey as lecturer of rhetoric in the local college. During his stay at Vevey he became a "bon chrétien" and adhered to the Reformed Church. While in the Republic of Bern, Castillon devoted himself both to the humanities and to mathematical studies. He provided an Italian translation of Pope's *Essay on Man*, entered into close contact with Gabriel Cramer, provided an edition of Newton's mathematical papers, published the correspondence between Leibniz and Johannes Bernoulli and edited the *Analysis infinitorum* of Leonhard Euler. In 1745 he was elected Fellow of the Royal Society, and in 1751 he was called to Utrecht University probably with the support of William IV. The fact that he had adhered to the Reformed Church was significant to his appointment. From 1751 to 1755 he was extraordinary professor of mathematics and physics, and from 1755 to 1763 ordinary professor. In 1763 he left Utrecht for Berlin where he became ordinary Fellow of the Royal Academy of Science, a colleague of Euler and, in later years, the director of the mathematical section of the Academy.

During the 12 years he spent at Utrecht, Castillon published a good number of mathematical and philosophical orations and dissertations, edited Newton's *Arithmetica Universalis*[18], published a French translation of

16. See the letter of Carlo Antonio Pilati to Ulysses von Salis-Marschlins of 27 September 1771, preserved in Coira, Staatsarchiv Graubünden, Familienarchiv: von Salis, Ms. D VI./CM, V. D7, 145-148.

17. Giorgio Spini, "Giovan Francesco Salvemini 'De Castillon' tra illuminismo e protestantesimo", in *I Valdesi e l'Europa* (Collana della Società di studi valdesi, 9) Torre Pellice 1982, pp. 317-350.

18. For instance: *Oratio de usu matheseos in vita humana habita a.d. IX. Decembris*

Vitaliano Donati's natural history of the Adriatic, and French translations of works by Locke and Campbell. In 1756 he also published an important book contesting Rousseau's *Discours sur l'origine de l'inegalité parmi les hommes*, claiming that the idea of the "bon sauvage" was an illusion, and defending the idea of progress through society and God's help.[19] For Castillon "nôtre espèce se perfectionne".

Jan Ingen-Housz is the only Dutch scientist of fame I know of who spent some time in Italy during the period under consideration.[20] A Roman catholic, with very poor prospects of a university career in The Netherlands, Ingen-Housz established his reputation as a physician and particularly as an inoculator. Appointed court physician in 1768 by the Empress Maria Theresa, he travelled to Tuscany in 1769 to inoculate the relatives of the imperial family. He spent about seven months in Italy: from June to December 1769.[21] In Florence he met Felice Fontana with whom he seems to have made a lasting friendship. This is testified not only by what is left of their correspondence,[22] but also by the fact that they carried out experiments together in Paris in the summer of 1777 and in London in 1778, and by many references made by Ingen-Housz to Fontana's published and unpublished investigations.[23] While Ingen-Housz was in Italy he bought several books

---

*MDCCLI*, Trajecti ad Rhenum: ex officina Joannis Brodelet, 1752; *Oratio in connexionarios habita a.d. XXIII. Junii MDCCLV*, Trajecti ad Rhenum: ex officina Ioannis Broedelet, 1755; *Oratio de legitimo studiorum nexu. Habita die XXVI. Martii MDCCLIX. Quum e Magistratu Academicu abiret*, Trajecti ad Rhenum: ex officina Joannis Broedelet, 1761; *Arithmetica Universalis ... Auctore Is. Newton, Eq Aur. Cum Commentario Johannis Castillionei ...*, Amstelodami: Apud Marcum Michaelem Rey, 1761 (2 vols.).

19. Jean de Castillon, *Discours sur l'origine de l'inegalité parmi les hommes. Pour servir de réponse au Discours que M. Rousseau, Citoyen de Géneve, a publié sur le même sujet*, Amsterdam: Chez J.F. Jolly, 1756.

20. J. Wiesner, *Jan Ingen-Housz, sein Leben und sein Wirken als Naturforscher und Arzt*, Wien 1905.

21. These dates are based on a diary kept by Ingen-Housz and preserved in Breda, Gemeentelijke Archiefdienst, Collectie Ingen-Housz, AFD. IV-16.7.

22. Up to now I have only been able to find three letters of Fontana to Ingen-Housz. One, dated 20 March 1772, is preserved in the Österreichische Nationalbibliothek of Vienna (4/63-1); another, dated 10 September 1773, in the Universiteitsbibliotheek of Amsterdam (119 Cq); and the third, dated 15 November 1773, in the Biblioteca Comunale of Forlì (Collezione Piancastelli). No original letter of Ingen-Housz to Fontana could be found. However, from a register of letters with short summaries of their content kept by Ingen-Housz from 1774 to 1793, and preserved in Breda, Gemeentelijke Archiefdienst, Collectie Ingen-Housz, AFD. IV-16.6., we learn that from 1774 onwards he wrote six letters to Fontana.

23. Peter K. Knoefel, "Famine and Fever in Tuscany. Eighteenth Century Italian Concern with the Environment", in *Physis*, XXI (1979) 7-35, on pp. 26-30.

and the complete Lucca edition of the *Encyclopédie*.[24] One of the first publications concerns a typically Italian research topic: the torpedo. In 1769 he studied the shocks of torpedos at Leghorn and reported his results in a letter to his British mentor Sir John Pringle, who communicated them to the Royal Society.[25]

Other Italians with whom Ingen-Housz corresponded, apart from Fontana, were Count Belgioioso, Beccaria, Landriani and Gianella.[26] We know that Ingen-Housz was briefly back in Florence in August and September 1772 to inoculate against smallpox other children of the imperial family, and of the court. In 1779 he illustrated in a Bakerian lecture Volta's electrical apparatus, the electrophore,[27] and in 1779 he provided the best description of the ultimate model of Fontana's eudiometer.[28] He returned to this subject in 1785.[29] What seems more significant, however, is the fact that he made use of Fontana's eudiometer in his own classical investigations on plant physiology.

It is difficult to assess the image that Italian natural philosophers of the second half of the 18th century had of Dutch contemporary science. It is well-known that in physics, during the Fifties and Sixties, the textbooks of 's Gravesande and especially of van Musschenbroek were both well-known and used in Italy; that works by Swammerdam and Leeuwenhoek enjoyed a good reputation among naturalists such as Felice Fontana, Lazzaro Spallanzani and Bonaventura Corti; and that in medicine authors such as Boerhaave, Gaub and Albinus were of constant reference. But what about the authors of the second half of the century? What did Italian scientists think of their Dutch counterparts? The documentation collected on this question is not sufficient to provide a definitive answer. The impression, however, is that they perceived a decline both at the level of university teaching and in scientific production. It should also be noted that the greater use of the

24. Breda, Gemeentelijke Archiefdienst, Collectie Ingen-Housz, AFD. IV-16.7.

25. "Extract of a Letter from Dr. John Ingen-Housz, F.R.S. to Sir John Pringle ... containing some Experiments on the Torpedo, made at Leghorn ...", in *Philosophical Transactions*, LXV/1 (1775) 1-4.

26. Breda, Gemeentelijke Archiefdienst, Collectie Ingen-Housz, AFD. IV-16.6.

27. "Electrical Experiments, to explain how far the Phenomena of the Electrophorus may be accounted for by Dr. Franklin's Theory of positive and negative Electricity ...", in *Philosophical Transactions* LXVIII/2 (1778) 1027-1055. On Volta's electrophore see: Giovanni Polvani, *Alessandro Volta*, Pisa: Domus Galilaeana, 1942.

28. Jan Ingen-Housz, *Experiments upon Vegetables* ..., London: P. Elmsly and H. Payne, 1779.

29. "Observations sur la construction et l'usage de l'eudiomètre de M. Fontana ...", in *Journal de physique*, XXVI (1785) 339-359.

national language in writing scientific works, instead of Latin or French as a medium of scientific communication, clearly frustrated those Italians wishing to read Dutch contributions of the period.

With the following words Carlo Antonio Pilati introduced that section of his two-volume book *Lettres sur la Hollande* published in 1780, which dealt with the sciences.

> "Quelque petit que soit ce pays, et quelque fort que soit le penchant général des habitans pour le commerce et pour les fabriques, cependant les arts, les belles-lettres et les sciences y ont été cultivées et portées beaucoup plus loin que dans bien des états de l'Europe qui ont beaucoup plus d'étendue et dont les habitants sont moins industrieux et moins occupés à gagner des richesses."[30]

His image of the Dutch scholar was built by contrasting it to stereotypes of the German, French and Italian:

> "les gens de lettres des sept provinces se distinguent de ceux des autres nations particulierement par trois choses: ils n'écrivent point d'abregés comme les Allemands: ils ne font point d'intrigues pour en imposer au public; et ne font point la cour aux femmes, aux grands seigneurs, et aux savans de la premiere classe pour s'en faire décerner une réputation que leurs ouvrages désavouent, comme ont coutume de faire quantité de faux esprits en France: enfin ils n'aiment ni les paradoxes, ni les futilités, qui sont les objets chéris de la plus grande partie des écrivains Italiens ...."[31]

Pilati noticed in The Netherlands of his time a general decline of the sciences.

> "Hors de Leyde on a depuis quelque temps un peu trop négligé la physique, et à Leyde même on néglige actuellement beaucoup trop l'histoire naturelle. Un ministre de l'Impératrice de Russie et un Ministre du saint évangile réussiront peut-être un jour à rendre en Hollande ce gout plus général. Le premier est M. le Prince de Callitzin [sic] ... L'autre est M. Martinet, pasteur de Zutphen, qui vient de publier un catéchisme de la nature en plusieurs volumes."[32]

Pilati also remarked that

> "il n'y a aucun pays peut-être où l'on trouve chez les particuliers autant de

30. [Carlo Antonio Pilati], *Lettres sur la Hollande*, op.cit. in note 6, vol. II, p. 232.

31. *Ibidem*, vol. II, pp. 251-252.

32. *Ibidem*, vol. II, pp. 254-255. Pilati meant the Russian prince Dmitry Alekseevich Golitsuin (often spelled Galitzin or Gallitzin) and Johannes Florentius Martinet, author of *Katechismus der Natuur*, 4 vols. Amsterdam: by de Wed. Loveringh en Allart (and from vol. II: by Johannes Allart), 1777-1779. Reduced editions of this work appeared in English (19 editions), German, French and Italian: *Catechismo della natura per uso dei fanciulli* ..., Firenze: presso Leonardo Ciardetti, 1837.

> cabinets d'histoire naturelle, et où dans le même temps l'on rencontre si peu de personnes qui cultivent la science même."[33]

In describing the decline of Leyden University, Pilati gave two main reasons. Firstly most students entered university unprepared, secondly they lacked "d'esprit républicain". They thus "abusent de leur liberté au point, qu'ils renversent dans leurs études tout l'ordre et toute l'harmonie des sciences."[34] He made one sole exception for the young people of Amsterdam, not only because they "font leurs études dans l'ordre où elles se suivent naturellement", but also because "les Amsterdamois sont de tous les Hollandois ceux qui ont le plus d'esprit républicain."[35]

Pilati was aware of the building up of provincial scientific societies,[36] but he did not master Dutch sufficiently to allow him to judge their contributions, and, as far as I have been able to establish up to now, it seems that they remained unnoticed to most Italian natural philosophers of the day with the notable exception of the *Societé Hollandoise des Sciences, établie à Haarlem* whose programmes were also sent out to a number of Italian scientists such as Paolo Frisi and Alessandro Volta. The problem of language constituted a permanent barrier, and only when Dutch scientific contributions were written in Latin or French did they reach an Italian audience.[37] It was, however, mainly by touring that some Italian natural philosophers came into closer contact with Dutch scholars and scientists.

An interesting feature of the life of Italian natural philosophers of the second half of the 18th century is that they travelled outside Italy more frequently and more freely than in the previous century and a half. This is an evident sign that the general atmosphere of the peninsula was loosening up. The main goals of their European tours were Switzerland, Paris, London and sometimes Germany. On their way they also tended to visit The Netherlands and contact scholars there. It should be borne in mind, however, that in the period 1750-1789 the foreigners with whom Italians had, from a scientific point of view, their most fruitful correspondence were mostly Swiss.

The holder of the chair in mathematics of the Collegio Romano, the Jesuit

33. *Ibidem*, vol. II, p. 256.

34. *Ibidem*, vol. II, p. 246.

35. *Ibidem*, vol. II, p. 247.

36. *Ibidem*, vol. II, pp. 356-357. On Dutch scientific societies of the time, see: W.W. Mijnhardt, *Tot Heil van't Menschdom. Culturele genootschappen in Nederland, 1750-1815*, Amsterdam: Rodopi, 1988.

37. For the use of Latin, French or Italian in the writing of scientific literature in Italy during the second half of the 18th century, see: Renato G. Mazzolini, "La questione del latino in una lettera di L. Spallanzani a Girolamo Ferri", in *Episteme*, X (1976) 313-326.

Ruggero Giuseppe Bošković, travelled throughout Europe and to Constantinople for five and a half years from the autumn of 1757 to 1763. He did so with the full approval of his Order, was provided with funds and was charged to carry out a number of diplomatic missions, some of which are not well known. In Paris and London he was very well received by the most distinguished scientists of the day. We know he was also in The Netherlands during the second half of January and the beginning of February 1761, and that he visited Rotterdam, Delft, The Hague, Leyden, Amsterdam and Utrecht.[38]

Another mathematician to visit The Netherlands during his tour to Paris and London in 1766-1767 was Paolo Frisi, one of the most eminent representatives of the Enlightenment in Lombardy. During his short stay in the United Provinces he met Rijklof Michael van Goens with whom he later corresponded till his death in 1784.[39]

In 1771-1773 the Bolognese mathematician Alfonso Bonfioli Malvezzi undertook a journey through Germany, The Netherlands, England, France and Switzerland. In The Netherlands he spent about 20 days from the middle of May to the beginning of June 1772. He visited Breda, Utrecht, Amsterdam, The Hague and Rotterdam. In Utrecht he contacted the physician Johannes Hahn and visited the University. In his diary he wrote that "la Salle de l'Anatomie, et celle des experiences Phisiques sont au dessous de l'ordinaire."[40]

38. Elizabeth Hill, "Roger Boscovich: A Biographical Essay", pp. 17-101, on pp. 47-49, in *Roger Joseph Boscovich S.J., F.R.S., 1711-1787. Studies of his Life and Work on the 250th Anniversary of his Birth*, edited by Lancelot Law Whyte, London: George Allen & Unwin, 1961. Ruggiero Giuseppe Boscovich, *Giornale di un viaggio da Costantinopoli in Polonia ... con una sua relazione delle rovine di Troia ...*, Milano: Giordano editore, 1966.

39. Frisi's diary of his tour is preserved in the Biblioteca Ambrosiana of Milan (Y 163 Sup.), and has been partially published by Franco Venturi in *Illuministi italiani*, vol. III: *Riformatori lombardi, piemontesi e toscani*, Milano-Napoli: Ricciardi, 1958, pp. 305-314 (but not the section of The Netherlands). Eleven letters of Frisi to van Goens, written from 1768 to 1784, are preserved in the Koninklijke Bibliotheek at The Hague (130 D 14), and one in the Universiteitsbibliotheek of Leyden (B.P.L. 246). Eight of them have been published by W.H. de Beaufort, "Brieven aan R.M. van Goens en onuitgegeven stukken hem betreffende", in *Werken van het Historisch Genootschap gevestigd te Utrecht*, n.s., XXXVIII/1 (1884) 482-494. Twelve letters of van Goens to Frisi are preserved in the Biblioteca Ambrosiana of Milan (Y 151, 189-204) together with letters of Rendorp and van der Aa to Frisi. See: *Ideologia e scienza nell'opera di Paolo Frisi (1728-1784)*, a cura di Gennaro Barbarisi, 2 vols. Milano: Franco Angeli, 1987, vol. II, pp. 505-507, 657, 662-663. On Frisi in The Netherlands see also: Mario Battistini, "Documenti italiani nella Biblioteca Reale dell'Aia", in *Giornale storico della letteratura italiana*, CIII (1934) 254-280, on pp. 255-257.

40. I owe the information on Malvezzi's diary to Professors L. Pepe and S. Cardinali who are presently editing it.

With the authorisation of and a subvention from the Grand Duke of Tuscany, Felice Fontana and his assistant Giovanni Fabbroni undertook a long journey abroad from 1775 to 1780.[41] The prime aim of their Grand Tour was to acquire scientific instruments and natural objects for the Royal Museum of Physics and Natural History in Florence. Fontana and Fabbroni spent most of their time first in Paris and later in London, where they carried out research and contacted a good number of scholars and instrument makers. They had both planned to visit the Low Countries[42] but, as far as I have been able to establish and contrary to what has been written so far, they were prevented from passing through Amsterdam on their way back to Italy, because the Grand Duke urged them to return as soon as possible.[43] In later years, however, Fontana corresponded between 1783-1789 with the physicist Jan Hendrik van Swinden.[44] Together with the well-known correspondence between van Marum and Volta, this is the most important and largest correspondence between an Italian and a Dutch scientist in the period under consideration here. The letters they exchanged run to about 40 and their historical significance lies mainly in the fact that they document how Fontana acquired a number of scientific instruments and especially natural objects for the Museum in Florence. It was through van Swinden and through Frederik Christian Meuschen that the Florentine Museum acquired specimens from the Museum Geversianum of Rotterdam in 1787.[45]

During his European tour, only partially subsidized by the government, Alessandro Volta spent about 20 days, between November and December 1781, in The Netherlands.[46] He went to Rotterdam, The Hague, where he was

41. Peter K. Knoefel, *Felice Fontana: Life and Works*, Trento: Società di Studi Trentini di Scienze Storiche, 1984, pp. 51-58.

42. Casimiro Adami, *Di Felice e Gregorio Fontana scienziati pomarolensi del secolo XVIII*, Rovereto: Ugo Grandi, 1905, p. xxxiii. See also: Renato Pasta, "La corrispondenza di Giovanni Fabbroni da Londra (1778-1779)", in *Critica storica*, (1980) 277-308, on p. 283, note 15.

43. Renato Pasta, "Alle origini del liberismo toscano: il contributo di Giovanni Fabbroni (1752-1822)", in *Annali della Fondazione Luigi Einaudi*, X (1981) 179-243, p. 188, note 13.

44. The letters of Fontana to van Swinden, together with copies of those of van Swinden to Fontana are preserved in the Universiteitsbibliotheek of Leyden (B.P.L. 755) and are going to be published in the second volume of the correspondence of Felice Fontana. For the first volume see: Felice Fontana, *Epistolario 1. Carteggio con Leopoldo Marc'Antonio Caldani 1758-1794*, a cura di Renato G. Mazzolini e Giuseppe Ongaro, Trento: Società di Studi Trentini di Scienze Storiche, 1980.

45. For the sale in September 1787, see: Gemeente Rotterdam, Archiefdienst, not. prot. 3238/776, where, however, Fontana's name is not mentioned.

46. Alessandro Volta, *Epistolario*, 5 vols. Bologna: Nicola Zanichelli, 1949-1955, vol. II, pp. 67-73.

invited by Prince Golitsuin, Amsterdam and Haarlem, where he visited the collection of natural history of the Dutch Society of Science directed by Martinus van Marum with whom he later corresponded.[47] Like other Italian visitors to The Netherlands, Volta was fascinated by the canals, the ships, the cleanliness of the towns and the number of collections of natural history.

In 1786 the Austrian Government supported and subsidized the abbot Barnaba Oriani in his European mission to visit astronomical observatories. As professor Tagliaferri has shown in his paper, Oriani also spent some time in Amsterdam where he met Jan Hendrik van Swinden.[48]

Directly appointed by the Austrian Government, the Milanese Marsilio Landriani, a physicist who later became a diplomat, undertook a long tour in 1787-1790 through France, the Austrian Low Countries, The Netherlands, England and Germany to study and collect information useful for "the progress of the sciences ... and the increment of manufactures and arts" in Lombardy.[49] On his return he delivered 14 reports — which have only recently been published — concerning his tour.[50] None of these reports concerns The Netherlands where he stayed at the end of 1788.[51] From other sources, however, we know that Volta provided him with an introductory letter to van Marum[52] and that instead of collecting information on Dutch manufactures he carried out physico-chemical experiments in Amsterdam with Adriaen Paets van Troostwijk and in Haarlem with van Marum with whom he later also corresponded.[53]

The above list is probably far from complete. Nonetheless I think it is sufficient to allow some remarks and put this data into a more general framework.

47. *La correspondance de A. Volta et M. van Marum*, publiée par J. Bosscha, Leiden: A.W. Sijthoff, 1905.

48. Mario Battistini, "Scienziati italiani in Belgio", in *Rivista di storia delle scienze mediche e naturali*, XII (1930) 212-214; Id., "Documenti italiani in Olanda: una lettera di Lazzaro Spallanzani ed una di Barnaba Oriani", in *Rivista di storia delle scienze mediche e naturali*, XXI (1939) 188-192.

49. *Relazioni di Marsilio Landriani sui progressi delle manifatture in Europa alla fine del Settecento*, a cura di Mario Pessina, introduzione di Aldo De Maddalena, Milano: Edizioni il Polifilo, 1981, p. XLV, note 4.

50. See note above.

51. Salvatore Pugliese, "I viaggi di Marsilio Landriani" in *Archivio storico lombardo*, s. VI, LI (1924) 145-185, on p. 172.

52. *La correspondance de A. Volta et M. van Marum*, publieé par J. Bosscha, Leiden: A.W. Sijthoff, 1905, p. 24.

53. Martinus van Marum, *Life and Work*, vol. 1, edited by R.J. Forbes, Haarlem: H.D. Tjeenk Willink & Zoon, 1969, pp. 234-235; vol. VI, edited by E. Lefebvre anf J.G. de Bruijn, Leyden: Noordhoff, pp. 186-190.

It is well-known that Italy had been and still was the focal point of the Grand tour. Generally we discuss the Grand Tours of the English, the French, the Germans and the Dutch, but not those carried out by the Italians. One of the reasons for this is that before the period I am considering, when Italians went abroad, they usually went for good. For over two centuries there had been a constant emigration of Italian artists, diplomats and intellectuals. Now, for the first time, there were Italian natural philosophers who were often financially supported to go abroad, to collect information, study, come back home and report on what they had seen. This fact has escaped the attention of historians and, I believe, is of some importance. But we cannot understand its meaning unless we turn our attention to the political scene and to the efforts of reform undertaken by some Italian States. It is only if we keep in mind such efforts that we can understand why Italians were now going abroad. Some rulers of the *ancien régime* in Italy became increasingly conscious of the importance of the sciences in the organisation and development of their States. The ideal of the useful sciences became the ideal of the rulers. They at times thus fostered the foundation of scientific or mixed societies and put an effort into reforming university curricula and acquiring scientific equipment.

If, on the other hand, we consider the facts mentioned above not in relation to the political scene, but in relation to the scientific movement as a whole, they point to a macrophenomenon that — as far as I know — seems to have escaped the attention of scholars of the Enlightenment. They point to the decline of the *Respublica litterarum* and to the slow but steady building up of national traditions within the sciences and their institutional organisations. At first glance the way in which this double process took place in Italy and in The Netherlands seems to have differed considerably. In Italy it seems to have depended mainly on the complex and unstable relationship between enlightened rulers and intellectuals, while in The Netherlands it was the building up of provincial societies. But this is only a historical hypothesis which requires specific investigation.

# THE ELEGANT ANATOMIST. THE ITALIAN MEDICAL CONNECTIONS OF EDUARD SANDIFORT (1742-1814)

*Antonie M. Luyendijk-Elshout*

Macies et nova febrium
Terris incubuit cohors
Semotique prius tarda necessitas
Lethi corripuit gradum

Horatius (Odes, I. 3.29-33)

("Decline and a new army of Fevers burdened the Earth.
and ineluctable Death, formerly distant
and very slowly drawing near, now quickened its pace".)

This passage in one of Horace's *Odes* is characteristic for the last three decades of the eighteenth century, when the medical world became aware of the social misery and the raging epidemics in Europe. In the spirit of the Enlightenment physicians became involved in social matters; treatment of the victims of raging diseases among men and cattle and the improvement of the conditions in the hospitals. Communication on medical information, formerly a matter of personal relations between outstanding physicians, was most urgent for university professors and general practitioners as well. The modernization of clinical teaching in Edinburgh and Pavia was of great importance for the new generation, which should become confronted with revolutionary concepts of diseases, which would change the medical world considerably after the French Revolution.

Around 1770 the situation in the established universities of Vienna, Leyden, Bologna and Padua, was dominated by men of excellence, who had

Fig. 43 Eduard Sandifort (1742-1814) (Rijksuniversiteit Leiden, Senaatskamer)

been witnesses to Boerhaave's doctrines and developed their methods along the lines of *experientia* and *ratione*.

Albrecht von Haller (1708-1777) was the undisputed leader in physiology. His experiments were repeated all over Europe. He corresponded with prominent scholars, such as Leopoldo Marc' Antonio Caldani (1725-1813)[1] and he was challenged by daring Italian scientists such as Tommaso Laghi (1709-1764), Felice Fontana (1730-1805)[2] and by Leyden students, who had watched the experiments taken by B.S. Albinus (1696-1770), who taught physiology at Leyden since 1747.[3]

The Leyden group was obviously impressed by the work of the Italian scientists. On August 8, 1757, Lambertus Bicker (1732-1801) presented his thesis *De natura humana, quae medicorum est* to the medical faculty, with Hieronymus David Gaub (1705-1780) as promotor. One week later, on August 15, his friend Iman Jacob van den Bosch (1731-1788) got his doctors degree on his thesis: *De vivis humani corporis solidis*, with B.S. Albinus as promotor.[4] We shall not discuss these remarkable studies in detail, but it should be mentioned that both students quoted the Italian studies on irritability and sensibility at large extent. Bicker even added Tommaso Laghi's letter to Cesareo Pozzi, professor of mathematics in Rome, to his thesis. In a footnote he explained:

> "I became acquainted with this letter by the kindness of the learned Musschenbroek, who received it.... Since the author is not so well known in Holland as he ought to be I have published it as a supplement to my thesis.... He has described several precious experiments on sensibility and irritability, done with great care and good judgment.... When he does not agree with Haller, he expresses himself in a very modest way."[5]

1. E. Hintzsche, *Albrecht von Haller, Marc' Antonio Caldani. Briefwechsel 1756-1776.* Bern/Stuttgart, Hans Huber, 1966.
2. Renato G. Mazzolini e Giuseppe Ongaro, "Carteggio con Leopoldo Marc' Antonio Caldani 1758-1794". In: *Epistolario di Felice Fontana* I, Trento/Bologna 1980.
3. H. Punt, *Bernhard Siegfried Albinus (1697-1770) 'On Human Nature'. Anatomical and physiological ideas in eighteenth century Leiden.* Amsterdam, B.M. Israel, 1983.
4. Lambertus Bicker, *De natura humana, quae medicorum est.* Leiden, apud Joannes Bos, 1757 and Iman Jacob van den Bosch *De vivis humanis corporis solidis.* Leiden, apud Samuel et Joh. Luchtmans et Nic. Moens, 1757.
5. Bicker, "De Natura", note 35, p. 37. Van den Bosch also mentions the letter, shown to him by van Musschenbroek. This letter should have been the one directed to '*Cl. Viro D. Caesareo Pozzi in romano lyceo matheseos professori*'. It was published in Bologna in 1756, and reprinted in *De Bononiensi Scientiarum et Artium Instituto atque Academia Commentarii* (opuscula), IV, 1757, pp. 208-218, as well as in '*Sulla insensitività ed irritabilità halleriana. Opuscoli di vari autori raccolti da G.C. Fabri*', II, Bologna, 1757, pp. 110-116. See Mazzolini, "Carteggio" (nt. 2) pp. 23, 125. Petrus van Musschenbroek (1692-1761) was professor of physics and mathematics at Leyden.

It was especially B.S. Albinus, who quarrelled with Haller on these matters and so any criticism from Italy was most welcome. The struggle on the irritability and sensibility of the brain, the meninges, the fascias and the tendons would reach its culminating point in the sixties. The disputes were endless, the experiments were gruesome and the quarrels were useless. Around 1770, the interest was fading, there were fresh fields for physiology, especially the experiments with electricity.

Albinus' unfailing search for details in anatomy was a guiding star for his students. This tradition was started by Giovan Battista Morgagni (1682-1771) whose careful and exact studies were well known in Leyden.[6] They became apparent in the interacting with the new physiology such as the structure and the functioning of the iris[7] and the detailed studies on the lymphatic system by Samuel Thomas Soemmerring (1755-1830)[8] and Paolo Mascagni (1755-1815).[9] The anatomists ranked with the best in the production of atlases: Albinus with his "perfect skeleton,"[10] Soemmerring with his "schöne Mainzerin" (beautiful woman from Mainz) (1797)[11] and Mascagni with his anatomical studies for artists, which were published after his death in 1816.[12]

Was the knowledge of the human body complete? In any case, the body of man could be reconstructed following the anatomical composition, which had been described by so many excellent scholars and which had been represented by so many skilful artists. Florence had its wax modellers, guided by Felice Fontana, and European travelers could soon admire the collection in the Imperiale e Reale Museo di Fisica e Storia Naturale founded by Peter

6. A.M. Luyendijk-Elshout, "Morgagni, Ruysch e Boerhaave. I tre moschettieri dell'anatomia sottile.' "*Scienza e Cultura*" Informazione dell'università di Padova, with English translation, Padova, 1983, pp. 301-320.

7. Renato G. Mazzolini, "The Iris in Eighteenth Century Physiology". *Berner Beiträge zur Geschichte der Medizin und der Naturwissenschaften*. Neue Folge, Bern/Stuttgart/Wien, 1980.

8. Samuel Thomas Soemmerring, *De cognitionis subtilioris systematis lymphatici in medicina usus*. Kassel, 1779 and *De Morbis vasorum absorbentium corporis humani sive dissertationis quae praemium retulit Societatis Rheno-Trajectinae anno 1794. Pars pathologica*. Frankfurt am Main. Varrentrapp & Wenner, 1795. This work was translated into Italian in 1813 and published in Milano under the title *Malattie de' vasi assorbenti del corpo umano ...*

9. Paolo Mascagni, *Vasorum lymphaticorum corporis humani Historia et Ichonographia*. Siena, Pazzini Carli, 1787.

10. B.S. Albinus, *Tabulae Skeleti et Musculorum Corporis Humani*. Leiden, Joa. et Hermannus Verbeek, 1747.

11. S.Th. Soemmerring, *Tabula sceleti femini juncta descriptione*. Frankfurt am Main, Varrentrapp & Wenner, 1797.

12. P. Mascagni, *Anatomia per uso degli studiosi di scultura e pittura*. Opera postuma, edited by his son Bernardino Mascagni, Firenze, G. Marenighi, 1816.

Leopold in 1775.[13] The Anatomy Museum in Leyden was to obtain two specimens, prepared by the Florentine artists, one human head with dissected parapharyngeal space from the hand of Clemente Michelangelo F. Susini (1754-1814) in 1804, and a dissected head with the venom-teeth of a viper, made for Fontana's studies on the action of poisons.[14] Peter Leopold's museum was meant for a more general public, he wanted to show the perfection of the human anatomy to artists, physicians and civilised laymen.[15]

But was anatomy as a discipline of the same importance for medicine as it had been since Vesalius' *Fabrica corporis humani,* published in 1543? In the early seventies, several scholars and learned societies had their doubts. Anatomists had done their utmost, each small vessel was localised, each group of muscles properly placed. Delicate nerves and lymphe vessels could be traced, the last group by injections with mercury. Specimens could be conserved better in terpentine, even the most delicate ones.

The tradition of the *anatomia subtilis*, forwarded by Malpighi a hundred years ago[16] was still alive, but gross anatomy was considered to be more or less sufficiently explored. It was time to pay more attention to pathological changes in the body, and to instruct the students in. the autopsy room, in order to compare the clinical picture with the post mortem, in other words: it was time to introduce the work of Giovanni Battista Morgagni *De sedibus et causis morborum per anatomen indagatis* which was published in 1761 in Venice, into the medical curriculum of the universities.

### *"Pathology new style" after Morgagni*

Morgagni's *Adversaria anatomica* had been introduced for the medical students by Herman Boerhaave in his lectures as early as 1711.[17] The connections between Morgagni and Boerhaave were very friendly, and this

13. Benedetto Lanza, Maria Luisa Azzaroli Pucetti e.a. *Le Cere Anatomiche della Specola*, Firenze, Arnaud, 1979.

14. W.J. Mulder, "Anatomical Waxmodels in the collection of Leiden". Biblioteca della *Rivista di storia delle scienze mediche e naturali* 20 (1977), pp. 427-432.

15. Erna Lesky, "Paolo Mascagni (1755 bis 1815) und die Restaurierung seiner 'Icones' ". *Österreichische Ärztezeitung* (1968), 4. Reprint.

16. Luigi Belloni, De la théorie atomistico-mécaniste à l'anatomie subtile (de Borelli à Malpighi) et de l'anatomie subtile à l'anatomie pathologique (de Malpighi à Morgagni). *Clio Medica* 6 (1971), pp. 99-107.

17. A.M. Luyendijk-Elshout, "Morgagni, Ruysch ..." *o.c.* (nt. 6). Boerhaave lectured for the first time on Morgagni's *Adversaria Prima* to the students in 1711, as can be learned from his manuscripts kept in the Kirov Museum at Leningrad. Copies of these are preserved in the Boerhaave Museum at Leyden.

tradition was continued by B.S. Albinus, who had a high esteem for his colleague in Padua.[18] However, Albinus was not interested in pathology, he detested anomalies of the human body and he only fulfilled his duties in surgery with reluctance.[19] Albinus died in 1770, and the tradition of *anatomia subtilis* seemed to come to an end. The next crew which entered the Dutch universities firmly stated the new policy for the faculty: *De anatomiae pathologicae utilitate et necessitate*, first presented by Paulus Graeuwen (1715-1779) in June 1771 to the audience in Groningen.[20] Graeuwen was a pupil of Albinus, but his interest was practical medicine and he praised Lancisi and Morgagni as guiding stars towards more certainty in medicine. One year later Eduard Sandifort offered his address to the Leyden audience: *De circumspecto cadaverum examine optimo practicae medicinae adminiculo.*[21] His address was more oppressive, he discussed pathology as the "Gubernatrix" of practical medicine, advocating Morgagni's empirical methodology. He praised Baglivi's knowledge of pulmonary diseases and suggested to direct *anatomia subtilis* towards the pathology of the parenchym.[22] Malpighi had prepared the way by his publications on the structure of the glands and the pulmonary tissue. He urged the students to combine the observations in the hospital with the autopsy of the patients, with a citation from the *sepulchretum anatomicum*, the well-known handbook by Theophile Bonet. But Morgagni excelled over Bonet:

> Morgagni whose work on the seat and the cause of diseases, worth of immortality, has more facts determined than all medical practitioners before him, not only by his innumerous own observations, but also by the explanation of the observations of his teacher Valsalva ...[23]

The hospital was the most important place to learn medicine. This, said Sandifort, was so rightly remarked by Morgagni, when he commemorated

18. Giuseppe Ongaro, *Il "Catalogo di Libri" di Giambattista Morgagni.* Edizione del testo e identificazione degli esemplari possedenti della biblioteca universitaria di Padova, a cura di Elisabetta Barile e Rosalba Suziano Padua, Ed. Lint, 1983. Albinus inscribed on the page of his *Historia Musculorum* (1734): "Eximo Morgagno observantiae ergo d.d. B.S. Albinus" (nr 128 in Morgagni's library).

19. H. Punt, "Bernhard Siegfried Albinus" *o.c.* (nt 3), p. 10.

20. Pauli Graeuwen, *De anatomiae pathologicae utilitate ac necessitate.* Groningen, Hajo Spandaw, 1771.

21. Eduard Sandifort, *De circumspecto cadaverum examini optimo practicae medicinae adminiculo.* Leiden, S. et J. Luchtmans, 1772.

22. *Ibidem*, p. 23. He quotes Baglivi's *Praxis Medicinae, Liber 1 - cap. 9.* Malpighi is especially mentioned for his letter to the Royal Society on the structure of the glands, written in 1688.

23. *Ibidem*, p. 28.

Valsalva in his *Commentariolum de vita et scriptis Valsalvae.*[24] Sandifort realised with his colleagues that clinical teaching was poor at Leyden and could in no way be compared with the facilities in Edinburgh! This serious want of an appropriate hospital should become one of the reasons why the medical faculty of Leyden lost its international splendour after the death of Hieronymus David Gaub. Gaub, an excellent pathologist, sent his *Institutiones Pathologiae Medicinalis*, Leidae Batavae, 1758, with a personal dedication to "his old friend Morgagni."[25] However, Gaub was mostly interested in theoretical medicine and no attention was paid to "the seat and causes of diseases" in the body until the arrival of Eduard Sandifort, as can be gathered from his inaugural address.

The Board of the University must have realised that a pathologist and a clinician had taken the chair of anatomy and that the Albinian tradition was fading. The famous collection of anatomical specimens was still under the guidance of Albinus' younger brother Frederik Bernhard Albinus (1715-1778).[26] But Sandifort's firm statement that autopsies would enter the program to teach the students both normal and pathological anatomy meant a watershed between *anatomy* and *anatomia subtilis.*

Sandifort would enrich the Museum with an impressive collection of pathological specimens, he would contribute to pathology by his publications in a most respectable way,[27] he would dissect about fifty corpses for the students,[28] he would catalogue the complete Leyden collection gathered from 1698 until 1790 into two volumes.[29] But his "elegant anatomy", after B.S.

24. In 1742, Johan Hasebroek, printer at Leyden, edited several works by Valsalva after an edition prepared by G.B. Morgagni in 1740. He dedicated this edition to the professors of the Medical Faculty, Herman Oosterdijk Schacht, B.S. Albinus and H.D. Gaub. It included Morgagni's biography of Valsalva. Valsalva recommends indeed observations in the hospital "Medicum qui suffecerit ad cognoscendum, sufficere etiam ad sanandum" but not as emphatically as Sandifort offers it to his audience.

25. Ongaro, "Catalogo di Libri", *o.c.* (nt 18) p. 74, nr. 1567.

26. Molhuyzen, *Bronnen tot de Geschiedenis der Leidsche Universiteit.* The Hague, Martinus Nijhoff, Vol. VI, 1923, Res. Cur. d.d. 27/8/1770. On November 17, 1771, B.S. Albinus and Sandifort had drawn up regulations for the supervision of the anatomical collections. Two professors of the medical faculty should be in charge of this: the anatomist and the physiologist.

27. E. Sandifort, *Observationes anatomico-pathologicae.* Leiden, P. v.d. Eyk et D. Vijgh 1779-1781. In four volumes.

28. Gerard Sandifort, *Museum Anatomicum Academiae Lugduno Batavae* Vol. III and IV, Leyden, S. and J. Luchtmans, 1827 and 1835. In Vol. III the author describes his father's biography: Introductio, pp. II-XII.

29. Eduard Sandifort, *Museum Anatomicum Academiae Lugduno Batavae.* Vol. I and II, Leiden, S. and J. Luchtmans, 1793. In these books the collections of J.J. Rau, B.S. Albinus

Albinus and more emphatically after Mascagni and others, only developed along sidelines. He never published in this field. According to his son Gerard Sandifort (1779-1848) his respect for B.S. Albinus was so deeply rooted, that he hardly dared to study *anatomia subtilis*. As we shall see, he nevertheless stimulated research in *anatomia subtilis new style*, as I would call "elegant anatomy". He even made a small collection of specimens. Sandifort considered *anatomia subtilis* as physiology. He felt obliged towards his teacher to present this field properly to the students, after the death of F.B. Albinus in 1778, when Sandifort became full professor of medicine. This he would continue even after his late fifties in 1798, when his son took over the autopsies and the instruction in normal and pathological anatomy. Sandifort's duties were then restricted to teaching physiology.[30]

*Encyclopedist of European medicine*

An elegant anatomist in his late fifties, a pathologist and gross anatomist in his midlife, and a practising surgeon and physician at the beginning of his career! Born in 1742 in Dordrecht, a small town along one of the branches of the Rhine, Eduard Sandifort, descendant of an English immigrant who died in 1652, matriculated at Leyden in 1758. After his study he settled down in The Hague to practise medicine. He was especially interested in surgery, and when he came upon a fatal case of a ruptured aneurysm he wrote his first study, which was accepted by the *Academia Curiosorum* and mentioned in their commentaries in 1766.[31] He was impressed by the wide scope of literature of which the *Curiosi*, under the guidance of Christian Gottlieb Ludwig (1705-1773) took care. In 1764 he decided to start a series of book reviews, like the learned societies of different countries offered abstracts to those physicians, who could not afford to buy medical publications from various countries. He used the same method as the editor of the *Commentarii de rebus in Scientia Naturali et Medicina gestis*, which were started in 1752. He wrote the abstracts in Dutch. Like the *Curiosi*, he added a list of titles also. He called his periodical *Natuur- en Geneeskundige Bibliotheek*, which he continued until 1776.[32]

---

and W. van Doeveren were described. See also A.M. Elshout, *Het Leidse Kabinet der anatomie uit de achttiende eeuw*. A scientific collection as a monument of cultural history. Leiden, University Press, 1952.

30. G. Sandifort, "Museum Anatomicum", *o.c.* (nt 28), p. X.

31. In the *Nova Acta Academia Naturae Curiosorum* T 4 (1760), p. 28.

32. E. Sandifort, *Natuur- en Geneeskundige Bibliotheek*, "in which work abstracts are presented of all new books, published outside our country dealing with Medicine and Natural History", 's-Gravenhage, Pieter van Cleef, 1765-1776.

In this "Library", Sandifort selected abstracts from medical books and periodicals, such as delivered by the European Societies and even contributions from America, for instance the *Transactions of the American Philosophical Society held at Philadelphia, for promoting useful knowledge* (1769-1771), Vol. I.[33] Needless to say that he paid ample attention to Morgagni's work, for instance the *Opuscula Miscellanea*, published in 1763,[34] and also to Morgagni's biography of Anton Maria Valsalva (1666-1723). Furthermore, Felice Fontana's studies on the motion of the iris were discussed,[35] Spallanzani's (1729-1799) work on reproduction[36] and various publications from other prominent Italian authors.

Sandifort gave a short survey of the history of the Bolognese society in vol. VI of his "Library", as an introduction to the *Commentarii* of the University of Bologna.[37] In this introduction he mentions one of the authors, Vincenzo Pozzi, who according to Sandifort had been travelling through France and the Netherlands in 1759. These commentaries were on a variety of subjects; on hyacinths, on silk-worms, diseased cattle, mushrooms, etc. The "Library" holds an impressive lot of information. Sandifort must have been very alert on new medical books, especially on work from Italy. Moreover, Sandifort stimulated his brother, a practising physician in The Hague, to translate into Dutch some surgical observations, made by Antonio Benevoli (1685-1756) in Florence and Gaetano Tacconi (1689-1782) professor of surgery in Bologna. The result of this work was also discussed in the "Library."[38]

Sandifort was also concerned with the publication of some very important Latin treatises in full length. Thus, the *Thesaurus Dissertationum* was composed between 1768 and 1778. Sandifort dedicated the first volume to the great encyclopedist Albrecht von Haller.[39] In the first volume he took Domenico Cotugno's (1736-1822) booklet on the inner ear.[40] Cotugno's

33. *Ibidem,* Vol. X (1773), pp. 483-510.

34. *Ibidem,* Vol. I (1765), pp. 230-247.

35. *Ibidem,* Vol. IV (1768), pp. 467-472.

36. *Ibidem,* Vol. V (1769), pp. 521-536.

37. *Ibidem,* Vol. VI (1769), pp. 468-517. He refers to "De Bononiensi scientiarum et artium Instituto atque Academia Commentarii". Bologna, 1767.

38. *Ibidem,* Vol. VII (1776), pp. 638-649. Jan Bernard Sandifort published these *Heelkundige Waarnemingen* in The Hague, at the widow Otto van Thol, in 1770.

39. E. Sandifort, *Thesaurus Dissertationem Programmatum aliorumque opusculorum selectissimorum ad omnem medicinae ambitum pertinentium.* Rotterdam, Henricus Beman, 1768. In three volumes. The last volume was published at Leyden by S. and J. Luchtmans and P. v.d. Eyk and D. Vijgh.

40. *De aquaeductibus auris humanae internae anatomica Dissertatio.* Published in Naples in 1760.

study *De ischiade nervosa* (Naples) was published in vol. II. In the third volume, he published Bartolomeo Patuna's (1738-1769) letter to Morgagni on extra uterine pregnancy[41] and Francesco Biumi's reports on autopsies, made in the Ospedale Milanese in 1765.[42] Sandifort published the illustrations, which were part of the treatises, with great care. He was very accurate, also on this point, as we shall see. For instance, several illustrations of authors from other countries were engraved by A.J. Polak, such as Cotugno's tables on the ear and some illustrations of German authors. Also dissertations from his own students were enlightened by copper engravings. The artists Abraham Delfos and F.F. Bylaart and R. Muys worked for Sandifort and his pupils over a long period.

In 1784, Sandifort published L.M.A. Caldani's work *Institutiones Physiologicae et Pathologicae* for the students. Haller had sent him the Institutions of Pathology, which he had received in 1772 from Caldani. Adriaan van Royen (1704-1779), Sandifort's colleague in botany, had brought the *Institutiones Physiologicae* from a visit to Padua. After consulting his colleagues of the Faculty, Sandifort decided to publish both books without any commentary, he only added an index to the book on physiology.[43] The last tribute which Sandifort paid to Italian authors was the publication of the *Opuscula Anatomica Selectiora* in 1788. In this book he collected three anatomical treatises, written by the Bolognese professor Germano Azzoguidi (1748-1814), by the Milanese surgeon Giovanni Battista Palletta (1747-1832) and the veterinary surgeon Giovanni Brugnoni (1741-1818) from Turin.[44] In the introduction to this work he complained about the difficulties in obtaining books from Italian authors, especially Azzoguidi's work on the structure of the uterus "frustra per plures annos a me desideratum". He was

41. *Epistola physico-medica continens historiam foetus sine involucris extra uterum inventis placenta intra uterum haerente, ad Cel. Morgagnum*. Vienna, 1765.

42. On Biumi's work, see note 50.

43. Sandifort explains all this in the preface. The book was published by S. and J. Luchtmans, on rather poor paper. Haller received two copies of Caldani's book on pathology, he may have forwarded one to Sandifort. See also Hintzsche, "Briefwechsel", *o.c.* (nt 1). Van Royen visited Padua probably in the mid-seventies.

44. Printed in Leyden by S. and J. Luchtmans. It holds the following studies:

G. Azzoguidi, "Observationes ad uteri constructionem pertinentes", Bologna, 1743.

G.B. Palletta, "Nova gubernaculi testis Hunteriani et tunicae vaginalis herniae congenitae descriptio", Milano, 1777.

Also: "Exercitatio de claudicatione congenita", Torino, 1785.

G. Brugnoni, "De testium in foetu posito, de eorum in scrotum descensu: de tunicarum, quibus hi continentur, numero et origine". The last study was published in the *Mémoires de l'Académie Royale des Sciences*, vol. 1784/1785, II part, Torino 1786.

even more angry about the inaccessibility of Palletta's work on the anatomy of the congenital hernia in the scrotum, which he had traced in the *Göttinger Anzeiger*[45] and which he finally translated himself from Italian into Latin. He published also Brugnoni's treatise on the same subject. Sandifort was directly involved in the discussion on the topography of the testis and its tunical wrappings,[46] since he studied inguinal hernia's in relation to topographical anatomy, an important issue for surgeons. He had published his *Icones herniae inguinalis congenitae* in 1781, with drawings made by A. Delfos and engraved by R. Muys, in order to demonstrate to the surgeons the importance of an abnormal situation in the inguinal canal.[47] In this work, he referred to Palletta's study. His interest in the uterus comes to light in the second book of his *Observationes anatomico-pathologiae*. He describes a public dissection of a pregnant woman with a series of drawings of a rather dramatic aspect.[48] In this study, the number and the accuracy of his references is remarkable. He was, of course, a well-read scholar, but these footnotes are complete documents on their own! This method, a meticulous survey of the existent literature, was an important feature of an enlightened physician! Sandifort referred to his colleague in Milano, Francesco Biumi, when he explained why he was so concerned with so many references:[49]

> "I am frightened of this bunch of criticisers of today, who everything, that has not come out of their own head, blame as much as they can, instead of valuing it. I feel like the philosopher and physician Biumi in Milano whose words, written in the preface of his "Observations in anatomy" are worth mentioning:[50]
>
> "There are" he says "troublesome people who judge other peoples writings too severely as rude, barbarous and without any learning. They consider their own work as splendid and famous, and they proclaim it! What I fear most in these weak voices is that no authors until today could escape these disparagements. But they expose themselves, and the authors will reprimand them, not in their inner circle, but with the help of edited books, instructed with well founded doctrines, to prove that they are false".

Sandifort had collected Biumi's reports on autopsies in this *Thesaurus*

45. He mentioned this in the introduction. I did not consult the reference.
46. Mazzolini, "Carteggio", *o.c.* (nt 2), pp. 62-76.
47. Eduard Sandifort, *Icones herniae inguinalis congenitae*. Leiden, P. v.d. Eyk et D. Vijgh, 1781, p. 11.
48. Eduard Sandifort, "Observationes", *o.c.* (nt 27). Liber II, cap. I, "De utero gravido", pp. 1-56.
49. *Ibidem,* Liber I, introduction.
50. Francesco Biumi, *Observationes anatomiae ac scholii illustratae*. T I, Milano, 1765.

of 1778. The Milanese school must have drawn attention in this period, not only from Sandifort, but also from Haller and other scholars.[51]

## *Anatomia subtilis new style, the "elegant anatomy"*

Leopoldo Caldani was most grateful to Sandifort for his effort to publish Caldani's two treatises in Leyden. In his letter of December 1, 1785 he informed Sandifort of a shipment from Padua of several books on anatomy and surgery.[52] Among those was Paolo Mascagni's *Prodrome d'un ouvrage sur le système des vaisseaux lymphatiques, contenant 24 planches in folio.* The work was published in 1784 in Siena, by "Vincent Pazzini Carli et fils". It was the overture of Mascagni's great work *Vasorum Lymphaticorum corporis humani Historia et Ichonographia* which appeared three years later at the same office.

This magisterial study on the lymphatic system would become the authoritative work in this field of anatomy.[53] It held an excellent introduction into the history of the exploration of the lymphatic system, with ample references to the Dutch scholars. Mascagni showed a great admiration for Antony Nuck (1650-1692), his predecessor in vascular injection with mercury.[54] Mascagni started his injections with mercury into the lymphatic vessels around 1777. He injected all organs, except the brain and he took care of proper drawings and engravings of the dissected areas. He also used a mixture of gelatin and cinnabar (mercuric sulphide) to study the micro-structures under a Dolland microscope.[55] He came to the conclusion that the lymphatic system was only connected with the venous capillaries and had to have a separate circulation from the arterial system. He denied arterial and venous lymphatic vessels, there were only lymphatic vessels with valves. In his profound study, he also mentioned the variations of the system. Therefore, he referred to the variety of the terminal of the thoracic duct into

51. Luigi Belloni, "La scuola ostetrica milanese dal Moscati al Porro". Publication from *VI Congresso della società di ostetricia e ginecologia del mediterraneo latino.* Milano, 2-4 May, 1960. Abstract.

52. F. Pierro, "Due lettere di Leopoldo M.A. Caldani (1725-1813) all'anatomico Leidense Eduardo Sandifort (1742-1814)". *Medicina nei Secoli a cura dell'Istituto di storia della Medicina dell'Universita di Roma* 7 (1970), 19-31.

53. Federico Allodi, *La storia e la conoscenza dei linfatici del cuore nelle opere di Paolo Mascagni (nel II centenario della nascita 1755-1955).* Firenze, Leo S. Olschki, 1955.

54. Mascagni, "Vasorum lymphaticorum", *o.c.* (nt 9), Introduction.

55. *Ibidem*, p. 7. See also Bruno Zanobio, "La microscopia dell'anatomico Paolo Mascagni", *Rendiconti – Istituto Lombardo, Accademia di Scienze e Lettere. Classe di Scienze Matematiche e Naturali, B, Scienze Biologiche e Mediche*, S. III, 94 (1960), pp. 221-228.

the subclavian vein and to the variety in position of the cysterna chyli, described by Eduard Sandifort "vir doctrina et eruditione pollens" and successor of the immortal Albinus![56] Mascagni's craftmanship was extraordinary and certainly equivalent to the art of his predecessors, Frederik Ruysch (1638-1731), B.S. Albinus and Nathaniel Lieberkühn (1711-1756). During the last decades of the eighteenth century several anatomists tried to perfectionize this art of injecting vessels. Not only accurate descriptions of layers, vessels and glands were important, but also beauty of the specimens.

In the Netherlands, a school of anatomy developed which may be called a school of elegant anatomists, characterized by beauty, transparency and perfectioned delicacy.

Such specimens have been preserved in Utrecht, in the collection of Jan Bleuland (1756-1838) and in Groningen, in the collection of Pieter de Riemer (1769-1831), a Bleuland pupil.[57] These specimens represent the decline of *anatomia subtilis*, the last attempt of the scholars to emphasize the importance of the fiber and the *vas minimum* in the formation of tissue. By high technical skill of vascular injection followed by inspection of tissue samples under the microscope, they tried like Mascagni to trace the course of the lymphe, the blood, the urine and the cerebro-spinal fluid. Furthermore, they were interested in the continuation of membranes, fascia's and their different density and structure. Finally, they studied osteogenesis and the parenchymal organs and glandular structures, such as lungs, liver, spleen and kidneys. They all used *analogia* as a method, they compared organs of different animals with human specimens.

Jan Bleuland was a student at Leyden between 1774 and 1780. He was deeply impressed by the Leyden Museum, by Sandifort's lessons and the collection specimens, which were shown to him by F.B. Albinus. Especially the skilfully injected organs drew his attention.[58] Bleuland became very friendly with Wouter van Doeveren (1730-1783), who encouraged him to mount specimens by himself. He applied the elegant method on specimens of normal and pathological structure. His studies were published in Leyden when he was already practising in Gouda, where he stayed until 1791. Among those was a profound study on the circulation of the lymphe in the

56. Mascagni, *Ibidem*, p. 132.

57. L.J. Endtz, *De Haghe-professoren. Geschiedenis van een chirurgische school.* Amstelveen, Specia, 1972, pp. 147-161 on De Riemer.

58. G.T. Haneveld, *Bijdrage tot de geschiedenis der pathologische anatomie: Utrecht in de eerste helft van de negentiende eeuw.* Amsterdam, Ronald Meesters, 1978, pp. 100-145 on Bleuland.

digestive tract.[59] However, Bleuland had presented his thesis to the Leyden medical faculty in 1780 on a different subject: a clinical-physiological study on deglutition and the oesophagus, regarding a case history he had observed under the guidance of van Doeveren. But the illustrations of the autopsy were made by two artists, under guidance of Sandifort.[60]

Sandifort's collection of "elegant anatomy" was never described. It was a small collection, of which only 52 specimens have been preserved. They are mounted in phials, filled with terpentine. This material hardens the tissues and makes the membranes transparent. The bloodvessels are injected with coloured material, perhaps the cinnabar/gelatin mixture, applied by Mascagni. They are different in colour from those made by Albinus and his contemporaries, who used a mixture of wax and sheep fat. There is a striking resemblance between Mascagni's injected specimens and the ones, present in this collection, such as a liver, where the lymphatic vessels have been filled with the mercury.[61] But there are also inguinal glands, ribs with intercostal membranes, delicate injected mesenteries, subcutaneous layers and submucous membranes of the intestines. Sandifort prepared a few specimens on comparative anatomy, such as the whiskers of a seal, mounted in alcohol, to show the innervating nerves, and the eye of a dolphin.[62] Eyes of whales and dolphins were usually dissected for demonstration of the iris and the chorioid layer to the students.[63] Several young human embryo's in the chorionic sack have been preserved, as well as pregnant uteri of animals. Except for a few varieties in ribs and a stuffed red baby cat with two heads, there are no pathological specimens. These were described by G.C.B. Suringar (1802-1874) and classed under his collection,[64] for instance a horseshoe-kidney of a child, which had been described by Sandifort in 1779.[65] The "elegant

59. J. Bleuland, *Experimentum anatomicum, quo arteriolarum lymphaticarum existentia probabiliter adstruitur.* Leiden, Abraham et Janus Honkoop. This study was dedicated to Wouter van Doeveren, Eduard Sandifort, Johan David Hahn and Nic. Georg. Oosterdijk, the members of the Leyden medical faculty. It is not mentioned by Mascagni.

60. J. Bleuland, *Specimen academicum inaugurale de difficile aut impedita alimentorum depulsione*, Leiden, apud Th. Haak and Jacob Douzy, 1780. The drawings were made by F.F. Bylaart, the engravings by R. Muys. Both artists worked for Sandifort.

61. Museum Anatomicum Leiden: Ag 22. Mascagni. "Vasorum lymphaticorum" *o.c.* (nt 9) Plate XVII.

62. Museum Anatomicum Leiden: Ag 34, Ag 36.

63. A.M. Elshout, "Anatomisch Kabinet" *o.c.* (nt 29), p. 42.

64. G.C.B. Suringar, *Pars supellectilis sive Catalogus Speciminium pathologico-anatomicorum*, Leiden, 1866.

65. E. Sandifort, "Observationes" *o.c.* (nt 27). Lib. 3, cap. 7, tab. 8, fig. 6. "De rene, numero unico, sed functione duplici". Suringar (see above) mentions this specimen in his catalogue on p. 20, nr. 108.

anatomy" has been preserved for its beauty and the skill of its creator, and is a distinguished collection in a modern museum of anatomy.

*Conclusive remarks*

Eduard Sandifort, an advocate of Morgagni's method in pathology, interested in documentation of European medicine, an "elegant anatomist" and an enlightened physician. He is best known in medical history by the cardiologists, for his accurate description of a congenital heart malformation which would become known under the name "tetralogy of Fallot,"[66] a quadruple defect of the embryonic heart, which causes the so-called blue baby. We may as well consider him as an accurate pathologist, whose illustrations of pathological specimens have highly contributed to a better knowledge of the "pathology of the solids", as Morgagni would have said. It is not a matter of rhetorics, when the French pathologist Jean Cruveilhier (1791-1874) praised him as the "father of pathological iconography."[67] Art and pathology were closely connected in his work, he was convinced that the diseased parts of the body should be documented in accurate drawings and engravings. During his active period as a physician and as a professor at the Leyden university, Sandifort took the trouble to collect publications from all over Europe to spread knowledge among his colleagues. As an enlightened physician, he took care of gaining information on variolation, cattle diseases and the pioneering work of Nils Rosèn von Rosenstein (1706-1773) whose work on pediatrics he translated into Dutch in 1768.[68] His interest in pathology led him to Caldani and the other Italian scholars, whose work he published in Latin editions. Unlike Petrus Camper, Sandifort was not a traveller. As far as is known from his biography, he never visited Italy, but he was kept informed by his colleagues and students. The *iter italicus* was very popular in the eigteenth century, albeit more in the spirit of romanticism than for educational values, as was the case in the seventeenth century.[69]

66. Lydian Russell Bennett, "Sandifort's 'observationes' chapter I concerning a very rare disease of the heart". *Bulletin of the History of Medicine* 20 (1946), pp. 539-570.

67. J. Cruveilhier, *Anatomie pathologique du corps humain.* Paris, chez J.B. Baillière, 1829-1835. Tome I, avant propos, p. III. "Au premier rang se place Sandifort, dont les belles planches en apprennent plus en quelques heures que des volumes de description, et qu'on peut considérer comme le père de l'iconographie pathologique".

68. Nils Rosèn von Rosenstein, *Handleiding tot de kennis en genezing van de ziekten der kinderen, ten nutte van het algemeen.* Translated by Eduard Sandifort, from a German edition, taken care of by J.A. Murray from Göttingen. It was published in The Hague in 1768.

69. A. Frank van Westrienen, *De Groote Tour. Tekening van de Educatiereis der Nederlanders in de zeventiende eeuw.* Amsterdam, Noordhollandsche Uitgeversmaatschappij BV, 1983.

Such a "wandervogel" was Wilhelm Xavier Jansen (1760-1793) who studied in Leyden and graduated under Sandifort in 1784. After his graduation he made a long journey through Italy. He described his journey in a series of letters sent to Sandifort.[70] Sandifort considered the very lively written letters of such importance, that he urged Jansen to make a German edition of the book. He even inserted some annotations.[71] It is obvious that his urge to spread knowledge had a firm grip upon his character.

When Johann Peter Frank (1745-1821) became professor in the academia Ticinensi (Pavia), he followed Sandifort's example for Italy.[72] Since his students did not read English and very little German, he composed his *Delectus Opusculorum Medicorum* in Latin. Like Haller, Sandifort and some other distinguished German encyclopedists he wanted to improve communication and information in the medical world. He showed his esteem for Sandifort's enterprise in the introduction of the first issue.

Sandifort's skill, accuracy and artistic qualities have contributed to his success as an anatomist. His description of severe malformations are austere and correct. His plates of pathological specimens are presented in such a way that they are beautiful and instructive at the same time. Therefore we may call Sandifort an "elegant anatomist", in a wider sense than the maker of skilfully injected specimens only. In his "elegant anatomy" he represented Morgagni, Albinus and Mascagni together.

70. W.X. Jansen, *Reize door Italien in den jare 1785 gedaan* (described in a series of letters to the learned Eduard Sandifort). Leiden, A.J. Honkoop, 1793.

71. G.A. Lindeboom, *Dutch Medical Biography. A biographical Dictionary of Dutch Physicians and Surgeons*. Amsterdam, Rodopi, 1984, p. 974. I was not able to consult the German edition, which was published in Düsseldorf in 1793/1794.

72. J.P. Frank, *Delectus opusculorum medicorum antehac in Germania diversis academiis editorum* ... in twelve volumes. Pavia, Petrus Galeatus, 1785-1793. Vol. I, preface.

## LIST OF FIGURES

**Addresses of the contributors**

Professor K. van Berkel, Instituut voor Geschiedenis, University of Groningen, Grote Rozenstraat 38, 9712 TJ Groningen, The Netherlands.

Professor H.J.M. Bos, Institute for the History of Science, University of Utrecht, Nieuwe Gracht 187, 3512 LM Utrecht, The Netherlands.

Dott. M. Cavazza, Dipartimento di Filosofia, University of Bologna, Via Zamboni 38, 40126 Bologna, Italy.

Professor H.F. Cohen, Vakgroep Geschiedenis, University of Twente, Postbus 217, 7500 AE Enschede, The Netherlands.

Professor A.M. Luyendijk-Elshout, Prins Bernhardlaan 60, 2341 KL Oegstgeest, The Netherlands.

Professor C. Maccagni, Centro C.N.R. per la Storia della Tecnica, University of Genoa, Via Balbi 6, 16126 Genova, Italy.

Dott. C.S. Maffioli, Institute for the History of Science, University of Utrecht, Nieuwe Gracht 187, 3512 LM Utrecht, The Netherlands.

Professor R. Mazzolini, University of Trento, Facoltà di Lettere e Filosofia, Via Santa Croce 65, 38100 Trento, Italy.

Dr A. McConnell, 46 Defoe House, Barbican, London EC2Y 8DN, England.

Dott. N. Morello, Istituto di Storia Moderna, University of Genoa, Via Balbi 6, 16126 Genova, Italy.

Drs L.C. Palm, Institute for the History of Science, University of Utrecht, Nieuwe Gracht 187, 3512 LM Utrecht, The Netherlands.

Dr C. de Pater, Institute for the History of Science, University of Utrecht, Nieuwe Gracht 187, 3512 LM Utrecht, The Netherlands.

Professor H. de Ridder-Symoens, Grote Baan 98, 9920 Lovendegem, Belgium.

Professor H.A.M. Snelders, Institute for the History of Science, University of Utrecht, Nieuwe Gracht 187, 3512 LM Utrecht, The Netherlands.

Professor G. Tagliaferri, Institute of Physics, Sezione Storia della Fisica, University of Milan, Via Celoria 16, 20133 Milano, Italy.

Professor P. Tucci, Institute of Physics, Sezione Storia della Fisica, University of Milan, Via Brera 28, 20121 Milano, Italy.

Dr G. Vanpaemel, EHSAL, Stormstraat 2, 1000 Brussel, Belgium.

# INDICES*

*A. Index of persons*



*These indices have been compiled by Lian M.J. Hielkema. They only refer to names mentioned in the text.

*B. Index of Places*

# NIEUWE NEDERLANDSE BIJDRAGEN TOT DE GESCHIEDENIS DER GENEESKUNDE EN DER NATUURWETENSCHAPPEN

Editors: M.J. van Lieburg, L.C. Palm, H.A.M. Snelders and R.P.W. Visser

No. 1.
Lindeboom, G.A.: Descartes and Medicine. Amsterdam 1979. 134 pp. Hfl. 30,—

No. 2.
Lindeboom, G.A.: Adriaan van den Spiegel (1578-1625) hoogleraar in de ontleed- en heelkunde te Padua. Amsterdam 1978. 125 pp. uitverkocht

No. 3.
Lieburg, M.J. van: Het medisch onderwijs te Rotterdam (1467-1967). Amsterdam 1978. 111 pp. Hfl. 18,—

No. 4.
Snelders, H.A.M.: Het Gezelschap der Hollandsche Scheikundigen — Amsterdamse chemici uit het einde van de achttiende eeuw. Amsterdam 1980. 111 pp. Hfl. 20,—

No. 5.
Brinkman, A.A.A.M.: De alchemist in de prentkunst. Amsterdam 1982. 69 pp. 4 to. Hfl. 25,—

No. 6.
Dumaître, Paule: La curieuse destinée des planches anatomiques de Gérard de Lairesse. Amsterdam 1982. XIV,107 pp. Hfl. 25,—

No. 7.
Deutsch-Niederländischen Beziehungen in der Medizin des 17. Jahrhunderts. Vorträge des Deutsch-Niederländischen Medizinhistorikertreffens 1981. M.J. van Lieburg und R. Toellner (Hrsg.). Amsterdam 1982. 108 pp. Hfl. 25,—

No. 8.
Antoni van Leeuwenhoek 1632-1723. Studies on the life and work of the Delft scientist commemorating the 350th anniversary of his birthday. Edited by L.C. Palm & H.A.M. Snelders. Amsterdam 1982. 212 pp. Hfl. 50,—

No. 9.
Berkel, K. van: Isaac Beeckman (1588-1637) en de mechanisering van het wereldbeeld. Isaac Beeckman (1588-1637) and the mechanization of the world picture (with a summary in English). 344 pp. Amsterdam 1983.
Hfl. 50,—

No. 10.
Fahrenheit's Letters to Leibniz and Boerhaave. Edited, translated and annotated by Pieter van der Star. Amsterdam/Leiden 1983. 195 pp.
Hfl. 50,—

No. 11.
Schraders, Catharina Geertruida: Memoryboeck van de Vrouwens. Het notitieboek van een Friese vroedvrouw 1693-1745. Bewerkt en ingeleid door Drs. M.J. van Lieburg. Met een verloskundig commentaar van Prof.Dr. G.J. Kloosterman. Amsterdam 1984. 218 pp. Hfl. 35,—

No. 12.
Visser, Robert Paul Willem: The Zoological Work of Petrus Camper (1722-1789). Het zoölogisch werk van Petrus Camper (1722-1789) (met een samenvatting in het Nederlands). Amsterdam 1985. 207 pp. Hfl. 45,—

No. 13.
Kracht en Stof. De introductie van moderne natuurwetenschappelijke denkwijzen in de geneeskunde, zoals blijkt uit Nederlandse medische vakbladen, 1840-1870. Verslag van een symposium, gehouden te Nijmegen op 21 september 1985, onder redactie van D. de Moulin. Amsterdam 1985. 127 pp. Hfl. 30,—

No. 14.
Deutsch-Niederländischen Beziehungen in der Medizin des 18.Jahrhunderts. Vorträge des Deutsch-Niederländischen Medizinhistorikertreffens 1982. Herausgegeben von R. Toellner und M.J. van Lieburg. Amsterdam 1985. 92 pp. Hfl. 25,—

No. 15.
Napjus, J.W.: De hoogleraren in de geneeskunde aan de hogeschool en het athenaeum te Franeker. Verzamelde opstellen. Ingeleid en voortgezet door Dr. G.A. Lindeboom. Amsterdam 1985. XIX,210,XXVI pp. Gebonden.
Hfl. 40,—

No. 16
Theunissen, Bert: Eugène Dubois en de aapmens van Java. Een bijdrage tot de Geschiedenis van de paleoantropologie. Amsterdam 1985. 302 pp.
Hfl. 50,—

No. 17.
Nineteenth-Century Scientific Instruments and their Makers. Papers presented at the Fourth Scientific Instrument Symposium, Amsterdam 23-26 October 1984. Edited by P.R. de Clercq. Amsterdam/Leiden 1985. 275 pp.
Hfl. 70,—

No. 18.
Lieburg, M.J. van: Het Bataafsch Genootschap der Proefondervindelijke Wijsbegeerte te Rotterdam 1769-1984. Een bibliografisch en documenterend overzicht. Amsterdam 1985. 283 pp. Hfl. 60,—

No. 19.
Hendrik Engel's Alphabetical List of Dutch Zoological Cabinets and Menageries. Second, enlarged edition. Prepared by Pieter Smit with the assistance of A.P.M. Sanders and J.P.F. van der Veer. Amsterdam 1986. X,340 pp. Bound. Hfl. 70,—

No. 20.
Hakfoort, Casper: Optica in de eeuw van Euler. Opvattingen over de natuur van het licht 1700-1795. Optics in the age of Euler. Conceptions of the nature of light, 1700-1795. (with a summary in English). Amsterdam 1986. 219 pp. Hfl. 49,50

No. 21.
Lieburg, M.J. van: Het Coolsingelziekenhuis te Rotterdam (1839-1900). De ontwikkeling van een stedelijk ziekenhuis in de 19e eeuw. Amsterdam 1986. XVIII,770 pp. 4to. Bound. Hfl. 160,—

No. 22.
Mother and Child were Saved. The memoirs (1693-1740) of the Frisian midwife Catharina Schrader. Translated and annotated by Hilary Marland. With introductory essays by M.J. van Lieburg and G.J. Kloosterman. Amsterdam 1987. 89 pp. Hfl. 25,—

No. 23.
Wiggen, G.J. van: In meer eerbare banen. De ontwikkeling van het tandheelkundig beroep in Nederland van 1865-1940. Amsterdam 1987. 369 pp. Hfl. 65.—

No. 24.
Mijnhardt, W.W.: Tot Heil van 't Menschdom. Culturele genootschappen in Nederland, 1750-1815. Amsterdam 1988. 430 pp. Hfl. 75,—

No. 25.
Lieburg, M.J. van: Bevorderlijk voor de kunst en nuttig voor de maatschappij. De geschiedenis van het Genootschap ter bevordering van Heel- en Verloskunde (1857-1929) en van het Amsterdams Geneeskundig Genootschap (1925-1988). Amsterdam 1988. 124 pp. Hfl. 25,—

No. 26.
's-Rijkskweekschool voor militaire geneeskundigen te Utrecht (1822-1865). Verslag van een symposium gehouden te Nijmegen op 26 september 1987 onder redactie van D. de Moulin met een lijst van kwekelingen, samengesteld door M.J. van Lieburg. Amsterdam 1988. 193 pp. Hfl. 45,—

No. 27.
A.I. Bierman: Van artsenijmengkunde naar artsenijbereidkunde. Ontwikkelingen van de Nederlandse farmacie in de negentiende eeuw. Amsterdam 1988. 312 pp. Hfl. 65,—

No. 28
Pieter Smit: Artis. Amsterdam 1988. 392 pp. 27 ill., 7 kleurenfoto's. Hfl. 45,—

No. 29
M.J. van Lieburg en H.A.M. Snelders: "De bevordering en volmaking der proefondervindelijke wijsbegeerte". De rol van het Bataafsch Genootschap te Rotterdam in de geschiedenis van de natuurwetenschappen, geneeskunde en techniek (1769-1988). Amsterdam 1989. 267 pp. Hfl. 60,—